Simulation-Gaming: On the Improvement of Competence in Dealing with Complexity, Uncertainty and Value Conflicts

Proceedings of the International Simulation and Gaming Association's 19th International Conference

Other Pergamon Titles of Related Interest

COLMAN
Game Theory and Experimental Games
The Study of Strategic Interaction

MOONEN & PLOMP
Eurit 86
Development of Educational Software and Courseware

SENDOV & STANCHEV
Children in the Information Age: Opportunities for
Creativity, Innovation and New Activities

STAHL
Operational Gaming
An International Approach

CROOKALL *et al.*
Simulation-Gaming in the Late 1980s

CROOKALL *et al.*
Simulation-Gaming in Education and Training

Related Pergamon Titles

Computers & Education

Computers & Operational Research

International Journal of Educational Development

Journal of the Operational Research Society

Language & Communication

Long Range Planning System

Simulation-Gaming: On the Improvement of Competence in Dealing with Complexity, Uncertainty and Value Conflicts

Proceedings of the International Simulation and Gaming Association's 19th International Conference

Department of Gamma-Informatics, Utrecht University, The Netherlands 16-19 August 1988

Edited by

JAN H. G. KLABBERS

WILLEM J. SCHEPER

CEES A. TH. TAKKENBERG
Utrecht University, The Netherlands

and DAVID CROOKALL
The University of Alabama, USA

PERGAMON PRESS
Member of Maxwell Macmillan Pergamon Publishing Corporation
OXFORD · NEW YORK · BEIJING · FRANKFURT
SÃO PAULO · SYDNEY · TOKYO · TORONTO

U.K.	Pergamon Press plc, Headington Hill Hall, Oxford OX3 0BW, England.
U.S.A.	Pergamon Press, Inc., Maxwell House, Fairview Park, Elmsford, New York 10523, U.S.A.
PEOPLE'S REPUBLIC OF CHINA	Pergamon Press, Room 4037, Qianmen Hotel, Beijing, People's Republic of China
FEDERAL REPUBLIC OF GERMANY	Pergamon Press GmbH, Hammerweg 6, D-6242 Kronberg, Federal Republic of Germany
BRAZIL	Pergamon Editora Ltda, Rua Eça de Queiros, 346, CEP 04011, Paraiso, São Paulo, Brazil
AUSTRALIA	Pergamon Press Australia Pty Ltd., P.O. Box 544, Potts Point, N.S.W. 2011, Australia
JAPAN	Pergamon Press, 5th Floor, Matsuoka Central Building, 1-7-1 Nishishinjuku, Shinjuku-ku, Tokyo 160, Japan
CANADA	Pergamon Press Canada Ltd., Suite No. 271, 253 College Street, Toronto, Ontario, Canada M5T 1R5

Copyright © 1989 ISAGA

All Rights Reserved. No part of this publication may be reproduced, stored in a retrieval system or transmitted in any form or by any means, electronic, electrostatic, magnetic tape, mechanical, photocopying, recording or otherwise, without permission in writing from the copyright holders.

First edition 1989

Library of Congress Cataloging-in-Publication Data
International Simulation and Gaming Association.
International Conference (19th: 1988: Dept. of Gamma-Informatics, Utrecht University)
Simulation-gaming: on the improvement of competence in dealing with complexity, uncertainty, and value conflicts: proceedings of the International Simulation and Gaming Association's 19th International Conference, Department of Gamma-Informatics, Utrecht University, The Netherlands, 16-19 August 1988/edited by
Jan H. G. Klabbers . . . [et al.]. — 1st ed.
p. cm.
1. Game theory — Congresses. 2. Uncertainty — Congresses. 3. Decision-making — Congresses. I. Klabbers, Jan H. G. II. Rijksuniversiteit te Utrecht. Dept. of Gamma-Informatics. III. Title.
HB144.158 1988 658.4'0352—dc20 89-8535

British Library Cataloguing in Publication Data
International Simulation and Gaming Association, International Conference, 19th; 1988, State University of Utrecht.
Simulation — gaming: on the improvement of competence in dealing with complexity, uncertainty and value conflicts: proceedings of the International Simulation and Gaming Associations 19th International Conference, Department of Gamma — Informatics, Utrecht University, The Netherlands, 16-19 August 1988.
1. Simulation games
I. Title II. Klabbers, Jan H.G.
001.4'24
ISBN 0-08-037115-9

Printed in Great Britain by BPCC Wheatons Ltd., Exeter

Contents

Foreword — ix
Acknowledgements — xi
Preface — xiii

Section One: Complexity, Uncertainty and Value Conflicts
Editor: Jan H. G. Klabbers

Jan H. G. Klabbers — 3
   On the improvement of competence
Jan Berting — 8
   "Structures, actors and choices"
Frances Mautner-Markhof — 24
   The reality, management and simulation of complex systems
Hiroharu Seki — 38
   Global modelling in Japanese political science
D. J. A. Kalff — 51
   Strategic decision making and simulation in Shell

Section Two: Organizational Change
Editor: Cees A. Th. Takkenberg

Cees A. Th. Takkenberg — 65
   Organizational change, implementation, social simulation, cultural change and leadership, model switching
Alan Coote and Maarten van Mens — 68
   Organizational change: workshop review
Bart van Linder — 72
   SWITCHER: an organization support system for improving reflective competence
Frans-Bauke van der Meer and Ton Roodink — 81
   Social simulation of organizing and organizational change
Maarten van Mens — 90
   PACT
Dolf Dekker — 98
   MIDAS: an awareness game on innovation management*
Alan Coote and Clive Loveluck — 99
   GHOSTS IN THE MACHINE: a computer-aided simulation/game to explore the relationships between strategic policies, tactical action and organizational cultures*

Section Three: Business Simulation
Editor: Willem J. Scheper

Willem J. Scheper — 103
   Business simulation: an introduction
Peter Schulein — 106
   Crisis gaming for research and training
W. C. Borawitz — 115
   Wargaming

vi   Contents

James M. Freeman   127
   Goal-setting and business gaming
T. Richard Whiteley and Anthony J. Faria   137
   A study of the relationship between student final exam
   performance and simulation game participation
John F. Lobuts, Jr.   146
   Dysfunctionalism in American management systems: management
   mania in corporate America
Erica Bates, Elizabeth Christopher and Barry Moore   155
   Australian rehearsal technique
Elizabeth Christopher   161
   Talking heads*
Richard D. Teach   162
   Designing an intercultural business simulation*
Ronald Brech   163
   The educational challenge of business simulations*
Precha Thavikulwat and Jimmy M. T. Chang   164
   Student- and instructor-oriented features in a business simulation*
David Crookall and Danny Saunders   165
   Adapting a wargame for multiskill tasks*

Section Four: Policy Exercise
Editor: Cees A. Th. Takkenberg

Cees A. Th. Takkenberg   169
   Policy exercise, (group) decision support
Steven E. Underwood   172
   Structured participation in technology assessment: the policy exercise
Mario Polic and Ivo Wenzler   180
   Project definition gaming/simulation exercise
Marian V. Sackson   189
   An expert system that simulates group decision making in a
   stochastic environment and exhibits learning
Henk G. Sol and Michael B. M. van der Ven   203
   Designing and evaluating decision support systems: a group
   DSS for international transfer pricing
P. W. G. Bots, F. D. J. van Schaik and H. G. Sol   210
   Gaming as an environment for testing the effectiveness of decision
   support systems*
Eduard Rădăceanu   211
   The value analysis of management information systems, problem
   analysis and gaming*

Section Five: Methodology
Editor: Jan H. G. Klabbers

Jan H. G. Klabbers   215
   Methodology: behavioural and social systems, design and evaluation of
   games/simulations, classification, taxonomy
Rainer Siebecke   219
   Problems of taxonomy for gaming in macro- and micro-economic subjects
Jan H. G. Klabbers and Barbara van der Waals   225
   From rigid-rule to free-form games: observations on the role of rules
Kiyoshi Arai   235

A simple method of scenario-making: two Japanese cases in community planning *Charles M. Plummer*	241
Design and evaluation of computer-based behavioural/social system simulations* *Tadeusz Selbirak*	242
Computer-based interactive procedure for analysing conflict resolutions within simulation and gaming*	

Section Six: Learning Environments and Communication
Editor: David Crookall

David Crookall Learning environments and communication: an introduction	245
Ken Jones Some dangers when using interactive events to improve competence	250
Donald Thatcher and June Robinson ME – THE SLOW LEARNER and some of its implications*	258
Michael J. Rockler Simulation/gaming: brain-compatible teaching strategy for the improvement of competence*	259
Peter A. Raynolds and Gennie H. Raynolds The "JOG your right brain exercise" at ISAGA 88	260
Alan L. Cudworth Thurston-Parkin – a case study in team communication*	269
Fred de Vries Student learning invoked by simulations embedded in a learning environment	270
D. Wells Coleman On modelling in CALL conversational simulations*	278
Hilbert Kuiper and Jos van der Arend Interactive simulation in a computer-based training environment*	279

Section Seven: Special Topics
Editor: Willem J. Scheper

Willem J. Scheper Special topics: an introduction	283

Environmental Planning

Bert de Vries Environmental planning: workshop review	286
Bert de Vries Learning about electric power planning: a gaming approach	292
Leopoldo Schapira Biosphere and underdevelopment	300
Frank R. Rijsberman and Gerrit Baarse Simulation of integrated rural development with IRDEM	309
Claude Bourlès CHERNOBYL, a game of negotiations under stress*	325

Health Care

E. M. Bronkhorst, G. J. Truin, et al. Improving dental planning through computer simulation	326

Louwrens ten Brummeler and Cor van Dijkum Physiotherapist's dilemma	332

Diplomacy

Paul W. Meerts Diplomatic games	340

Gambling

Danny Saunders and Dave Turner Competency in gambling games	348
Alan L. Cudworth SURVIVAL: a case study in risk taking	359
R. Iain F. Brown Gaming, gambling, risk taking, addictions and a developmental model of a pathology of man-machine relationships*	368

*Applies to chapters where abstract only appears.

Foreword

This volume records the proceedings of the International Simulation and Gaming Association's 19th International Conference, which took place at Utrecht University, The Netherlands, from 16 to 19 August 1988.

Seven sections are contained in this volume. The first section on complexity, uncertainty and conflicts deals with theoretical and methodological issues. It is the introduction to the conference theme "On the improvement of competence". The following sections cover broad areas: organizational change, business simulation, policy exercise, methodology, learning environments, and special topics like, environmental planning, health care, diplomatic games and gambling.

Overlooking the sections and papers, especially against the background of the first section, I have the strong impression, that we are at the cross-road, where several disciplines are looking for new directions in coping with issues that do resist one-sided views. When we talk about complexity and uncertainty, we actually have to admit that it is our own lack of competence that prevents us from comprehending our contemporary world adequately. Ideologically we have been fabricating images of societies that may have been suitable until recently. Nowadays, it seems necessary to release ourselves from (out-dated) ideologies and to start searching for a new order and establishing new dynamic societal equilibria utilizing new insights gained from various disciplines. This applies for example to looking for new patterns of international relations, to managing complex systems in a wicked world, to coping with threats like AIDS, that challenge our current conceptions about drugs, crime and health care. A science will be necessary, that deals with self-referential (autological) processes, that is susceptible for a coherent search into multiple reality, that generates variety instead of killing it, that generates basic knowledge about self-organizing and self-destructing systems. This knowledge has to be produced from the inside, through communicating, negotiating and co-operating between the actors involved. For this science a spectator's position is no longer feasible. Gaming-simulation offers an adequate theoretical, methodological and empirical basis for dealing with these issues. The papers presented in this volume offer a wide spectrum of ideas, suggestions and solutions. Those who wish to explore the field of gaming/simulation further are encouraged to read copies of the 1986 and 1987 Conference proceedings of ISAGA. They are entitled *Simulation-Gaming in the Late 1980s*, respectively *Simulation-Gaming in Education and Training*, and published by Pergamon Press, Headington Hill Hall, Oxford OX3 0BW, England.

I should like here, on behalf of ISAGA and the 1988 Conference delegates, to take this opportunity of thanking my colleagues of the programme and organizing committee, and the workshop convenors for their involvement and support in putting together this fine meeting. Contributors to these Proceedings must be thanked as well.

My gratitude to my co-editors should be recorded, to Willem Scheper, Cees Takkenberg, and David Crookall for their help in putting together these Proceedings. Michèle Norton of Pergamon Press provided valuable editorial support.

A few words about ISAGA: The International Simulation and Gaming Association was established in Birmingham in 1972 with the purpose of creating an international forum for the exchange of ideas and information on simulations and games. The main objective of ISAGA is to promote the use of simulation and gaming as a learning, training and research medium.

It publishes a quarterly newsletter, and its official journal is *Simulation & Games: An International Journal of Theory, Design & Research* (quarterly, Sage Publications). It also holds an annual international conference, with the proceedings published by Pergamon Press.

Current ISAGA officers are as follows: The 1988/1989 President is Jan Klabbers, who is General Secretary as well; President-elect is Hans Gernert; the Treasurer is Danny Saunders; and the *ISAGA Newsletter* Editor is David Crookall. ISAGA has also regional secretaries in various parts of the world (information from the General Secretary). Requests for information should be sent to Jan Klabbers, Department of Gamma-Informatics, Faculty of Social Sciences, P.O. Box 80140, 3508 TC Utrecht, The Netherlands. Applications for membership applications should be sent to Danny Saunders, Department of Behavioural and Communication Studies, The Polytechnic of Wales, Pontypridd CF37 1DL, UK. Potential contributors to the *Newsletter* should write to David Crookall, Dept. of English, the University of Alabama, Tuscaloosa, AL 35487, USA.

Utrecht, The Netherlands
December 1988

JAN KLABBERS

Acknowledgements

A large number of people involved with the Conference should be thanked. First thanks go to the members of the Conference Honorary, the Programme and Organizing Committees.

Honorary Committee

Chris Brand	Maxim Training Systems Ltd, UK
Klaas Bruin	Teacher Training Institute Ubbo Emmius, The Netherlands
Vladimir Burkov	Institute for Management Problems, USSR
Elizabeth Christopher	University of Sydney, Australia
Arnaldo Cecchini	DAEST-IUV, Italy
Alan Coote	Polytechnic of Wales, UK; Editor Simulation/Games for Learning
David Crookall	The University of Alabama, USA; editor ISAGA Newsletter
Richard Duke	University of Michigan, Ann Arbor, USA
Paolo Frignani	University of Geneva, Switzerland
Hans Gernert	Humboldt University, GDR, ISAGA President-Elect
Cathy Greenblat	Rutgers University, USA; editor Simulation & Games
Kiyoshi Arai	Kinki University, Japan
Hubert Law-Yone	Technion, Israel Institute of Technology, Israel
Dennis Meadows	University of New Hampshire, USA
Jacob Ngwa	Regional Pan African Institute for Development, Cameroon
Giorgio Panizzi	CNITE, Italy
Florosito Q. Pimentel	PHILSAGA, Philippines
Charles Petranek	Indiana State University, USA
Walter Rohn	Deutsche Planspiel Zentrale, FRG
Danny Saunders	Polytechnic of Wales, UK
Leopoldo Schapira	Universidad Nacional de Cordoba, Argentine
Richard Switalski	University of Warsaw, Poland
Richard Teach	College of Management, Atlanta, USA

Programme Committee

Jan Klabbers, chairman
Willem Scheper, secretary
Hotze de Jong, technical and logistic management
Catja Klabbers, computer support

Organizing Committee
Cees Takkenberg, chairman
Cay van Bremen, congress manager
Cor van Dijkum, treasurer
Jacqueline Dijns, secretary
Marianne Dierckx, catering
Hans Oosthoek, adviser

Address ISAGA 88
Department of Gamma-Informatics
Faculty of Social Sciences
Heidelberglaan 1
P.O. Box 80140
3508 TC Utrecht – The Netherlands

Support
Special gratitude must be expressed to the following organizations for their generous support of ISAGA 88:
- Algemeen Burgerlijk Pensioenfonds
- ASPA BV
- Bergdrukkerij Amersfoort
- BSO/Nederland BV
- Comshare BV
- Coopers & Lybrand Associates
- Hewlett Packard Nederland BV
- IBM Nederland BV
- Mercedes-Benz Nederland BV
- Pergamon Press
- Rijks Universiteit Utrecht
- Special Promotions
- TES Nederland BV
- Verlinden Wezeman, Ernst & Whinney
- Wang Nederland BV.

Preface

During the conference participants have been actively involved in a process of learning-by-doing, that is through participation they have found out for themselves how gaming/simulation may help them in dealing appropriately with complex issues. As this field is multidisciplinary, participants had the opportunity to meet people from very different fields of science and areas of practice. The conference was a forum where games and simulations have been demonstrated in an informal atmosphere during fifteen parallel workshops. Although the issues to be covered during the conference were serious, issues such as management, policy formation and policy making, environmental planning, delivering quality services, education, learning, user-oriented game design, and gambling, they nevertheless have been dealt with in a stimulating, open and playful atmosphere.

In order to run a conference like this we needed various microcomputer configurations. We were very glad to receive equipment from several computer firms like, Hewlett Packard, IBM, Comshare. They enabled us to provide the session convenors and participants with adequate support. Without their assistance it would not have been possible to run a number of workshops on the basis of learning-by-doing.

Finally, we have received too many papers, totalling 600 pages. Consequently we were forced to cut about 250 pages for contract reasons. We have decided to include the abstracts of the papers we have cut in order to show the variety of topics that have been discussed during the conference. In doing so we hope to have produced a well-balanced volume.

Section One

Complexity, Uncertainty and Value Conflicts

Editor: **Jan H. G. Klabbers**

On the improvement of competence

Jan H. G. Klabbers
Utrecht University, The Netherlands

We are living in a world that is rapidly growing more and more complex. Consequently we find ourselves in the position of having to cope with problems that pass our comprehension. In the flux of problems we are confronted with, from a global scale down to the level of local communities and families, it is almost impossible to find events that can be dealt with in isolation. Most contemporary political issues reveal entangled events, processes and consequences that are difficult to understand. The acceleration of technological development and the impact it has on governing industrial societies, on managing large multinational corporations and small medium size enterprises drive many governments and boards of directors to despair because they find themselves entangled in inextricable cobwebs. For example it seems impossible to stop our ecology from deteriorating. It seems impossible to control the costs of health care, while at the same time the quality of service is declining. It seems impossible to tune human resource planning adequately in a rapidly changing technical environment, to find adequate strategies for dealing with crime, and to deal properly with international relations. These issues show us that governments, institutions and corporations are becoming less competent in dealing with complex problems and in coping with high levels of uncertainty. Thus more and more we are becoming aware of our collective incompetence. Incompetence can be the result of inability to do a job or to perform a task properly. It can be a consequence of bad thinking, a result of stupidity or bluntness, or of the way we have been shaping our societies and causing a loss of flexibility and adaptability. Here we touch upon the contrast between ideology and competence respectively the role of political power, stressing single-mindedness, versus the potential role of science, generating variety. But as anyhow problems and issues are becoming increasingly complex, how can we improve our individual and collective competence in steering and self-steering our societies, organizations and institutions? Can we learn a lesson from our mistakes? Can we become better educated instead of more educated?

What has gaming-simulation to offer, and who might gain from its utilization?

We think of several answers, each of them highlighting various characteristics of gaming theory.

From a practitioner's point of view, gaming and simulation have proved to be a powerful combination of methods and ideas in dealing with complex and unique issues and with value conflicts between various parties (stakeholders). Gaming-simulation provides a language for combining the social-human domain with the physical; technological and economic domains and provides a shared language for communication between the social and natural sciences. With the utilization of gaming-simulation we can improve our abilities to deal with multiple realities, any of them being legitimate from the point of view of each stakeholder (actor). Through gaming-simulation we can learn to look for just solutions to complex problems without destroying their variety, i.e. we can learn to converge on solutions through shared knowledge and a will to understand and act.

Epistemologically, gaming-simulation refers to systems, to structure, patterns, and form, to states and processes, to order and chaos. When applied to social systems, gaming theorists should take two coupled perspectives into account. Both are crucial in understanding the double nature of the theory, which is puzzling many scientists who are not accustomed to dealing with more than one perspective. The first perspective is the spectator's standpoint. It is the position taken from outside the social system involved. The second one is the insider's, participant's perspective. The outsider/spectator is trying to understand regularities in the ongoing processes within the system. Insiders/participants mutually give meaning to their interactions. In doing so they develop and use a shared language and shape order. Outsiders lack this shared language. What might look chaotic from the outsider's viewpoint, who may not understand the signs that are being used, may be very transparent for the insider. As gaming-simulations are viewed as models for social systems these notions apply to human organizations as well. Regularities, patterns and structure detected by spectators of a gaming-simulation may be of quite different order than the meaning of collective structure enacted by the participants. Both the inside and outside positions have to be embedded in gaming theory, otherwise basic features of gaming-simulations are being left out of consideration. Consequently our understanding of social systems suffers, decreasing our competence in dealing with them.

Elsewhere I have stated that gaming and simulation belong to two different academic cultures (Klabbers, 1987). Simulation being predominantly concerned with description of general characteristics and ultimately control of reality, fits well into the positivist philosophy of the natural sciences. Gaming is more receptive to making sense of reality and to meaning processing between human beings. It has more hermeneutic (interpreting) characteristics. Simulation allures the gaming theorist to the outsider's position, while gaming attracts more the insider's position i.e. "knowing-in action". Apparently, reflections on improvement of competence in coping with complex social systems will have to take into account both perspectives, either implicitly or explicitly. The keynote speakers reflect on the state-of-the-art of simulation and management of complex social systems.

Jan Berting reviewing sociological theories in "Structures, actors and choices", points out that already within the realm of the social sciences there are four academic cultures instead of two. He notices two major oppositions in the social sciences. The first opposition is between a rationalist conception of (social) science and a historicist conception. The second opposition is between explanations which are based on methodological individualism and those which take a structuralist point of view. The resulting two by two matrix generates four academic cultures. Berting describes them vividly. He looks for possibilities for bridging the rifts between those positions and aims at solving some of the problems of the social sciences and humanities via the field of gaming and simulation. He evaluates opportunities provided by gaming and simulation for solving problems connected with the longstanding opposition between social sciences and policy-oriented or applied sciences. Berting favours process models "in which the interaction between observers, policy makers and the interest groups plays a pivotal role". This is not only an epistemological statement, it is an ethical statement as well, for "in a democractic society this (interaction) is an antidote to the tendency to use human beings as things".

Frances Mautner-Markhof discusses general issues of management and simulation of complex systems. She reflects among others on creation of order out of chaos, stability requirements, innovation, the role of negotiations, limitations of knowledge of complex systems. She considers simulation as the third reality. Although her theoretical viewpoint presumes an outsider's perspective, she keeps an open eye for the insider's position when she talks about the role of negotiations. Accordingly, she takes a boundary position of switching between both perspectives searching for "the right standpoint for seeing and judging the behaviour of complex systems, as well as for their management and simulation". She warns that "the simplification process which inevitably occurs in modelling and analysis may be factoring out the essence of complexity – and thus of the system itself". This is especially important when taking a spectator's perspective. In this view "complexity refers to information that we lack, but need to know – that is, it is related to what we observe when we do not have sufficient information to see the underlying patterns". What seems to be chaos from one point of view, may become order when changing position. Complexity seems to be in the beholder, who does not understand to what extent those complex systems show a sensitive dependence on initial conditions. – Self-organization – that is, creating order out of disorder, implies selective and coherent amplification of certain random fluctuations leading to the system's evolution to a higher level of organization and dynamic stability. Because of this shift from organization to self-organization Mautner-Markhof moves from a spectator's position to a boundary position, balancing between an outsider's and insider's perspective. She bases herself on Prigogine's work when stating that complex systems generate internal disorder (viewed from the outside, maybe experienced from the inside), disorder which is to be balanced through the introduction of order via a constant influx of energy, matter and information. This enacted influx is the result of internal communication, control and feedback, co-operation, competition i.e. dynamic interactions among its component parts. When considering social systems, it is obvious that the central component parts are (aggregate) actors – that is, individuals, groups, organizations, institutions, nation-states. From insider's point of view, actors generate order out of chaos and show accordingly self-organizing capacities. From the boundary position it is possible both to understand this internal process of creating order, and at the same time to be able externally to detect regularities during those phases that the system is in a metastable state. During shifts from one metastable state to another, a spectator expresses his lack of information and knowledge by stating that the respective system is chaotic.

Mautner-Markhof pays much attention to dynamic stability and co-operative change through reliability, sufficiency and confidence in communication and the role of negotiations to achieve these goals. She mentions that negotiations are essential mechanisms for maintaining dynamic stability. Support systems like gaming-simulations, based on information technology, may play a key role in "achieving the optimum balance between systemic options and constraints". Finally she discusses simulation and gaming as a third reality for coping with complexity.

Seki reviews global modelling from a Japanese political science point of view. He describes the development of concepts used to model the world system and international relations starting with Guetzkow's Internation Simulation. He mentions that post World War II Japanese studies of the world system were totally oriented to self-criticism of the pre-war Japanese militarism. On the basis of Guetzkow's model a simple simulation model had been developed, which was different from simulations conducted in the US during the 1960s. Seki distances himself from the American abstract scientific approach, "which had no correspondence with any transformation of the given International System". The weak infrastructure of social science departments during the sixties, seventies and eighties has hampered progress in the field of Japanese global modelling and simulation. The development of an adequate database for international relations was extremely limited as well. Seki stresses that the post-behavioural revolution and the related positivist attitude of American trained Japanese political scientists were sometimes contradictory to the Japanese peace culture. Here he stirs a tender chord. He points out that political science, international relations and peace studies are intrinsically difficult to be explored from what I have called a spectator's perspective. Seki states that the "subjective and autonomous human nature should become the central core of the discipline". Does he favour an insider's perspective?

Seki proposes the building of an infrastructure of computer networks on a global scale for exploring alternatives to solve the global problematique. He suggests the use of culture bound expert systems as interfaces between personal computers.

Rapid development of microcomputers and related software has spurred the use of computers in daily life and opened a new era of game culture. Game software is being developed in Japan to generate very successful territorial competitive games. Seki proposes reviving the early international simulation studies using the international network as a gaming facility for intercultural exchange.

Finally, Kalff addresses an experimental programme in Shell for supporting strategic decision making. Insiders (general managers) develop jointly an image of their social system via modelbuilding and simulation. In doing so the outsider's position often taken by planners is abandoned, for managers become true planners, facilitators of the actual decision-making processes. Strategic problems nowadays have become wicked. In terms of Mautner-Markhof, managers face disorder and they have to create dynamic stability and improve self-organizing capabilities within their organization. Kalff mentions seven important characteristics of the wicked world of large institutions. Subsequently he sketches some implications from his observations on strategic decision making, that are meant to bring forward order via identifying interdependencies, increasing commitments and a wider spectrum of information to select from. The joint building of simulation models is considered very promising indeed for meeting many of the requirements of coping with adverse, turbulent circumstances. A necessary condition for success is a generic language used by the managers for mutually developing a simulation model, cutting down manager's scarce time investment to acceptable levels.

"Structures, actors and choices"

Jan Berting
Erasmus University, Rotterdam

EXTRACT: Many authors who are working in the domain of gaming and simulation have emphasized that the worlds of gaming and simulation can be regarded as two different cultures which, however, seem to converge in recent times. Confronted with this statement, we have to ask ourselves several questions: What do we understand by "cultures" and which are their main characteristics in the context of gaming and simulation? What is the nature of the differences between the two cultures? Is it possible to supersede these differences? Which are the implications of a convergence: will one of the two cultures be integrated into the other or are there indications for the development of a new paradigm which will replace the "two cultures"?
KEYWORDS: Paradigms, culture, simulation, gaming, choice, constraint, universality, specificity, rationality, historicism.
ADDRESS: Faculty of Social Sciences, Erasmus University, Rotterdam, Burgemeester Oudlaan 50, 3000 DR Rotterdam, The Netherlands.

Games, Simulations and Plays

Looking at the many activities which are going on in the field of gaming and simulation, it is evident that confusion prevails concerning the nature of the "culture of gaming" and the "culture of simulation", and with respect to the meaning of the concepts of game, play and simulation.

Almost 60 years ago, the famous historian Huizinga made a remark in his well-known analysis of the concept of play (game) which still has a modern tinge: "The whole of the category of game (play) remains curiously unstable. This category is very difficult to deliminate, e.g. it is almost impossible to distinguish it from those of myth and of cult" (Huizinga, 1950).

By the way, I would like to remind you that Huizinga mentions "*earnestness*" as one of the most important characteristics of play. Without earnestness, he says, a play is not any longer a play. ". . . because . . . of contradiction, the play must be earnest, to be a play! - Tournament or match, game of chance, game of intelligence or of pure fate, a play of chess and gambling, as types of struggle they have almost everything in common, tremendously deeply rooted as they are in human nature" (Huizinga, 1950).

By playing a game people remove themselves from everyday life. The game creates a world of its own, which fascinates and captivates the players, which are embedded in a specific network of interactions between them. The game, when played, leads to an outcome or "dénouement".

Moreover, Huizinga points out, it is not primarily an imitation but, in the first place, a creation; it shapes what seemed to be shapeless, it is action and drama. In a certain way, every play is a cult, its rules and actions are "sacred".

Finally, its constituents - rhythm, repetition, cadence, closed structure, chord and harmony - create a style, and, on the ethical level, order and faith. Also these fruits ripen in the garden of games. Playing a game presupposes *association*, in the best sense of the word (Huizinga, 1950).

When playing a game, the participants (re)create a social world of their own, with a specific social structure in which their interactions take place according to "sacred" rules (culture). Like on the stage, the playing of a game is carefully delimitated: there is a beginning, a development, and an end after which "normal life" resumes its normal course. But unlike the play in the theatre, the players of a game are *participants*; there is not a formal division of labour between those who are "inside the play" and an audience, as is the normal situation within the theatre.

Moreover, the players do seem to have more freedom as they don't have to follow a specific plot and to abide to the characteristics of their role as described in the scenario.

We will have to say more about the "culture of gaming" later on. Although Huizinga's ideal typical description of game and play may be considered to be inadequate in several respects when dealing with gaming in modern times, it appears from what we have said that it is a good starting point for the discussion about similarities and divergences between gaming and simulation. In contradistinction to gaming, simulation has several characteristics which are – or seem to be – at odds with gaming: it is directed at the representation of complex processes which are taking place in the "real world", it is system-oriented, nomothetic, highly abstract, universalistic and approaches the actors as "trivial machines" (Klabbers, 1988), to mention a few differences. Are those differences indeed related to the "culture of simulation", or will it appear from our analysis that this dichotomy between gaming and simulation is obscuring rather than elucidating? Which are the convergences and divergences that are taking place in the field that is covered by gaming and gaming theory, designing of scenario's, simulation, dramaturgy, sociodrama, decision theory, modelling, development of simulation games, multi-actor simulation, (symbolic) interactionism, theory of social action within general systems theory (Baumgartner, 1978), "figuration models"[1], (Elias, 1969), "les jeux de coopération" (Crozier and Friedberg, 1977) and "l'intervention sociologique" (Touraine, 1978)?

In order to get a grip on this problem I will utilize a heuristic model which I developed in order to elucidate some major problems which are related to the fact that sociology is a multiple-paradigm science. As we will see, this model may also be used to clarify problems which are related to the division of labour between different social sciences. In fact, the model deals with a deep and longstanding rift that can be discerned within the social sciences between two cultures. One culture is connected with a rationalist view on social and cultural life; the other presents a historicist view, emphasizing the role of actors' meanings and of the context-boundedness of all observations.

Moreover, within both cultures researchers are confronted with the opposition between methodological individualist or interactionist approaches and structuralist ones.

With this contribution we try to achieve several goals:
(a) a systematic analysis of the similarities and differences between the "cultures of gaming and of simulation" and a brief discussion of the possibilities to bridge the rifts between them;
(b) an analysis of the opportunities provided by gaming and simulation to solve some of the problems of the social sciences and humanities. To what degree are the solutions which are available in the field of gaming and simulation to bridge the rift between "the two cultures", applicable to the social sciences as such?
(c) an evaluation of the opportunities provided by simulation and gaming be applied to solve the problems connected with the longstanding opposition between social sciences as scientific disciplines and as policy-oriented or applied sciences. This opposition pertains strongly to the divide between the quantitative variable-oriented approach and the qualitative case-oriented (comparative) approach (Ragin, 1987).

The Main Oppositions on Which the Major Paradigmatic Differences in the Social Sciences are Based

In order to achieve our goals we have to make a detour which is inevitable: "reculer pour mieux sauter". Without going into their origin, we start with the observation that the main differences between contemporary paradigms in the social sciences are related to two major oppositions. We will treat these oppositions briefly, as we elaborated on them already elsewhere (Berting, 1979).

The first opposition is between a *rationalist* conception of (social) science and a *historicist* conception, while our second opposition is between explanations which are based on *methodological individualism* and those which take a *structuralist* point of departure (see Scheme I).

The structuralist and methodological individualist approaches may be found both within the rationalist and the historicist conceptions of social and cultural life (see Scheme I).

The *rationalist* conception of science, Bendix says: ". . . seeks to develop methods which will ultimately make the right understanding compelling to all who seek truth. In this view, understanding is the work of reason emancipated from all forms of unreason like emotions and partnership" (Bendix, 1984). The historicist search for the truth asserts ". . . that for all who seek truth the right understanding can be made compelling only for a time – even with the best available methods. Historical conditions change, and that means the facts and their contexts, as well as the scholars with their interest and methods" (Bendix, 1984). In this view understanding remains always conditioned, because the observer and society, subject and object, are aspects of history and undergo change, Bendix states. A logical conclusion, based on this view, is that our knowledge of social life is not cumulative – as is so strongly stressed in the rationalist conception – or that it is only cumulative as far as the principia media do not change.

In opposition to the rationalist conception, in which the social scientist tries to discover the nature of social reality and to reduce the variety of social life's forms to simple basic structures (e.g. causal models) which explain social life, in the historicist conception human beings do not only reflect an essentially unchangeable order, but are able to create new social orders, based on new interpretations of the meaning of social and cultural life. Human beings are not the product of the determining forces of the main orders of complexity, but are "non-trivial machines", real actors with discretionary powers – on an individual and collective level – which transcend the limits of the rationalist determining forces.

Our second opposition – between structuralist and methodological individualist approaches – is considered by many social scientists as representing the deepest gulf separating them. Structuralists assert that all interesting and important social phenomena can be explained by the operation of supra-individual factors, i.e. at a level of analysis involving social structures and/or aggregations of some form (Markowsky, 1987). Structuralist approaches often represent social holistic types of explanation[2].

In contradistinction to the structuralist, the methodological individualist asserts that the most important and interesting social and cultural problems can be explained in terms of the operation of intra-individual factors (e.g. the capacity of rational choice) in relationship with interindividual factors (e.g. social interaction and its unintended consequences or other social phenomena emerging from interaction processes).

This brief presentation of the major oppositions in social sciences has to be considered as instrumental to the discussion of (a) the relationship between culture and behaviour in the context of the major paradigms and of (b) the solutions which have been proposed in order to bridge the rifts between the major oppositions.

The *rationalist* conception of science is related to a perspective on social life in which cultural phenomena (value patterns, images of society, ideologies, individual preferences etc) are primarily treated as *dependent* variables. Differences on the cultural level are explained by variables which are external to human beings as conscious creatures. Human beings are not considered to be relatively autonomous goal-seeking actors. Generally, the representatives of the rationalist approaches (AX/AY) are rather distrustful of what people say they believe. In the rationalist view the search for objective standards, detached from emotion, affection, tradition and interests, means that a person's statements about the world can be trusted (only) if they are submitted to established rules and values deemed legitimate by a professional community[3].

SCHEME 1.1   Major oppositions in social sciences

	A Rationalist Conception of Science ("Nature")	B Historicist Conception of Science ("Culture")
X Methodological individualist approaches/interactions		
Y Structuralist approaches		

Structures, actors and choices 11

In this respect the differences between the structuralists and the methodological individualists within the rationalist conception are important to note. We will give a few examples of both approaches.

(AY) The structuralist approach of Lévi-Strauss is clearly antisubjectivist and antihistoricist, the ideas, belief and other manifestations of the conscience of the social actors not being *as such* legitimate objects for scientific analysis. As Lévi-Strauss states: "La conscience, qu'il soit individuelle ou collective, est trompeuse vis-à-vis d'elle même, par conséquent, si l'on veut atteindre des réalités plus solides, il faut descendre en dessous du niveau de la conscience . . ." (Lévi-Strauss, 1986).

According to him, the different aspects of social life can only be studied by the methods of, and with the help of concepts, which are similar to those employed in linguistics. The social and the cultural world is a world of signs, of different semiotic codes, such as myths, ideologies, customs and marriage rules. This signification of social and cultural phenomena *cannot be found in the consciousness of the actor*, but has to be discovered at another level than that of conscious representations. The hidden codes of the social and cultural world are like the syntax of a language, which structures speech, but is not a set of rules to which a person, while speaking, conforms in a conscious way. The code "determines" the structuration of social and cultural facts. Put differently, the codes or structures are to be regarded as models, constructions of the social scientist, which permit him to understand the nature of social or cultural phenomena (Lévi-Strauss). In this approach men are represented as acting, in an unconscious way, in accordance with covert structures, and their "subjective" interpretation in terms of their values and meanings is rejected as being: ". . . illusoir car appartenant au règne de l'imaginaire, qui est toujours 'méconnaissance' " (Herman, 1983).

In a quite different way, most Marxist approaches do give us a structural interpretation of social life in which the structure of the production relations plays, as we all know, a dominant role. Here we see a strong tendency to present a rather mechanical interpretation of the relationship between objective structures and the consciousness of individuals and classes within these structures: individuals as members of a social class have either a class conscience, an interpretation of social and cultural life which is in accordance with their position in the class structure or, when their interpretation is not in line with their objective position, they are supposed to have a false class consciousness, presumably as a consequence of the manipulation by the dominant class or of the lagging behind of traditional ideas and religion in comparison with the development of the technological and economic forces[4].

In line with their position in scheme I (AY) as rationalist-structuralists, the authors to which we referred in the preceding subparagraph view social life either as a system of "codes" which are enacted by individuals whose own consciousness is considered to be misleading ("trompeuse" as Lévi-Strauss says) or as an organized system that "produces" culture. In the latter case "culture" is seen as a *dependent* variable (Marx, Kerr).

In both cases individuals seem to reflect a given order. They do not play a major role in the development of economic and social life.

(AX) Turning to the rationalist-methodological individualist approaches (AX), we perceive a quite different conception of social life. In the AX-approaches, individuals as *rational* actors play a dominant role. Nevertheless, the AX-approaches tend to treat the cultural level in much the same way as the adherents of the AY-perspective. To elucidate this point we turn to the works of Pareto, which can be regarded as paradigmatic for the rationalist-individualist perspective. Pareto states that in social life two types of behaviour can be discerned. The first type consists of logical acts which are characterized by a correspondence between the relationship means-goals as it exists in the conscience of the actor (subjective relationship) and the relationship means-goals as it can be observed in real life (objective relationship). In those cases where this correspondence is absent, behaviour is considered to be nonlogical. Logical (or instrumental-rational) behaviour is based, according to Pareto, on reason (as is the case when we are acting as good scientists); nonlogical behaviour springs from human emotions and sentiments. Those sentiments are considered to be rather fixed components of the personality structure. The sentiments are related to verbalizations in society (cultural products) to which Pareto refers as "derivations" (quickly changing verbal surface-phenomena in social life) and "residues" (constant elements which can be discerned behind the extreme variety of verbal expressions. See Lévi-Strauss's "codes"). It is the task of the social sciences to *unmask* the nonlogical character of those verbalizations (ideologies, religious expressions in social life) (Pareto, 1968).

Modern variants of this approach tend to neglect the analysis of "nonlogical behaviour" and emphasize the role of human beings as rational actors. So rational choice theory states that men have some choice-making discretion under all circumstances and that they will use rational strategies to reach their goals.

This individualistic approach is closely tied to Enlightenment with its emphasis on individualism, rationalism, universalism and cosmopolitism (Haarscher, 1986). Rational individuals, when confronted with a conflict between a collective interest and an individual interest, will always be guided by their self-interest. This is a universal axiom of the theory. These theorists ". . . combine a concern for individual action with an appreciation for structural constraints that these actors face" (Hechter, 1983).

Although in this perspective the actor's role is emphasized, it nevertheless becomes evident that the freedom of the individual actor is a rather limited one as – given the preferences which guide his actions, and the instrumental rationality on which his actions are based – only one course of action can be considered as "rational", at least from the perspective of the observer. Moreover, considering the variety of cultural institutions, it seems to be rather naïve to reduce all this to "conditions" and "individual preferences", and to neglect the analysis of collective (BY) and individual (BX) interpretations.

(BX) In the historicist-individualist perspective we find many authors who adopt an actor-oriented or phenomenological approach. They are primarily interested in the ways in which social life is "produced" or "created" in daily life. They often take their starting-point in the works of Weber, Mead or Schütz. Among them are those who see as their main task to show what it is like to live in one particular setting to those who inhabit another (and vice versa) (Giddens, 1987; Willis, 1977).

Other scientists devote their attention to the interaction processes in which social life is produced and reproduced, to the ways in which people cope with unstructured social situations (e.g. in symbolic interactionism). In those approaches the perspective(s) of the actor, as a real participant in social life, is dominant. Often this type of analysis is restricted to the micro-level of society, but this need not be so, as follows from Touraine's "théorie de l'action", with its emphasis on the idea that society is the result of (collective) cultural creations and conflicts between social movements (Touraine, 1978).

In this approach social scientists are not seen as detached observers of social movements. The observer is a participant who is helping the leaders of a social movement in their self-analysis of the movement's goals, of its opposition and its place in the societal struggle for control. Touraine's "intervention sociologique" places the observer in a process of intensive interaction with the participants. In this process both parties, the "observers" and the "observed", learn more about the nature of social reality while acting and as such they are being changed. The distinction between the observer and observed tends to fade away[4].

The approaches in BX accentuate human freedom, the opportunities for individual and collective choice, unrestrained by the deterministic ideas of the past or the "restrictions of underlying codes". Behaviour is, in this perspective, primarily "cultural", "symbolic". But while emphasizing human autonomy, it tends to neglect the limitations in social life which are the object of analysis in the AY- and AX-approaches.

(BY) The fourth position in Scheme I, BY, is connected with the idea of the primacy of cultural systems as language, science, philosophy, religion, the fine arts, ethics, law, and the vast derivative systems of applied technology, economics, and politics.

The cultural systems are regarded as consistent wholes which are realized when grounded empirically in the material culture and in the behaviour of the members or bearers of these systems, according to Sorokin (Sorokin, 1986).

Related to this view are those authors who state that societies are primarily to be regarded as moral entities, characterized by a common system of values. The integration of society is considered to be dependent upon a collective conscience, including collective representations, which represents ". . . la plus haute réalité, dans l'ordre intellectuel et moral, que nous puissions connaître par l'observation, j'entends la société", as Durkheim says (Durkheim, 1968).

In structural functionalism, the core is also the integrating function of the cultural system. According to Parsons, the institutionalized patterns of values are collective representations that define the desirable types of social action. These are correlative with the conception of types of social systems by which individuals orient themselves in their capacity as members of society (Parsons, 1968).

Structures, actors and choices

After having presented the four perspectives which are implied by Scheme I, we are in a better position to comment on the meaning of some major differences between them. In the next paragraphs we will make a few remarks with respect to the ways in which social scientists tried to overcome those oppositions.

The following points are relevant for our discussion about gaming and simulation:

Culture

In Scheme I AX and AY are related to a "materialist" conception of culture, and BX and BY to an "idealist" one. In the first conception culture is "produced" by the economic and social order and by "human nature". In the second conception culture is the "informing spirit" of a whole way of life which is manifest over the whole range of activities within a social order. Williams points out that each position implies a broad method. In the idealist approach this is the *illustration* and *clarification* of the "informing spirit", as in national histories of styles of art and kinds of intellectual work, which is manifest in relation with other institutions and activities, the central values and interests of a group. In the materialist approach it is the *exploration* from the known or discoverable character of a general social order to the specific forms taken by its cultural manifestations.

According to Williams a new kind of convergence is coming to the fore in contemporary work, in which "cultural practice" and "cultural production" are being considered as major elements in the constitution of a social order. In this new convergence culture is seen as ". . . the *signifying system* through which necessarily (though among other means) a social order is communicated, reproduced, experienced and explored" (Williams, 1981).

Conceptions of Man

The four perspectives represented in Scheme I correlate with rather different conceptions of man. In the rationalist-methodological individualist perspective man is understood as an individual whose behaviour is primarily determined by his biological nature as conditioned by his social environment or as a rational being, making rational choices which are based on his own interest under social conditioned circumstances.

In Pareto's case both conceptions of man are interlocked. In rational choice theory rational man is primarily represented as part of a system of interacting rational men. In this conception the outcomes of preceding interactions of the participants in the system produce constraining conditions (social interactionism).

In the historist-individualist perspective the conception of man is as a *conscious producer* of social and cultural life, as a being that creates new meanings and values and by doing so changes social life. Moreover, he is represented as a being that does not apply rules in a mechanical way but, on the contrary, is able to find new interpretations, applications and solutions in interaction with other persons (symbolic interactionism).

In the structuralist approaches the conception of man is subordinated to the conception of the totality (structure, system). This means that often the flexibility of the human being to adapt to (the changing) conditions of the system is emphasized, together with the processes which produce conformity to the system (e.g. the role of socialization in structural functionalism (BY). In the AY approaches men are either the bearers of hidden codes which they "enact" without being conscious of doing so, or they are regarded as parts of a social system whose characteristics are derived *from the system as such*.

Universality-specificity

It follows from our presentation that the A-side of Scheme I is associated with universality and the B-side with specificity of social and cultural life, with the context-boundedness of all social phenomena. At the same time, the A-side approaches tend to emphasize determinism or *rigidity* as a general characteristic of social life. The B-side emphasizes flexibility or freedom and the possibility that new types of social life will arise which are not determined by preceding states. While the A-side stresses continuity in social change, the B-side points out that history is characterized by "breaks" which are related – *inter alia* – to new interpretations of social and cultural life (e.g. the rise of Protestantism or of Socialism and their social consequences).

System-oriented – actor-oriented

The X-side of Scheme I is associated with actor-oriented approaches and the Y-side with system-oriented approaches. However, it follows from our analysis that we must be careful to distinguish between type A actor-oriented approaches (e.g. Boudon) and type B actor-oriented approaches (e.g. Touraine). In the same vein, the system-oriented approach of the A-side (e.g. Lévi-Strauss) is far removed from the B-side system-oriented approach (e.g. Parsons).

Observers-participants

It is quite understandable after what we have said that the A-side of Scheme I is linked to approaches in the social sciences in which the perspective of the observer is dominant with respect to the perspective of the participants (the "observed"). The B-side approaches emphasize that it is important to know the "inside view" in order to be able to understand social life (for BX, e.g. symbolic interactionism, for BY, e.g. the specific patterns of culture).

Moreover, the observer's perspective is tied to a variable-oriented approach. In contradistinction to this, the researchers who emphasize the role of the participants' interpretations, values etc. in social life, are inclined to adopt a case-oriented approach. This distinction has, of course, important consequences for the nature of the knowledge produced within both "cultures" (A- and B-sides), although on the methodological level this distinction between "quantitative" and "qualitative" methods may be overcome (Ragin, 1987).

Strategies to Overcome the Rifts Between the Two (or Four) Cultures in the Social Sciences

Many social scientists have been – and still are – engaged in developing strategies to overcome the rifts between the perspectives or "paradigms" which we discussed in the preceding paragraph. In which way is it possible to bridge differences between approaches pertaining to the rigidity of structures, the autonomy of individual and collective actors, in both social and cultural life, and the nature of the outcomes of human choices?

Within the context of this contribution we can only refer very briefly to the different solutions that have been proposed. As we will see, almost all of the solutions refer to the combination of only two of the four different positions which we distinguished in Scheme I.

Solutions proposed to resolve the tension between:

AX-BX: Blalock, confronted with shortcomings of rational choice theory, states that many social situations are "fuzzy". As a consequence of this the danger looms large that individual behaviour that deviates from what is expected by the researcher on the base of his rational-choice model, will be stigmatized as "irrational" (or, in Pareto's term, as nonlogical). He suggests reducing the problem by including a greater number of behaviour alternatives in the model and, moreover, to analyse in depth discrepancies between the "objective" situation and individual behaviour. "Is this (behaviour JB) irrational or a mistake, or simply rational behaviour based on a different set of values than those assumed by the investigator?' Blalock remarks (Blalock, 1984).

Boudon's solution to bridge the AX and BX approaches goes farther than the inclusion-of-variables strategy of Blalock. Boudon conforms, in line with the AX-position, to the conception of man as a rational actor. According to him the universality of this action paradigm is firmly grounded, although it is far from accepted by everybody. For Boudon macro-social phenomena are a function of a set of individual acts. Those acts themselves are dependent upon the situation in which the actors are embedded. This position includes the interpretation of the situation by the participants themselves (BX). In its turn, this situation is a function of macrosocial conditions. Rationality of behaviour can only be judged with respect to the structure of the situation in which the actors find themselves.

Although Boudon tends to neglect the role of nonrational, emotional, affective and traditional elements in human behaviour, his approach offers some promises for the development of a new synthesis between the rationalist and the historicist actor-oriented approaches. Less promising is the relationship between the AX-approaches and the macrosocial and macrocultural approaches (AY/BY). The latter are considered by Boudon as being largely an "échec" (Boudon, 1984).

AY-BX

We have made the remark that the AY-approaches tend to be rather rigid and mechanical in their dealing with processes in social life. This leaves little room for the actor-oriented elements in social life. This relationship between structural determined positions of groups and classes and their types of consciousness is elaborated upon by Therborn who stresses the dialectical character of ideology. Ideologies, he says, subject people but they also qualify them to take up roles, including the role of possible agents of change. Ideologies qualify subjects by telling them, relating them to, and making them recognize what exists, what is good, and what is possible[6]. This dialectical relationship between subjection and qualification emphasizes the role of actors in social life, and as such takes in some elements of the BX-approaches. Nevertheless, it is quite evident that in Therborn's approach the taking up of roles by the actors remains dependent on the objective conditions of their lives (Therborn, 1980).

In these efforts to integrate two approaches, either one of the two is "absorbed" by the other, or the two approaches are regarded as complementary to each other. Finally, several social scientists tried to solve the problems to which we are referring in a more ambitious way: the absorption of the "alien" paradigms into an overarching theory or system (Parsons, Etzioni, Elias and most Marxian authors)[7].

The Culture of Gaming and of Simulation

In our introduction we referred to the two cultures of gaming and of simulation. After our discussion of the main paradigmatic differences in the social sciences and of the efforts to overcome them, we are, I hope, in a better position to discuss the developments in the field of gaming and simulation.

In the literature concerning gaming and simulation the same type of remarks can be observed which abound in the social sciences when the consequences of the present multiple paradigm situation are being discussed. So Klabbers asks, referring to simulation, "Can we stick to formal methods, time-invariant solutions and rational behaviour for well-defined issues, or on the contrary do we have to cope with ill-defined problems for which solutions are context dependent?" (Klabbers, 1987).

Law-Yone states that the structure of a game hides an ideology. He regards as basic requirements of games "simplicity", "fairness" and competition. For simulation the basic requirements are "structural isomorphism" and "reproducibility" (Law-Yone, 1987). These remarks point into the direction of our distinction between the rationalist and the historicist approaches. In fact, Klabbers says that simulation is generally considered to belong to the realm of the so-called hard sciences, while gaming, especially the role-playing variety, is having more affinity with the social sciences (Klabbers, 1987). However, we have seen in our preceding analysis that the social sciences are themselves divided in this respect. Moreover, we find also a lot of references in the gaming and simulation literature which deal with the methodological individualist-structuralist divide, such as differences between actor-oriented approaches in which purposive action of individuals plays a central role and system-oriented approaches in which ". . . causal or power relations are maintained in equilibrium, either stable or dynamic, by some form of autonomous control relationships" (Beniger, 1986). As gaming and simulation also deal with social and cultural life - be it often in a rather detached way - it is no wonder that they have to deal with the same types of fundamental theoretical oppositions as the "traditional" social sciences.

We will now proceed to classify different approaches within the field of gaming and simulation on the basis of Scheme I.

AX (rational-individualist):
1. decision theory and rational choice theory. Here we refer to the abstract types of decision-

theoretical research, in which theorists investigate the logical consequences of different rules of decision making by "ideally rational agents" or explore the mathematical features of different descriptions of rational behaviour (Resnik, 1987).

BX (historicist-individualist):
1. gaming
2. experimental/descriptive decision theory
3. sociodrama (Moreno)
4. "intervention sociologique" (Touraine)

AY (rationalist-structuralist)
1. simulation
2. formal modelling (including models of system development)
3. scenarios

BY (historicist-structuralist):
1. role-playing (theatre)
2. construction of ideal types (Weberian)

We will not try to disentangle all of the problems which are indicated by this classification in this short contribution. It is necessary to continue with painstaking research in this domain in order to find adequate solutions for the problems which are raised concerning (a) the relationships between the four cultures, (b) the relationships between the four cultures and (social) reality or realities to which they may refer; (c) the relationships between gaming and simulation, the social sciences and policy problems which have to be solved within democratic societies.

We will briefly refer to a number of the problems which are related to a, b and c.

The relationships between (a) the "four cultures" and (b) the four cultures and "reality"

Gaming

It follows from our analysis which is based on the major oppositions in the social sciences (Scheme I) that the reconciliation of the "culture of simulation" and the "culture of gaming" is a goal which is extremely difficult to attain - if at all - on a general level. Nevertheless, some opportunities for convergence seem to be present. In the first paragraph we referred to Huizinga's analysis of the concept of play (game) and his conclusion that this whole category "remains curiously unstable". However, looking at our analysis it appears that the concept of play or game refers to activities which fall into different categories of Scheme I. We distinguish between:

1. An approach in which actors can resolve a problem in a rational way by calculating the outcomes of their (inter)actions, given the issue and the conditions of the *play* or, put in a more general way, of the *problem* to be resolved.

    This approach can be carried back to one of the main exponents of Enlightenment, Condorcet, who had the endeavour to rationalize the decisions of the law and in political life by the application of the methods of the exact sciences. He drew the attention to the fact that in voting the logic of the collective decision is not similar to the individual decision (the Paradox of Condorcet)[8].

    This approach emphasizes predictability of outcomes, reproducibility and the universality of decision making (the process is ideally "culture free"). The approach is outcome-oriented and as such a priori. It abstracts from most of the psychological and practical features of a game (Resnik, 1987) while applying mathematical procedures. The participants take the decisions which lead to certain outcomes. Paradoxically, this does not imply that those actors are to be regarded as "non-trivial machines". As rational persons they have to comply to what logics dictates under the given circumstances. When they deviate from what is a logical way of action - as seen by the (outside) observer - these deviations stem from unreason, emotions or nonlogical theories (ideologies). Moreover, an analysis of decision making may show that the actions which are logical from the perspective of the individual actor - there is a logical link in the mind of the actor between the goals to be achieved and his application of means - do have so many unintended "perverse effects" within the *system* of interactions that the outcomes of the interactions are quite different from what they were intended to be.

Structures, actors and choices   17

In both cases it is expected or hoped that the results of the systematic analysis of decision-making processes will result – in line with the ideals of Enlightenment – in a rationalization of actors' behaviour. The analyst instructs the actors how to improve their potentiality for rational decision making, using a systematic body of science-based knowledge (decision theory, game theory).

2. An approach which emphasizes, in contradistinction to the former, the ways in which the participants are being changed, and change themselves in the process of playing. As such this approach is *process-oriented*, not primarily outcome-oriented. The game played by the participants is a virtual world, separated from "real" life as a game, nevertheless related to real life processes because in the process of playing the game the participants may change in such a way that they can cope better with the vicissitudes of social life. The play uses the creative spontaneity of persons, their ability to learn taking the role of others and stimulates cathartic processes. As examples of this we refer to Moreno's psychodrama and sociodrama, where the play is to be regarded as a means to structure social relationships. The play provides the participants with an extensive gamma of opportunities to express their internal life in the dramatization of situations and the expression of emotions. Moreover, the play gives the participants the pleasure of confrontation with unexpected representations or imagery, of incongruent juxtapositions and quite a lot of effects which are related to the fact that the players are, to a certain extent, liberated from the social control of everyday life. The play signifies for the players "le plaisir dérobé au Surmoi" (Kestemberg and Jeammet, 1987). This conception of game (play) is not variable-oriented, but stresses the importance of the play as a figuration created by the players themselves (see Huizinga). The players are not confined a priori to the role of rational actors. In this conception of play the pivotal role is stressed of communication between the participants, of flexibility, self-reflection and self-reference.

3. An approach which states that the best way to analyse social life is to regard it as a game. This approach is exemplified by the analysis of organizations by Crozier and Friedberg. For them, the game is a concrete mechanism ". . . grâce auquel les hommes structurent leurs relations de pouvoir et les régularisent tout en leur laissant – en se laissant – leur liberté" (Crozier and Friedberg, 1977). The game regulates cooperation between the participants within an organization.

The participants are not just adapting to (formal) roles in the organization. Role behaviour is only an extreme case. In most cases the participants have opportunities to choose and they will use these to choose their own behaviour or *their* "role behaviour" in line with the characteristics of the game as they see it, with their own affective, cognitive and cultural capacities. The strategies of the participants vary according to the capacities of the players, the specific figuration of the strategic field and the rules of the game. As such the organization can be regarded as a structured whole of potential strategies pertaining to both the organization's internal and external relationships. The game reconciles constraint and freedom. The player does not have to adopt a fixed role but must adopt, if he intends to win, a rational strategy which is a function of the nature of the game. Moreover, he has to respect the rules of the game, at least for the time being.

Crozier and Friedberg state that the decision theory (our first approach) would be superior to their own if one would have at his disposal all of the necessary information, if there would be no ambiguities, if all participants would have the same values and if all resources would be equally available to them. However, the actors never know exactly what they want and they adopt new goals during the process of decision making itself and, moreover, "Ce sont les pressions, contrepressions et négociations qu'impose l'adjustement mutuel partisan qui les feront sortir dans l'action" (Crozier and Friedberg, 1977).

In contradistinction to the a priori character of the first approach, Crozier and Friedberg state that the rationality of decisions can only be ascertained a posteriori. This position is related to their observation that rationality is not interculturally stable, and that players may succeed in changing the rules of the game.

So this approach seems to be a plea for case-studies, using this game-model as an instrument for this type of analysis. It remains rather close to social reality in comparison to the two approaches which we mentioned earlier. It may be that the results of a series of those case-studies, in which rationality of actions is revealed a posteriori, will contribute to the first approach and vice versa.

The approach of concrete organizations as games may also provide insights into the ways in which parties or participants become prisoners of their own strategies and roles. Organizations may become blocked: the participants and the institutions don't have the capacity to create new systems of exchange between the parts when the conditions are changing (Mucchielli, 1983). The analysis of the blockade may show the opportunities to reinforce the creative capacity of the system and its participants.

4. An approach in which the conception of play leaves the actors, in comparison with the three other approaches, almost no room for decisions of their own. It is the designer of the play who pre-empts almost all of the decisions by fixing the conditions of the play, its structure, roles, plot and its dénouement. The participants or actors are part of the play as an instrument with which the designer (author) tries to communicate with his audience in order to instruct, to shock, to criticize, etc. But as we all know, the actors in a play never comply just mechanically to the prescriptions of their role. They express themselves in their role when enacting it, they have opportunities to negotiate with the stage-manager and the other actors and they respond to cues coming from the audience. This means that some elements of our second conception of play are present. Nevertheless, in this "functionalist" conception of play an important element is the adaptation of the actors to structural exigencies of the play.

Our analysis has revealed four different conceptions or "models" of game/play. The first one (decision) is to be situated in AX (Scheme I), the second and the third (drama and strategy) in BX and the fourth (stage) in BY. Those conceptions of play/game imply different conceptions of man (as we outlined earlier): in the first one man is caught in the iron cage of rationality, in the second man is a free, self organizing unity, in the third he is both free to choose *and* constrained by the fact that he *has* to play the game if he does not want to be a loser and in the fourth he is almost vanishing as an individual, being largely reduced to an adapting entity.

It is interesting to note that all of the four conceptions emphasize the *systematic* character of the game/play, although they differ with respect to the nature of the system: resultant of a set of actions of rational individuals (1), a specific configuration of actors as a creation of themselves (2), a set of individual and collective strategies (3) and a set of roles (4). But although the four models have, on a general level, the idea of system in common, none of them pays attention, in a systematic way, to long-term developments of systems, not to speak to the transformation of systems. The third model (strategy) does not exclude such an analysis, but Crozier and Friedberg do not undertake this task on a theoretical level. The first model (decision) provides a theoretical base for such an analysis. This strategy has been elaborated by Boudon in his *La place du désordre* (Boudon, 1984) and may moreover provide a bridge between 1 and 3.

In order to be able to handle the problem of long-term developments as consequences of human action Elias has suggested to develop "play models" or "didactic models", being mental constructions which may help us to understand the unintended long-term consequences of interactions, especially the changing power balances between interest groups within human figurations (Elias, 1969). This analysis of the unintended long-term consequences as flowing from the ways the players influence each other reciprocally is cognate to the analyses of Boudon and Crozier and Friedberg. The contributions of those authors can be regarded as providing a solid base for the development of games which take into consideration long-term consequences of action and the transformation of action systems.

Simulation

As we already remarked in the first paragraph, simulation is in certain respects quite different from the four models of play/game. Simulation tries to bring together in a dynamic model the main variables which are considered to represent the systemic character of a part of social reality. Simulation has to *re-present* this reality. This signifies that simulation has as an important task the amelioration of the isomorphy between the model and the "real system" by improving the quality of indicators and by being on the guard against the possibility of system alterations (Scheme I, AY). Simulation is in principle related to the real world, not to virtual worlds, it is outcome-oriented, it abstracts from the participants to social life or adopts a construct of man of the AX-type. As such simulation is related to decision theory, which implies that it is feasible to study the effects on the system of the interactions of "rational" man and the effects of system changes on the actions of the participants by using computer simulations, as suggested by

Markowsky. But such solutions leave out almost everything that we said about the BX and BY approaches, especially concerning the self-organizing potentialities of actors and systems. After what has been said about the nature of the 4 cultures it must be possible to adopt types of simulation and decision theory in which self-reference and self-organizing properties of units are accounted for or to introduce types of games which take into account, in a systematic way, the (changing) rigidities of social and cultural life. In this way we can avoid conceptions of man and society in which men are regarded either as "free" units and society as totally malleable or as determined by forces out of human reach (e.g. technological determinism).

Indeed, it is very interesting to note that the strong emphasis on unilinear reductionism and on predictability is growing weaker. Brown speaks in this context of a "crisis in the Newtonian paradigm", while referring to Morin who remarks that if the universe were subject to the predictability of determinism, nothing new could ever exist. Moreover, Brown quotes Prigogine and Stengers who regard living systems as "the supreme expression of the self-organizing processes that occur" (Brown, 1988). This being so, they say, physics must pay special attention to the evolution of these living, autonomous, self-organizing systems existing at far from equilibrium conditions. In the same vein Luhmann rejects all traditional solutions to the problem of social order and states the centrality of "autopoiesis" or the "self-referential" constitution of social systems. Those self-referential systems are able to observe themselves "By using a fundamental distinction schema to delineate their self identities they can direct their own operations toward their self identities" (Luhmann, 1988). Social systems (organizations, societies, interaction systems) are not reducible to their constituent elements, i.e. they are emergent.

Those publications indicate the direction of our research that intends to solve problems which are related to the rifts between the four cultures. On the theoretical level an enormous task is still to be done. This task contains the systematic analysis of the concept of self-reference and self-organization and of the nature of the (possible) relationships between these concepts and the problems which we have raised in our contribution in combination with the conception of culture, of system and of man, with the universality-specificity debate (including the distinction between variable-oriented approaches and case-studies), with the divide between system- and actor-oriented approaches and with the distinction between observers and participants.

Gaming and Simulation: Their Contributions to the Social Sciences and to the Solution of Policy Problems

In the preceding paragraphs we have presented a sketch of the main oppositions we are confronted with in the social sciences and of the efforts to overcome them. This analysis has been used to elucidate some of the problems which come to the fore in the relationships between the culture of gaming and the culture of simulation.

The "interface" between gaming and simulation on the one hand and theoretical developments in the social sciences on the other, offers a promising perspective in which co-operation between the two "cultures" leads to a lot of advantages, both on the theoretical and practical or pragmatic level.

When we have a look at the social sciences it is clear that those sciences, and especially sociology and social or cultural anthropology, are extremely descriptive or "data-oriented" on the one hand or speculative on the theoretical level on the other. Gaming and simulation offer a challenge to those sciences to develop a stronger orientation toward the *scientific* identification of problems and toward the understanding or explanation of the dynamics of social and cultural life. What Belshaw wrote recently about anthropology applies also to sociology: ". . . we have not made nearly enough use of the games of abstract model building and of playing with the logic of alternatives and modifications to the systems embedded in our monographs" (Belshaw, 1988).

In the second place gaming and simulation may open new avenues to the solution of the theoretical problems which we raised in the preceding pages, by applying their methodology to a systematic analysis of the interplay between systemic constraints and choices (e.g. by using computer simulation/gaming). On the other hand, gaming and simulation could take advantage of developments in the social sciences with respect to the relationship between nomethetic, variable-oriented research and case-approaches (Ragin, 1987).

In the third place, related to the first and the second ones, gaming and simulation may help to overcome several of the "nonlogical" effects of the division of labour between the social sciences by enhancing the solution-oriented approach which may stimulate multi- and interdisciplinary collaboration without falling into the trap of (extreme) reductionism.

Finally, by emphasizing the role of actors as "non-trivial machines", gaming and simulation may contribute to the reinforcement of a development in the social sciences in which the role of the observer of social life, using "top-down models", is weakened in favour of an approach in which the "inside view of life" comes (also) to the fore.

We must also be aware that gaming and simulation have something to learn from the social sciences, as is, I hope, demonstrated in the preceding pages. In this connection special attention should be paid to the fact that simulation tend to be *synchronic* and that the models of gaming are inclined to stick to a very short time perspective. The models of social science in which long-term processes have a dominant place could, perhaps, be used for designing games or simulations?

The changes in the domain of simulation and gaming, and in the domain of the social sciences, will have a strong impact on the relationship between social scientists and policy- or decision makers, as follows from what we said in the beginning of this paragraph. Those changes influence the use of top-down or hierarchical models, in which the observer analyses social processes and provides the policy maker with data or "knowledge" which he tends to apply within his own frame of reference.

In many cases this top-down model will have to be replaced by process models in which the *interaction* between observers (scientists), policy makers and the interest groups which are likely to experience the impact of measures to be taken, plays a pivotal role. In a democratic society this is an antidote to the tendency to use human beings as things. Almost 40 years ago, Norbert Wiener, the father of cybernetics, wrote his *The Human Use of Human Beings* (Wiener, 1956). In this book he expressed his deep concern about the possibility that the new communication technologies might be used to control human beings as things, to transform human societies into the model of the ant community (the Fascist Model). Such a development would be, he says, a tremendous loss as ". . . variety and possibility are inherent in the human sensorium – and are indeed the key to man's most noble flights – because variety and possibility belong to the very structure of the human organism" (Wiener, 1956, p.52). Wiener shuddered at the very thought of the use of machines, including bureaucracies, constructed for the making of decisions, which do not possess the power of learning. But he also rejected the idea of a machine which can learn and can make decisions on the basis of its learning, but which is in no way obliged to make decisions "as we should have made them, or will be acceptable to us". His warning, still urgent, is: "The hour is very late, and the choice of good and evil knocks at our door" (Wiener, 1956, p.186). It seems to me that the recent developments in our fields of interest may contribute to the making of the right choice. Such a choice will imply, *forcément*, conflict and struggle.

Notes

1. Elias, N., *Die höfische Gesellschaft*. Neuwied und Berlin: Luchterhand, 1969, pp. 313f.; p.79 ff.; id., *Was ist Sociologie?*. München: Juventa Verlag, 1970, Chapter III. Elias applies his "figuration models" to analyse complex processes of changing interdependencies between actors.
2. Piaget discusses also the non-holistic variant of structuralism (Piaget. J., *Le Structuralisme*. Paris: PUF, 1972 (Que sais-je? no. 1311), p.9–10.)
3. Bannister points out that in American sociology the emergence of objectivism coincided with a growing interest in efficiency, adjustment, and social control. These slogans, he says, ". . . in turn reflected a growing concern with order over freedom, and with how society shapes the individual rather than vice versa" (R. C. Bannister, *Sociology and Scientism. The American Quest for Objectivity*, 1880–1940. Chapel Hill and London: The University of North Carolina Press, 1987, 235–236).
4. We have also to include in Scheme I, AY, the proponents of the industrial convergence thesis, a thesis which is put forward in many variants. See: C. Kerr, *The Future of Industrial Societies. Convergence or Continuing Diversity?*. Harvard University Press, 1983, p.65.
5. This development can be characterized by what Gellner has named the "Pirandello-effect", referring to Pirandello's experiments with the elimination of the separation between stage and public. When the spectators become (inter)actors, the play seems no longer to be "a

spectacle but a predicament". See: E. Gellner, "The Scientific Status of the Social Sciences". *International Social Science Journal*, Vol. 36, no. 102, 1984, ref. 7.
6. G. Therborn, *The Ideology of Power and the Power of Ideology*. London: New Left Books, 1980, p.16-17. The interpretation of the relationship between structural condition, culture and behaviour can be enriched, within the Marxian tradition, by the ideas developed by Weber in the field of the sociology of religion (BX) and by Gramsci's dynamic view on the relationship between dominant conceptions and norms and the ways in which popular culture is negotiated by the subordinate classes. Here the role of "common sense" or the "spontaneous philosophy" of people comes to the fore. (K. Thompson, *Belief and Ideology*. Chichester: Ellis Harwood Ltd/London & New York: Tavistock Publications, 1985, p.101).
7. Some other attempts to bridge are: B. Markowsky, "Toward multilevel sociological theories: simulations of actor and network effects". *Sociological Theory '87*, Vol. 5, 1 (Spring 1987), p.101-117. (AX-AY); P. L. Berger en T. Luckmann, *The Social Construction of Reality: A Treatise in the Sociology of Knowledge*. New York: Anchor Books, 1967. (BX-BY); P.M. Blau, *Exchange and Power in Social Life*. New York, London and Sidney: John Wiley and Sons, Inc., 1964. (AX-BY); P. Singelmann, "Exchange as Symbolic Interactionism: convergences between two theoretical perspectives". *American Sociological Review*, Vol. 37, 1972, p.414-423. (AX-BX); P. L. van den Berghe, "Dialectic and Functionalism: Toward a Theoretical Synthesis", *American Sociological Review*. Oct., 1963, pp. 695-705.
(AY-BY); In a latter stage he rejected both types of theory altogether, and considered sociobiology as a bridge between approaches within AX, while rejecting AY, BX and BY. P. L. van den Berghe, *Man and Society. A Biosocial View*. New York, Oxford, Amsterdam: Elsevier, 1975. id., "Bringing Beasts Back In: Toward a Biosocial Theory of Aggression". *American Sociological Review*, Vol. 39, 6 (Dec. 1974), p.777-788.
R. Dahrendorf, *Class and Class Conflict in an Industrial Society*. London: Routledge & Kegan Paul, 1959 (first German edition: 1957).
8. J. de Condorcet, *Essai sur l'application de l'analyse à la probabilité des décisions rendues à la pluralité des voix*, published in the "Corpus des oeuvres de philosophie en langue francaise": *Condorcet, Sur les élections et autres textes*. Paris: Fayard, 1986, pp.7-177. The position of Condorcet is treated comprehensively in: E. Badinter and R. Badinter, *Condorcet (1743-1794). Un intellectuel en politique*. Paris: Fayard 1988, 1988, esp. chapter IV "Au Service des Lumières" (1777-1785), pp.143-198.

References

Badinter, E. and Badinter, R. 1988. *Condorcet (1743-1794) un intellectuel en politique*. Paris: Fayard.
Bannister, R. C. 1987. *Sociology and Scientism. The American Quest for Objectivity. 1880-1940*. Chapel Hill and London: The University of North Carolina Press.
Baumgartner, T., Burns, T. R. and Devillé, Ph. 1978. Actors, games and systems: the dialectics of social action and system structuring. *In*: Geyer, R. F. and Zouwen, J. van der (1986).
Belshaw, C. 1988. Challenges for the future of social and cultural anthropology. *International Social Science Journal*: 116.
Bendix, R. 1984. *Force, Fate and Freedom. On Historical Sociology*. Berkely, Los Angeles and London: University of California Press.
Beniger, J. R. Control theory and social changes; toward a synthesis of the system and action approaches. 1986. *In*: Geyer, R. F. and Zouwen, J. van der (1986).
Berger, P. L. and Luckmann, T. 1967. *The Social Construction of Reality. A treatise in the Sociology of Knowledge*. New York: Anchor Books.
Berghe, P. L. Van den. 1978. Dialectic and functionalism: Toward a Theoretical Synthesis. American Sociological Review **43**.
id.1975. *Man and Society. A Biosocial View*. New York, Amsterdam: Elsevier.
id. 1974. Bringing Beasts Back In: Toward a Biosocial Theory of Aggression. *American Sociological Review* 39:6.
Berting, J. 1988. Paradigms, Culture and Behaviour. The relevance of the main paradigms in

social sciences to the relationship between culture and behaviour. *Report to Unesco. Paris.* Paris: Unesco.
id. 1979. A Framework for the Discussion of Theoretical and Methodological Problems in the Field of International Comparative Research in the Social Sciences. *In* Berting, J. and Geyer, R. F. (1979).
id. and Geyer, R. F. 1979. *Problems in International Comparative Research in the Social Sciences.* Oxford: Clarendon Press.
Blalock Jr., H. M. 1984. *Basic Dilemmas in the Social Sciences.* Beverly Hills: Sage Publications.
Blau, P. M. 1964. *Exchange and Power in Social Life.* New York: John Wiley and Sons, Inc.
Boudon, R. 1984. *La place du désordre. Critique des théories du changement.* Paris: PUF.
Brown, C. W. 1988. A new interdisciplinary impulse and the anthropology of the 1990's. *International Social Science Journal* **116**.
Crookall, D. *et al.* (ed). 1987. *Simulation-Gaming in the Late 1980s.* Oxford: Pergamon.
Crozier, M. and Friedberg, E. 1977. *L'acteur et le système.* Paris: Editions du Seuil.
Dahrendorf, R. 1959. *Class and Class Conflict in an Industrial Society.* London: Routledge and Kegan Paul.
Durkheim, E. 1968. *Les formes élémentaires de la vie religieuse.* Paris: PUF.
Elias, N. 1969. *Die höfische Gesellschaft.* Berlin: Luchterhand.
Gellner, 1984. The Scientific Status of the Social Sciences. International Social Science Journal **36**: 102.
Geyer, R. F. and Zouwen, J. van der. (eds.) 1986. *Sociocybernetic Paradoxes.* Leiden/London/Boston: Martinus Nijhoff.
Giddens, A. 1987. *Social Theory and Modern Sociology.* Cambridge: Cambridge Polity Press.
Haarscher, G. 1986. *Philosophie des droits de l'homme.* Brussels: Edition de l'Université de Bruxelles.
Hechter, M. 1983. *The Microfoundations of Macrosociology.* Philadelphia: Temple University Press.
Herman, J. 1983. *Les languages de la sociologie.* Paris: PUF.
Huizinga, J. 1950. Over de grenzen van spel en ernst in de cultuur. *In* Huizinga (1950).
id. 1950. *Verzamelde Werken V.* Haarlem: Tjeenk Willink en Zonen.
Kerr, C. 1983. *The Future of Industrial Societies. Convergence or Continuing Diversity?.* Cambridge, Mass.: Harvard University Press.
Kestemberg, E. and Jeammet, P. 1987. *Le Psychodrame Psychoanalytique.* Paris: PUF.
Klabbers, J. H. G. 1987. A User-Oriented Taxonomy of Games and Simulations. *In* Crookall, D. (1987).
Klabbers, J. H. G. 1987. Methodological issues and taxonomies in simulation. *In* Crookall, D. (1987).
Klabbers, J. H. G. 1988. Spelen op Onzekerheid. *In* Klabbers, J. H. G. (1988).
Klabbers, J. H. G (ed.) 1988. *Kennen en organiseren van informatie. Over machines en actoren.* Deventer: Van Loghum Slaterus.
Law-Yone, H. 1987. The production of SPACE: experiments with a marxist game. *In* Crookall, D (1987).
Lévi-Strauss, C. 1986. Interviewed by P. Simonnet. Un anarchiste de droite. *L'Expres.*
id. 1967. *Structural Anthropology.* New York: Anchor Books.
Luhmann, N. 1988. Tautology and Paradox in the Self-Descriptions of Modern Society. *Sociological Theory* **6**:1.
Markowsky, B. 1987. Toward multilevel sociological theories: simulations of actor and network effects. *Sociological Theory* **5**:1.
Mucchielli, A. 1983. *Les jeux des rôles.* Paris: PUF.
Pareto, V. 1968. *Traité de sociologie genérale.* Genève: Librairie Droz.
Parsons, T. 1968. *Societies. Evolutionary and Comparative Perspectives.* Englewood Cliffs, N.J.: Prentice-Hall.
Piaget, J. 1972. *Le Structuralism.* Paris: PUF.
Ragin, Ch. C. 1987. *The Comparative Method. Moving Beyond Qualitative and Quantitative Strategies.* Berkely, Los Angeles, London: University of California Press.
Resnik, M. D. 1987. *Choices.* Minneapolis: University of Minnesota Press.
Singelmann, P. 1964. *Exchange and Power in Social Life.* New York: John Wiley and Sons, Inc.

Sorokin, P. A. 1986. *Sociological Theories Today*. New York: Harper and Row.
Therborn, G. 1980. *The Ideology of Power and the Power of Ideology*. London: New Left Books.
Thompson, K. 1985. *Belief and Ideology*. London, New York: Tavistock Publications.
Touraine, A. 1978. *La voix et le regard*. Paris: Editions du Seuil.
Wiener, W. 1956. *The Human Use of Human Beings. Cybernetics and Society*. New York: Garden City.
Williams, R. 1981. *Culture*. New York: Fontana Paperback.
Willis, P. 1977. *Learning to Labour. How Working Class Kids Get Working Class Jobs*. Westmead, Farnborough Hants: Saxon House.

The reality, management and simulation of complex systems

Frances Mautner-Markhof

International Atomic Energy Agency, Vienna

ABSTRACT: This paper on the reality, management and simulation of complex systems addresses: the characteristics and stability requirements of complex systems, the mechanisms by which complex systems create order out of chaos or fluctuations; co-operation and competition; innovation and control for the management of complex systems: the role of negotiations, information and technology; limitations on the knowledge and analysis of complex systems; simulation as the third reality: its necessity and constraints.
KEYWORDS: Complex systems; self-organization; chaos, pattern, dynamic stability; innovation; negotiations; information; technology; simulations.
ADDRESS: International Atomic Energy Agency, P.O. Box 200, A-1400 Vienna, Austria.

Introduction

I should like to address a topic I believe to be of relevance to the development of simulation and gaming approaches, and to their effectiveness in the understanding and management of complexity. This topic is the reality, management and simulation of complex systems. Real systems are becoming increasingly complex and there are fewer simple or simplifiable systems which are topics of current research or practice. Thus there is both a trend and necessity in research in the social and natural sciences to discover how to meet the need to encompass and deal with more and more of the totality, diversity and unpredictability of complex systems. In this work a multidisciplinary approach is essential – as is a cross-fertilization of relevant research and ideas between the social and natural sciences.

It is now becoming clear in many areas of intellectual activity that there is a need for a new paradigm to deal with complex systems, which is capable of representing better the characteristics and requirements of complexity and of newness or innovation.

Nothing is more important in life than finding the right standpoint for seeing and judging events, and then adhering to it. One point and *only one* yields an integrated view of all phenomena, and only by adhering to that point of view can one avoid inconsistency.

This statement of Clausewitz (1979), and his work, are based on a systemic point of view in which events, processes and environment are interdependent and interactive, and in which their history and time development are essential.

The topic of this paper concerns a new paradigm which could provide a more consistent and coherent framework for guiding, implementing and evaluating not only research, but also actions related to complex systems. This requires trying to find the right standpoint for seeing and judging the behaviour of complex systems, as well as for their management and simulation.

Why should we need to understand how to manage complexity? Not only are systems becoming increasingly complex and decreasingly simplifiable, but also complex human systems are not designed or modelled so much as managed. Disaggregation of complex systems into components which are more accessible for analysis, and subsequent linear superposition of these parts no longer suffice for the adequate and coherent comprehension, representation and management of complex systems. We must deal with – and are dealing with – nonlinear dynamic systems far from equilibrium, consisting of functionally dependent parts and processes. Not only is the whole complex system greater than the sum of its parts, it is usually different, and in critical ways. The point here is that the simplification process which inevitably occurs in modelling and analysis may be factoring out the essence of complexity – and thus of the system itself.

Complexity refers to information that we lack, but need to know – that is, it is related to what we observe when we do not have sufficient information to see the underlying patterns; thus the role of information is critical. For these reasons it is necessary to understand and maintain the essential characteristics of a complex system, so that this order or pattern can be observed to the extent possible.

When dealing with non-linear dynamic systems we must also take into account their chaotic behaviour. Chaos here is taken to mean a sensitive dependence on initial conditions. What has been identified as simple order or as disorder may in fact be neither. Reality is more complex and less "chaotic" or disorderly – in fact there is indeed an order to be found in and achieved through disorder and only through disorder.

Reality is the process of realization – i.e. the event, which becomes "the ultimate unit of natural occurrence" (Whitehead, 1956). The key is the identification of enduring patterns in these events/processes.

The paradigm I am discussing refers to the self-organization of complex systems, which has sometimes been referred to as order through fluctuations (Nicolis and Prigogine, 1977), or order out of chaos or disorder. This paradigm has as its basis a conceptual framework and standpoint for elucidating and explaining how order or pattern emerges and evolves, what endures and why.

It is clear that the evolution of a complex system depends on its capability for adaptation and innovation – i.e. for flexibility and the creation of newness. For managing complex and interdependent social systems, the main mechanisms will depend increasingly on information, technology and negotiations.

Order Out of Chaos and Fluctuations

I should like to discuss in more detail the manner in which order can emerge and endure in a complex system, order here being not static but an enduring dynamic pattern.

In principle, complex systems have an intrinsic capacity to deal with large unpredictable change or surprise. Nonlinear systems far from equilibrium are capable of reacting to certain perturbations or fluctuations by successive reorganizations onto new branches or bifurcations of the system's development, characterized by a higher level of functional and structural order and complexity. Not all systems do this – those which can or do not will not survive.

It has been argued by von Neumann that in a system's development and behaviour there is a kind of threshold complexity below which the system exhibits regular, stable behaviour, but above which entirely new modes of behaviour can appear which are qualitatively very different from the behaviour of the less complex system, and which are precipitated by disturbances or perturbations in the system. This idea is closely related to Prigogine's concept of order out of fluctuations.

I should now like to take these ideas a bit further. This will require discussing briefly some of the key characteristics and requirements of complex systems.

Self-organization, or order out of disorder, takes place essentially through the selective or coherent amplification of certain random fluctuations or disturbances leading to the system's evolution to a new branch of its development associated with a higher level of organization and dynamic stability (Nicolis and Prigogine, 1977). Thus a complex system survives and evolves only through this capacity to create order out of randomness or unpredictability, and to deal with surprise and chaos. Self-organization corresponds to the low-entropy ordered states to and through which nonlinear dynamic systems evolve. These are the kinds of systems which have up to now been either oversimplified to permit analysis and/or quantification, or considered entirely random and intractable in their behaviour. It is such systems with which we are concerned.

Complexity itself is not only an objective condition but also a measure of the information content, as opposed to the transmitted information, of the system. It is related to the number of different ways in which the system could be ordered structurally and functionally, which corresponds to the various possible states of the system.

Complexity is thus related to a lack of knowledge about the system, and decreases in accordance with an increase in knowledge of how the basic components of the system are put together and interact. This corresponds to information which is known about the internal constraints between the system's constituent parts and/or operative principles of interaction. Atlan (1983) has noted that this characteristic is sometimes measured by a redundancy function, which is what knowledge of one part of the system can tell about the other parts of the system and is closely related to mutual information.

Organization is both a state and a dynamic process. These patterned processes are what make the system distinct – the elements of the system are bound together by a certain unity of purpose, environment and function. Self-organization and chaos recognize explicitly the importance of the mutual interaction between a complex system and its environment for dynamic systemic stability.

The process of self- or re-organization implies that a complex system has the potential for transformation of what appears as disorder (which is associated with the information needed but not known about the system, and is produced by stochastic or random processes) at one level of organization into what will appear as functional and/or structural complexity at a different, higher level of organization.

In this way both order and new meaning or newness can be achieved in the process of the system's dealing with instabilities. So we see that, for a complex system, there can be no such thing as static structural stability.

Maintaining Dynamic Stability in a Complex System

A necessary and important characteristic of complex systems is their relative robustness or stability with respect to small changes or pertubations, and their sensitivity and capability of adapting to large disturbances or changes.

A complex system far from equilibrium will correspond to a metastable state of order. A system in this state can be stabilized and maintained only through the introduction of order via a constant influx of energy, matter and information from the environment outside of the system – to balance the production of internal disorder (Nicolis and Prigogine, 1977). These systems require also communication, control and feedback mechanisms to utilize effectively the input from the environment and to combat potential perturbations which could lead to instabilities. Complex systems will develop with time in such a way as to dissipate both order and energy – i.e. they are characterized as dissipative structures with increasing entropy.

In their work, Nicolis and Prigogine (1977) have shown that, for such systems, this higher level of input of order from, and coupling with, its environment helps to dampen the inhomogeneities accompanying fluctuations in the system. So the a priori probability that a fluctuation will arise and attain macroscopic proportions is less likely to be realized in a complex system. Thus, susceptibility to instabilities caused by fluctuations is less, the more complex the system.

Another characteristic means for maintaining dynamic stability – that is, for ensuring the capacity for innovation, flexibility and adaptability – is the mechanism through which an optimal balance between systemic diversity and redundancy, or between heterogeneity and homogeneity is maintained. Diversity in a complex system is associated with the number of possibilities or potential states available to the system, and thus with the amount of information or entropy of the system. This in turn is related to the degree of unpredictability – a system with only one state available has minimum information content and entropy, complete predictability and no capacity for innovation or adaptation.

The survivability of a system is clearly related to its ability to innovate, to create new options or possibilities. The adaptive potential of a complex system is related not only to its total potential for diversity or non-repetitive order, but also to the redundancy of the system (known constraints among parts).

Co-operation and Competition

A complex system is characterized by dynamic interactions among its component parts, including co-operation, competition, communication, feedback and control. Nicolis and Prigogine (1977) note that the interactions of the entities or components of a complex system can exhibit a competitiveness which leads either to the elimination of some of the entities from the system or else to "a sort of dynamic equilibrium . . . enabling the coexistence of widely differing entities". In this way, functional, time-dependent, evolving order replaces static equilibrium. The history of the system can be considered to be made up of successive phases of relatively predictable evolution along a particular branch, separated by periods of instability during which the future of the system is determined by its reaction to unpredictable events which push it onto one branch or another of organization. The key point is that, as a result of the competitive and co-operative interactions of the components of the system, which can lead to successive instabilities, complexity and stability can emerge in a system.

One of the criteria for systemic stability is a dynamically stable balance between the options, diversity or possibilities available to a system, and the constraints or redundancy which characterize and provide static stability to, and usually limit, the system.

At any given time a complex system is to be associated with a certain level of information content or entropy, corresponding to various possible states or options available. In a dynamically stable system, these options cannot be arbitrarily large but must achieve some optimum balance with systemic constraints. Crises are associated with abrupt and/or large changes in the number of possible states or options and/or in the number of constraints. These changes are accompanied by a decrease or loss of control.

Dynamic stability can be achieved by a co-operative change (creation/congruence/limitation) of options of some or all of the main parts of the system and thus of the possible states available to the system, and/or of mutual knowledge or constraints. Patterns, orders or options which are imposed (or are attempted to be imposed) by one component part of the system on the other(s) cannot achieve this result, as there is no common interest or co-operation, and thus no real congruence or *agreed* change to either increase or limit the potential states of the system. Rather, such situations cause potential instability or crises which could lead to systemic breakdown. In this way, co-operation permits a mutually beneficial result not obtainable by one part of the system alone.

Cooperation must be based on reliability, sufficiency and confidence in communication. It is important to establish, through co-operation, such a balance between options and constraints as to ensure that the system does not become destabilized by too much heterogeneity on the one hand nor overly rigid by too many internal constraints on the other. To avoid destabilizing the system it is essential that co-operative modes of functioning and interaction dominate the competitive modes early enough to avoid crisis and prevent a loss of control. These interactions will reach a level where either co-operation generates more co-operation, or competition/crisis drives the system to more competition/crisis.

Innovation and Control for the Management of Complex Systems: The Role of Negotiations, Information and Technology

Innovation and control for the management of complex systems will have to depend increasingly on information, technology and negotiations.

From the original Greek meaning of crisis, which is turning point, we see that complex systems evolve through a succession of crises, turning points or discontinuities. By means of self-organization through a succession of turning/bifurcation points or crises, complex chaotic systems, which exhibit a sensitive dependence on initial conditions, can evolve to states of higher order or organization. Whether, how and what crises arise will depend not only on the specific political, economic, technological and cultural conditions prevailing within the system, but also on the environment in which the system is embedded. More specifically, there are not only political, economic and technological risks but also systemic risks, which are associated with the system's mode of organization, environment and perceptions of and ability to deal with risk.

In general, a system tends more to crisis and instability the more changes occur in a given time, or the shorter the time for a given change in the balance between systemic options and constraints. These changes are exacerbated by a breakdown in co-operative modes and an increase in competitive modes of interaction between the system's main entities. The tendency towards crisis and instability can be mitigated by controlled and agreed changes in the possible options or states available to a system and/or all or some of its parts, by changing constraints to achieve an appropriate new balance with options, and/or by extending the time during which such changes occur.

Negotiations are essential mechanisms for maintaining dynamic stability – through their potential to develop new options and constraints on a co-operative basis and to enhance the system's capability to deal with the implications of uncertainty, unpredictability and conflict, i.e. with risks, crises, and instabilities.

Negotiations are important means for dealing co-operatively with the rapidly increasing complexity, uncertainty and uncontrollability related to change and risk in complex systems, and for developing innovative approaches (in the political, legal, institutional, technological and economic areas) for the associated issues and problems. Effective negotiations require sufficient, regular and reliable transmission of information among the negotiating sides, which in itself represents an important confidence-building measure. Negotiations processes and the resulting agreements must seek to contain sufficient flexibility to deal with unpredictability and change, and must therefore be not only reactive but also anticipatory (Mautner-Markhof, 1988).

Information, information technology and technology generally are the other main mechanisms for achieving the dynamic stability necessary for the adaptation and survival of modern complex systems, for example the systems of international trade and finance, economic systems and sectors and the strategic system. Information and technology are not only the main areas and causes of innovation and change – and the means of their diffusion within complex systems – but also the primary mechanisms of control of the system itself, including policy formulation, implementation and assessment. As such, information and technology – and the negotiations relating to their development, use and control – are becoming the key determinants of the options, constraints and thus dynamic stability of a complex system. For example, this can be seen on the one hand in the expanding freedom in international financial markets associated with the increasing number of financial instruments made possible by information technology, and on the other in the increasing interdependence of and constraints on national economic and financial policies and practices due to the transboundary scope and effects of information technology.

For a complex system to be dynamically stable, it must have appropriate means for information collection, communication, control and feedback. These are a function of the state of the information and technology flow and utilization within the system. Information and technology will also provide the regulatory, communication and feedback mechanisms which a system requires to introduce and utilize effectively the input of order from the environment and to combat potential perturbations which could lead to instabilities.

Information technology and information will be essential for creating, evaluating and implementing options and will predominate in the diffusion and control of both the positive and negative effects of innovation. Information and technology will thus be key factors in achieving (any changes in) the optimum balance between systemic options and constraints.

If we now turn to the implications of innovation, we see that any innovation implies risk, and no innovation also implies risk – and in fact innovation itself usually comes as a response to a crisis or risk – actual or perceived. Therefore, it is essential to study how change and innovation can be guided, the associated risks, and whether or not they are acceptable and/ or controllable. Connected with this is the need to understand the cultural-based differences in perceiving and dealing with risk.

The management of complexity – as well as of crises and risks – depends on understanding, guiding and controlling innovation and change in a complex system. The means for innovation and control increasingly demand access to and command of information, knowledge and technology. Those who have this will be able to influence irreversibly the evolution and survival of complex systems.

Limitations on Knowledge and Analysis

How can effective options and constraints in a complex system be developed co-operatively? How are sufficient, reliable information and knowledge to be obtained? Clearly there are limits to obtaining objective knowledge and information and to their utilization. In this connection it will be useful to discuss briefly the benefits and successes, as well as limitations in the use of analytic methods and models to understand and structure complexity and to evaluate efficiently certain alternatives.

There are many aspects of complex problems, processes and systems which analytical approaches such as classical decision theory are not equipped to handle. Here the role and meaning of rationality and logic is essential.

The model of the decision maker (or group acting as a unit) who lays out goals and uses the precepts of the logic of choice to select the best means to reach these goals is not immediately, if at all, applicable to situations involving two or more decision makers or negotiators with different Objectives . . . when (in these cases) there is nothing in the model that requires them to agree on how to order consequences or probabilities; each may be rational (i.e. consistent) in holding quite divergent views (Majone, 1989).

Simon (1978) has pointed out that limitations on rationality will cause actual behaviour to fall far short of the objective rationality presupposed in the process of maximizing utility in decision making, in at least three ways:

(1) rationality requires a complete knowledge and anticipation of the consequences that will follow on each choice. In fact, knowledge of consequences is always fragmentary; (2) since these consequences lie in the future, imagination must supply the lack of experienced feeling in attaching value to them. But values can only be imperfectly anticipated; (3) rationality requires a choice among all possible alternative behaviours. In actual behaviour only a very few of all these possible alternatives ever come to mind.

Thus complexity limits the use of objective rationality.

Therefore, a distinction can be made between approaches based on assumed strictly rational behaviour on the one hand, and the problem-solving adaptive approaches which recognize the limits to rationality on the other. Knowledge on the applicability and limitations of the various analytical methods and models is necessary to understand and structure problems more effectively and to evaluate more efficiently complex alternatives. The problem-solving approach is more relevant when dealing with systems, processes, events and issues which are highly complex and interdependent, and where there are limits on the capacity to process and disseminate information, even assuming all relevant information can be assembled – especially within the required time.

Complexity is the ultimate constraint. As Gottinger (1983) has pointed out,

> . . . there is one major link missing which is to be considered one of the fundamental properties of any model-building in the large, namely that one must come to grips with the structural constraint of complexity, that is, the information processing limits arising in the control of dynamic systems.

Complexity imposes constraints on the power to take decisions, on computability, selectivity and controllability. This limits proper functioning and thus rationality. The uncertainty inherent in complexity cannot be treated adequately in terms of probabilities only. There are thresholds of complexity beyond which individuals and entities are unable to choose and reveal cognitive limits.

Thus, while the ideal objectives of entities within a system, and of a system itself, may be the achieving of preferred goals, actual objectives in real-world, real-time problem-solving situations (for example, negotiations, crises etc) are usually the attainment of satisfactory or sufficient outcomes. In many cases the actual objective is or has to be the sheer survival of the system in question.

Many factors and facets of real complex systems are not susceptible to quantification in any rigorous or even meaningful sense. Thus, the danger exists of focusing increasingly sharply and with greater analytical and mathematical rigour and computational power on an ever-narrower or insufficient number of aspects of highly complex problems and systems. To simplify or omit unquantifiable or uncontrollable factors, regardless of their importance, in order to deal in a quasi-rational, quantitative, objective manner with complex issues and problems by simplifying the number or nature of parameters to a point where they become susceptible to manipulation and computation is to doom such activity to increasingly less relevance and impact on the actual state of affairs.

The importance of the mutual interaction between a system and its environment has been stressed. Not only are the system and its environment interdependent, but a system is in key respects observer-dependent. Thus, it is essential to identify and elaborate, to the extent possible, the main cultural, sociological, political and other factors which influence the perceptions of and in a system. Mechanisms or methods developed to support negotiations and/or the management of complex systems – including simulations and gaming – should therefore aim to achieve multiuser consistency, e.g. through cross-cultural evaluation by multidisciplinary groups of users from various cultural backgrounds and political/economic environments. In this manner the impacts of perceptions and cultural differences can be dealt with more effectively.

Not only is the perception of a system influenced by cultural, sociopolitical and other observer-dependent factors, but the capacity for innovation and the attitudes towards co-operation/conflict also depend on these context-related factors. So reality and risk in a complex system, as well as the potential for dealing with crises or instabilities through innovation and co-operation, will depend on the perceptions and environments of those involved.

The logic and inexorability of mathematics do not carry over to the cultural, sociological, political or economic domains. Each of these has its own functional equivalent of logic or rationality, but not in the strict mathematical sense. This interior "logic" of a discipline or area – or of a culture – i.e. the rules and procedures by which meaningful statements can be made and tested – must be discerned and utilized, not bypassed. In this way, a more reliable and effective process can be developed for gaining and utilizing information and knowledge concerning a specific system, its parts, processes and patterns.

Concerning logic and rationality Clausewitz (1979) has observed that

If we were to think in purely absolute terms, we would proclaim with inflexible logic that, since the extreme must always be the goal, the greatest effort must always be exerted. Any such pronouncement would be an abstraction and would leave the real world quite unaffected, since subtleties of logic do not motivate the human will. . . . It follows that [conflict] is dependent on the interplay of possibilities and probabilities, of good and bad luck, conditions in which strictly logical reasoning often plays no part at all and is always apt to be a most unsuitable and awkward intellectual tool.

Simulation: The Third Reality – Its Necessity and Constraints

The manner in which the processes and patterns of complex systems are to be represented and understood – that is, the reality of complex systems – also will determine the possibilities and limitations for analysis and simulation. In conditions where it is not possible to know all necessary information and to anticipate all possible circumstances – that is to say in most real situations – analytical methods will be of limited use. They can be most helpful in defining questions which can best or only be answered by simulation – in particular computer simulation. In this way also, information technology will be an essential tool for the management of complex systems and for providing advanced support systems.

Simulation and gaming, broadly defined, represent the forward edge of a much needed *rapprochement* between the natural and social sciences. Their importance and responsibilities are large and growing. As multidisciplinary activities which aim to bridge the gap between theory and practice, they are taking on an increasingly essential role both in research and in practical applications in the political, economic, scientific, technological, security and other areas.

The new role which simultations and gaming – and especially computer-based simulations – are taking on has to do with what may be called the emergence of simulations as a third reality. They are now being relied upon increasingly to lead as well as to test theory, and to replace as well as to aid experiments, e.g. through numerical experiments (Ulmer, Häfele et al., 1987). The basic reason for this is that systems and the reality they comprise are becoming increasingly complex and intractable to conventional methods of theory and experiment, and cannot be simplified sufficiently for traditional analytical approaches without losing the essence of complexity and reality. The research and discoveries on the behaviour of nonlinear dynamic systems, called chaos, is a prime example. Others include computer simulations to test complex technological and other systems for safety and reliability, and to analyse various security problems and policy options.

Simulation deals intimately with complexity – and in fact owes its *raison d'être* to complexity. In taking on its new role as a third reality, it must recognize and accept the responsibility for defining the areas and limits of validity of its methods and approaches, identifying assumptions and conditions explicitly, and including to the maximum extent possible all relevant factors. Operationally, this implies taking specifically into account the dynamic interactions and interdependence of the system with its environment, and the culture- or observer-dependent aspects or perceptions of a system's processes, patterns and history.

Computer systems themselves can and do exhibit strange and unpredictable behaviour. Thus, in the management and simulation of complex systems, the application of computers and the inherent logic they utilize demand a thorough awareness of the strengths, limits and pitfalls of their use and the concommitant approaches, and of (over) reliance on them. The factors and constraints to be taken into account include an understanding of:

– The real effects and implications of chaotic behaviour, that is, of sensitive dependence on initial conditions

The reality, management and simulation 35

- The characteristics and requirements for the dynamic stability of a finite, self-organizing complex system – these must be apparent in and treated adequately by the structure of the simulation
- What it is that does and should endure in such systems (these are usually patterns and processes, not static entities)
- The reality and role of information content and transition – that is, of entropy and communication
- The limits associated with computational complexity
- The time-dependent effects and factors in a complex system
- The role of friction (in the Clausewitzian sense) in a complex system – related to unpredictability, surprise and chaos
- The uses and limitations of classical analytical methods.

Simulation – in functioning as a quasi-third reality, supporting and leading both theory and experiments, developing, testing and evaluating real practises and options – requires information as raw material. More than this, it must have the technology to process information, to search for and identify underlying patterns, and to make more transparent not only the history or past but also the future possibilities and probabilities of a complex system. Information technology can also be used to evaluate the impacts and effectiveness of the use of simulation and gaming. Computer systems can become not only aids for the memory but also for innovation, and thus a partner in discovering order in disorder and in managing complex chaotic systems.

It is understood that nonlinear dynamic complex systems can exhibit all of the attributes of an idealized random process. That is, as Jensen (1987) has noted,

. . . chaotic dynamic systems are mathematical models which "read" initial conditions . . . the unpredictability of chaotic dynamic systems arises from the fact that slight changes in initial conditions correspond to large, unpredictable changes in systemic development.

More generally, and of more operational relevance, if nonlinear models describing the evolution of complex systems can exhibit chaotic behaviour, "then it may not be possible to predict the behaviour of these systems or their responses to external forces, since errors or perturbations will grow exponentially."

The computational limitations of complex dynamic systems place added importance, as well as responsibility and constraints, on the use of simulation and gaming approaches to understand and assess the likely outcomes of the development of a complex system.

This is related to another important characteristic associated with the unpredictability of chaotic dynamic systems – namely that their time evolution or future development is computationally irreducible. In other words,

. . . there is no faster way of finding out how a chaotic system will evolve than to watch its evolution. The dynamical system itself is its own fastest computer. So, because of the complexity and unpredictability of chaos, direct numerical simulations are, and are likely to remain impracticable, although one can compute probabilities for the outcomes of these processes (Jensen, 1987).

Conclusion

Perhaps the main point of this discussion on the reality, management and simulation of complex systems is to gain a clearer idea of what newness, chaos and surprise are – how they impact and limit the representation, computability and dynamic stability of a complex system. From the underlying continuum of possibilities, certain actualities emerge. Newness or innovation arises at those points where actualities correspond to discontinuities in the system's development.

In a way which I hope has become clearer through this presentation, one cannot, strictly speaking, manage innovation. One can, I believe, manage complex systems so that innovation is fostered which can maintain dynamic systemic stability – through the collection, creation, communication and feedback of information; through identifying and maintaining the enduring patterns and processes which constitute the essence of a system, and through co-operative mechanisms such as negotiations to deal with and anticipate risks, crises and instabilities by creating an appropriate/optimal systemic balance between options and constraints, or diversity and redundancy.

Thus, maintaining the vitality and dynamic stability of a complex system means devising those conditions in which sufficient and sufficiently controlled diversity can exist to permit new options to be created, leading to innovation which takes the system further along its development to a higher levels of organization and complexity.

Ultimately, a complex system buys order at the expense of increasing complexity and missing information. It gains dynamic stability through its capability to deal with inherent unpredictability by means of diversity and innovation. And it achieves control and capability to avoid/manage crises and risks through attaining and maintaining an optimal balance between systemic diversity and redundancy, options and constraints.

Just as a complex or chaotic system is its own fastest computer, because there is no faster way of finding out how it will evolve than to watch its evolution, in the same way there is no better way to understand and deal with the reality of a complex system than to let this reality determine, and become apparent through, an approach or paradigm which can encompass the emergence and development of the system's underlying patterns and processes – that is, the essence of complexity.

References

Atlan, H. 1983. Natural complexity and self-creation of meaning. Proc. Symposium on Complexity, Montpelier, France, 9, A–3.

Clausewitz, K. von. 1979. *On War*. Princeton: Princeton University Press. Original work in German: *Vom Kriege*, Ferdinand Dümmler, Berlin, 1832–4 (Ullstein, Frankfurt/M, Berlin, Wien, 1980).
Gottinger, H. W. 1983. *Coping with Complexity*. Dordrecht: D. Reidel Publishing Co.
Jensen, R. V. 1987. Classical Chaos, *American Scientist* **75**:168.
Majone, G. 1989. *Evidence, Arguments and Persuasion in the Policy Process*. New Haven: Yale University Press.
Mautner-Markhof, F. 1988. International Negotiations: Mechanisms for the Management of Complex Systems. *Cooperation and Conflict* **XXIII**:95.
Nicolis, G. and Prigogine, I. 1977; *The Self-Organization of Non-Equilibrium Systems*. New York: Wiley.
Simon, H. A. 1978. *Administrative Behaviour*. 3rd Edn. New York: The Free Press.
Ulmer, K., Häfele, W. and Stegmaier, W. 1987. *Bedingungen der Zukunft*. Stuttgart-Bad Cannstatt: Frommann-Holzboog.
Whitehead, A. N. 1956. *Science and the Modern World*. New York: Macmillan.

Global modelling in Japanese political science

Hiroharu Seki

Ritsumeikan University, Kyoto

ABSTRACT: In this paper on global modelling in Japanese political science the concept of Japanese models of the world system is related to American efforts in this realm, especially the international simulations studies initiated by Harold Guetzkow in the fifties. The first big scale person-computer simulation was exercised in 1969. It focused on the Asian International System in the 1990s. The impact of the post-behavioural revolution of American political science on Japanese studies is described. Subsequently the (necessary) infrastructure for model building is sketched. Advances in computer technology and software in Japan have been beneficial for social science departments, which lacked the necessary infrastructure to carry out international relations studies adequately, especially database construction was hampered considerably.

Finally recent developments of personal computers in Japan and their potential use for global model building are discussed. The higher-order consequences of the innovation of mini- and personal computers has drastically changed the opportunities for global modelling and computer-networking. This is illustrated by some recent examples of Japanese territorial competitive games.

KEYWORDS: Global modelling; world system; person-computer simulation; database construction; post-behavioural revolution; Japanese language processing; game-software; computer-networks; process of gaming.

ADDRESS: Faculty of International Relations, Ritsumeikan University, Kyoto, Japan.

Concept of Models of World System in Japan

Post-World War II Japanese studies of world system was totally oriented to self-criticism of the pre-world war Japanese militarism at least among intellectuals and academicians. The Japanese Peace Constitution which was drafted and initially supported by the American Occupation Authority was the strong basis of this way of thinking by Japanese intellectuals and academic liberals. Later, so-called realist thinkers of international politics who ardently tried to advocate the importance of the US-Japan security relationship against communist threat or the pre-war Japan's trajectory of development appeared and criticized peace-oriented Japanese intellectuals and academic liberals as ideologically naïve and stupidly ignorant of the real characteristics of international politics. According to some realists, Japan has become forcefully a half-sovereign nation state during the occupation period. This half-sovereign characteristic was rather institutionalized by the New Constitution. Article 9 of the Constitution states that the Japanese people forever renounce war as a sovereign right of the nation and the threat or use of force as a means of settling international disputes. This type of criticism against Japanese Constitution always asserted that the origin of the Constitution was the idea by the American Occupation force and occupied Japan had no sovereignty. It is ironical that this criticism against the Peace Constitution was raised when the sovereignty of every nation states was faced with the accelerating crisis of the nuclear arms race and the crisis of global militarization. However, under the US-Japan Security relationship in the era of the first Reagan Administration, Japan's military build-up and US-Japan military co-operation were promoted by the international pressure of the reopening of the cold war or of the collapse of the detente between the US and USSR. It was the Pacific region where Japan was forced to increase conventional marine forces in the intimate collaboration with the US nuclear marine force against the USSR.

In the process of the military build-up of Japan, right-wing intellectuals emphasized the Soviet belligerency and on the other hand realist intellectuals explained the necessity of military build-up in terms of balance of power or in terms of the nuclear deterrence doctorine recently expressed in the elegant forms of MAD (Mutual Assured Destruction) strategy. Support to this argument was step by step expected to increase because specialists of international studies became gradually conservative by the general atmosphere of specialist's culture in the Japan Association of International Relations. However, the original intellectual tendency of international studies strongly inspired by the Atomic bomb experience in Hiroshima and Nagasaki, by the Japanese Constitution, and by the world peace research development, was never dead even after the period of reopening of the cold war (the second cold war) began. Younger generations of peace-oriented intellectuals specialized in international studies continued to consolidate the PSAJ (the Peace Studies Association of Japan, established in 1973) and other peace research infrastructures since the middle of 1960s. Early effort of this movement were directed to person-computer simulation models of international relations slightly different from the original Northwestern model (Guetzkow, 1963).

It was in 1969 that the first big scale person-computer simulation which tried to forecast the Asian International System in the 1990s, was exercised. It was partly based on the Computer Center of University of Tokyo. The model used in this person-computer simulation was a type of model which tried to combine the original Guetzkow-type model with a simple but new trade and loan matrix model modified by an asymmetrical factor of cultural influence. A 2-day's exercise of this simulation clearly illustrated to us that the phenomenological change in the Asian International System in the last phase of this person-computer simulation was extremely vivid in the sense of radical drop of the US power conflict resolution between the North and South Koreans, and continuous conflict in Vietnam. It was in this sense that our person-computer simulation was considerably different from several person-computer simulations conducted in the US in the middle of the 1960s. The latters were more oriented to abstract scientific experiment which has no correspondence with any transformation of the given International System. The total analysis of Japanese exercise was not published except a partial report by the Annual of Japan Political Science Association (1969) in Japanese in 1971. In the US, unhappily, studies of international relations by person-computer simulations have become unpopular because of Richard Chadwick's severe and strict criticism of several exercises with Northwestern models while basing himself on a statistical validation study. Increasing cost of use of large computers including database construction and the impact of post-behavioural revolution of political science to international studies have also influenced the decline of international studies by means of person-computer simulation. In Japan, I also tried to freshen database construction for the model building of person-computer world system simulation. But, that has become also difficult because of the relative weakness of computer software and the increasing costs of use of computers combined with gaming exercises.

Impact of the post-behavioural revolution of American political science has also been strong in the sense that behavioural type of study could only be accepted within the narrow framework of solid analysis such as in the fields of opinion surveys or in the fields of election studies. In case of simulation studies, a simple all-computer world model such as Club of Rome types of simulation was used for forecasting the limits of growth. On the contrary, person-computer simulations for the study of the referent international system were discredited and gaming-type simulations were exercised only for the classroom learning of international relations.

This was the general situation also in Japan in the 1970s and in the 1980s. In spite of the above-mentioned setback of person-computer simulation exercises for the study of the world system, I especially want to bring to attention the marked improvement of the validity of Japanese exercise of person-computer simulation which was conducted in 1969. It should be compared with the validity estimated by Richard Chadwick in his study of Northwestern exercises. See (diagrams 1, 2, 4) in comparison with (diagrams 1, 3, 5).

Infrastructure of Model Building

Generally speaking, development of computer technology in Japan was considerably backward during the 60s, 70s and the former part of the 80s. Backwardness of computer technology in Japan was particularly illustrated in computer terminals, in the 60s, software applications in the 70s and network applications for the social sciences in the 80s. It was for this reason that most of the talented students, who wanted to study computer applications to the analysis of political and international phenomena should have been trained in America. This unfavourable situation still continues among Japanese political science circles as far as computer application curriculum is concerned. Japanese political scientists were too quickly influenced by the trend of post-behavioural revolution in the US. Ironically most of Japanese political scientists still stand to be pre-behavioural although they advocated the necessity of post-behavioural revolution. This situation of political science in Japan is still to be an institutional obstacle to the further advancement of person-computer simulation studies of global system.

In the early days, use of computers by political scientists and international studies specialists imposed considerable high costs. This is still the case because the infrastructure of social science departments is extremely backward except for a very few cases compared with natural science and engineering departments. For many years in the 60s and 70s and in the early 80s, even large-scale computers (IBM or HITAC etc.) were not fitted to input and output of characters of Japanese language. Only present-day personal computers are capable of handling Japanese characters. Then, time sharing computer terminals were not available for political scientists for the same time span in Japanese universities. It implies that the above-mentioned conditions imposed not only high costs for the use of computers for creative purposes, it also forced political scientists to abandon big scale person-computer simulation projects.

In any case, 1969 person-computer simulation exercise, conducted in the Computer Center of University of Tokyo, to forecast the Asian Pacific

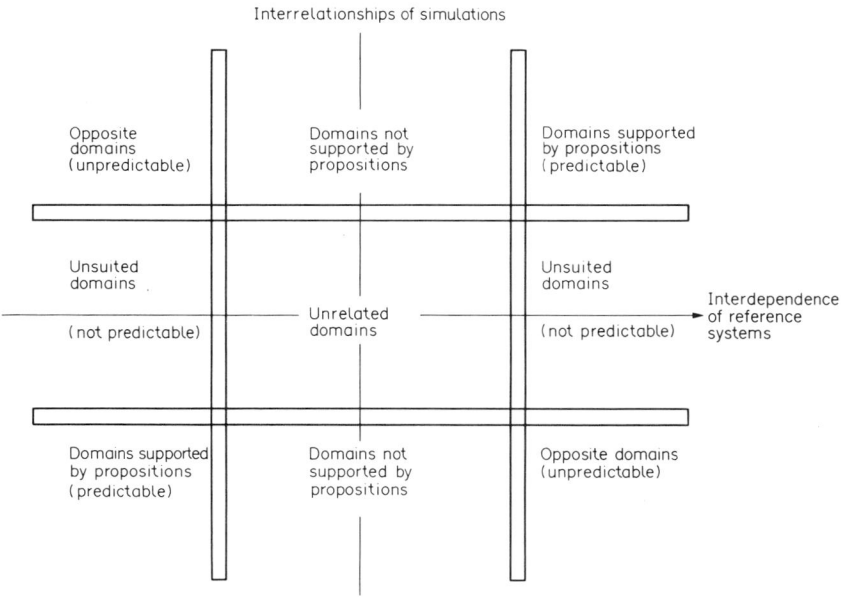

DIAGRAM 1

Correlation Coefficient Matrix
Aggregate Data

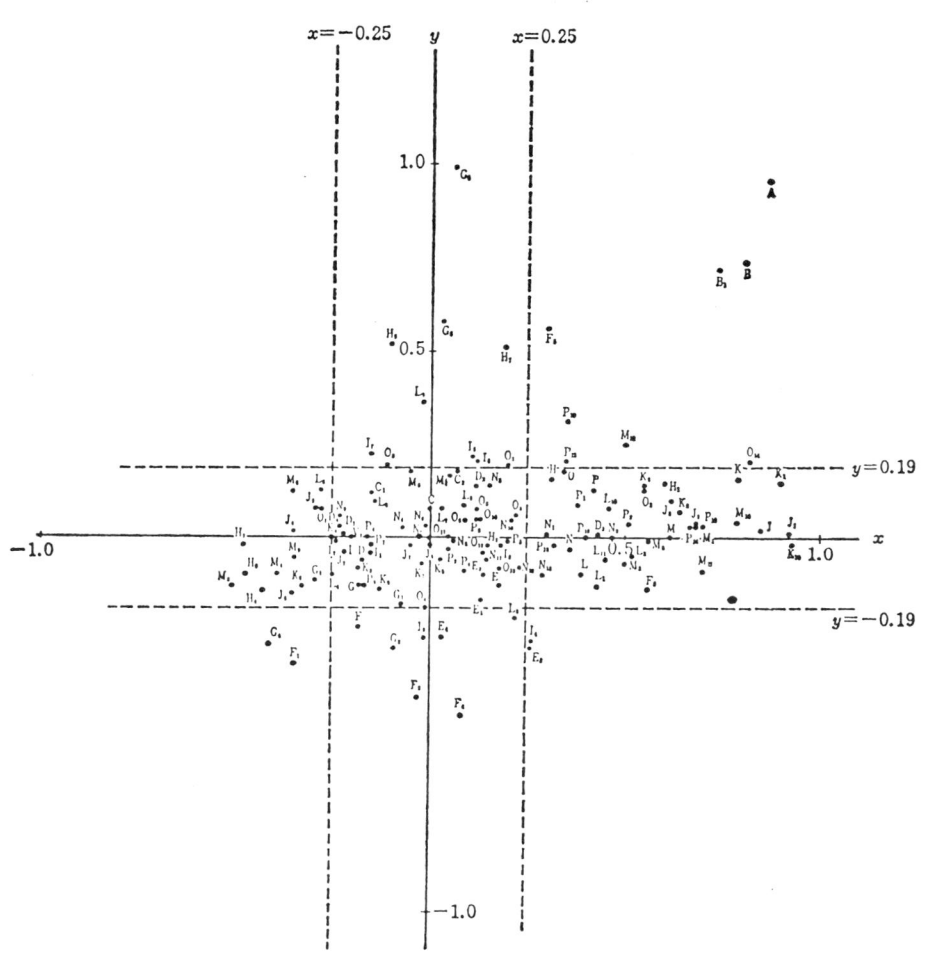

Guetzkow Model
DIAGRAM 2

International System of the 90s, could not be repeated with the further revised model and it was decided to postpone the total project for a while.

Database construction for the international and global system continued. Even in the US it did not develop so easily because of both increasing costs for additional data and weakness of theories involved. Rummel's DON (Dimensionalities of Nations Project) data at University of Hawaii was criticized by a group of specialists of international relations in the US as well as in Europe. Till the end of the 1970s, many scholars involved in this type of study recognized that costs to improve previous efforts for database construction involved more fund raising because of the exponentially escalating nature of

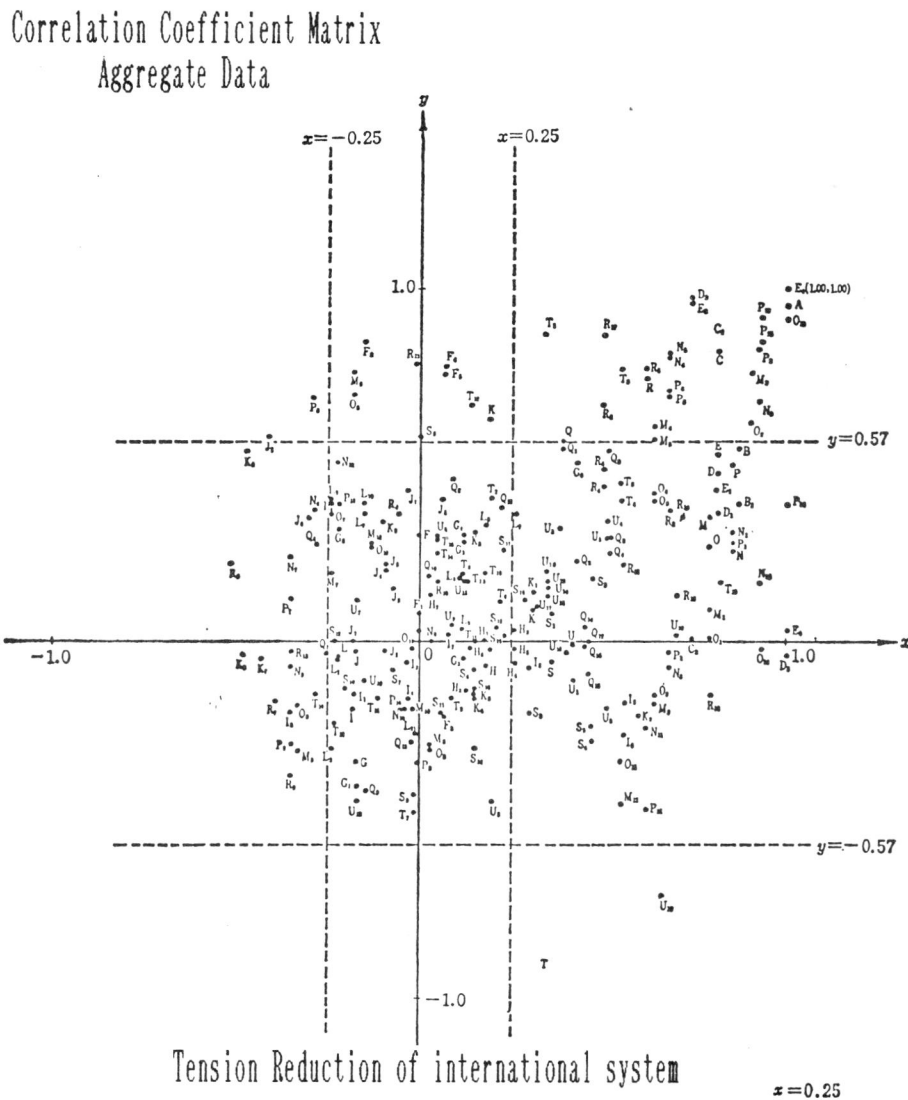

DIAGRAM 3

Table of different interdependencies
in simulations in the Guetzkow model

Number of different interdependencies	Significant interdependencies in reference system		No significance in simulation system	Unsignificant interrelationships in reference system	
	Significance in simulation system			Significance in simulation system	No significance
	Same direction	Opposite direction			
Between 1–10	6	2	8	13	16
Between 1–10, 11–17	0	0	33	4	33 — 56
Between 11–17	4	0	10	0	7

DIAGRAM 4

Table of different interdependencies
in simulations of tension reduction of international system

Number of different interdependencies	Significant interdependencies in reference system		No significance in simulation system	Unsignificant interrelationships in reference system	
	Significance in simulation system			Significance in simulation system	No significance
	Same direction	Opposite direction			
Between 1–13	6	1	25	4	42
Between 1–13, 14–22	16	2	54	3	24 — 76
Between 14–22	4	1	19	2	10

DIAGRAM 5

costs of further data accumulation. Even in this situation, survey research or election studies were easily conducted. So-called SPSS or other types of statistical package programmes were quite useful for the social scientists. However development of a database in the field of international relations was extremely limited and any effort to develop big scale simulation models in this field of international studies and global studies did not start again. The only exception is the all computer global simulation, called GLOBUS.

Most of Japanese political scientists or specialists of international relations capable of using computers were influenced by the American culture, because they stayed and studied in the US for many years. Their way of thinking and their value priorities tended to be Americanized particularly through the interpretation of international history and the deep impressions caused by Hiroshima and Nagasaki were dismissed. Their attitudes to the Peace Constitution or to peace studies was sometimes negative. Their empirical basis and methodology is also sometimes contradictory to peace culture. Some of them even criticized peace studies as subjectively biased regarding criterion of strict objectivity. But the fact that the survival of total humankind has become the first priority among peace researchers and among some of global thinkers is strictly objective. It is subjectively recognized as well.

We must recognize that such disciplines as political science, international relations, peace studies and theories of states face intrinsic difficulties as subjective and autonomous human nature should become their central core. Scientific empirical studies tend to ignore this core value within the theory because of the fundamental weakness of the theory itself. We must again recognize that the subjective, autonomous and creative nature of human personality is becoming more and more important once we pick up political, economic, technological and scientific leaders as future planners of the global transformation for human survival. It is almost impossible to deny that their subjective consciousness, imagination and creativity could surely transform the present international system to a global developmental peace system. Otherwise, the total human history would be very difficult to understand and we might end in the previously unimaginable terrible catastrophe.

In the development of political science in the US, the empirical basis of methodology has become very strong. It is surely appreciated as characteristic for American Political Science. However, as Professor Ricci's recent work named as "The Tragedy of Political Science" points out, empirical science in the study of politics was a failure because it could not identify the core part of truth of political reality in the US. Studies did not mention any aspect of the global system. Once we pick up the global problematique, focus on it and think about how to solve it, political science is becoming more realistic and true. It is now the most adequate time to do our utmost for increasing variety of global modelling in exploring alternatives for solving the global problematique. The most important and imminent task for this area of research is unquestionably building and maintaining infrastructure of computer networks of global scale for this specific purpose. In due process, it simultaneously would be realized that a better atmosphere for the revival of person-computer simulation focused on the study of the global system would become more feasible. Because of the global nature of such a newly formed network, it could become a potential source of culture-bound expert systems enhancing communication and learning among different cultures. It is an ironical truth that Gorbachev's PERESTROIKA has been highly appreciated by American intellectuals as extraordinarily creative to be repeated among other cultures. Thinking about the risk of accidental nuclear war, dialogue between Reagan and Gorbachev for true arms control and arms reduction should have been started. Once a global network could be formed for the study of person-computer simulation of the global system, the highest level of infrastructure would be created through which better learning and take-off for mutual integrated-global development would be sincerely explored.

Development of Personal Computers in Japan and its Potential Use for Global Model Building

Since the beginning of the 1980s, rapid development and innovation of mini – and personal – computers drastically changed the world of users of computers. This revolutionary process is still going on by the increasing numbers of common persons' familiarity with such computers. Of course it is restricted to word processing, game play or data processing for daily life with which layman can become easily familiar. However, this situation of the new computer world should not be neglected when we want to construct the world of application of computer-networks. In Japan, such revolutionary change was particularly silent in the field of Japanese language processing. Development and innovation of softwares for Japanese language processing were really revolutionary because previous images in this world was restricted to processing English or Romanized characters of Japanese by traditional typewriters. On the other hand, various types of computer games, mostly produced in the end of 1970s, opened a new era of game culture particularly for children. This was illustrated by the sudden rise of popularity with playing so-called "Invader Games". We should not define such tendency simply as militarization of the children's world because efforts to develop more sophisticated varieties of games were step-by-step successful in the software enterprises. Intellectual youth including peace-oriented intellectuals have become interested in playing such sophisticated games. University students, or even young university professors were also included in this category.

One of the pioneering game-software companies called "KO-EI" was extremely successful to produce and develop the warlord territorial competitive games. The most eminent one was called "Nobunaga's Ambition" for which English language edition was also published. The original English language edition of the "Nobunaga's Ambition" was designed as a competitive game among 17 warlords of 16th-century Japan whose territory was set in the middle part of Japan between Osaka and Niigata. Characteristics of this type of restriction, fictitious enough to play as a simple game than reality displays, is already well known among skilful specialists of person-computer simulation. Validation studies of Northwestern type of models which were conducted in the middle of the 1960s became sometimes very difficult to compare with the referent system by this fictitious character of territory itself. KO-EI's later edition of "Nobunaga's Ambition" was successful in

Global modelling in Japanese political science **47**

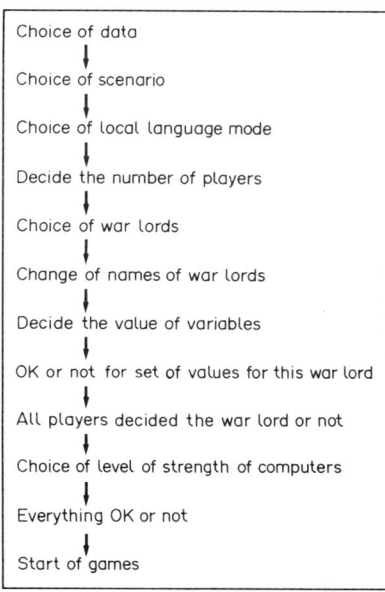

DIAGRAM 6

overcoming this restriction because it set 50 warlords of the 16th century within total Japan. The game itself set the broad repertory of variables such as age, health, ambition, fate, charming character for leadership and I.Q. of each individual or warlord as well as money, rice production, loan, control of towns, rice for military use, irrigation, loyalty of people, wealth of people, number of military soldiers, loyalty of military, military training, equipment for each soldier etc. Players can set the scenario such as choice of loading data for their own characters, choose to fight or not, use of local language, computer level of strength of opponent players, number of players, choices of players to take a role, change of names of warlords etc. through each phase of HEX displays. This process is summarized in the diagram 6.

Many Japanese people generally like to read historical biographies of eminent warlords or, otherwise, want to see TV dramas and video editions based on such historical narratives. It is not denied that popularity with such stories was a result of the conservative intellectual atmosphere to which revival of historicism in post-war Japan greatly contributed. "Nobunaga's Ambition" received the unexpected popularity and the KO-EI Company has become one of the most successful venture businesses. Range of variables in "Nobunaga's Ambition" is far broader than the "Balance of Power" game produced by Apple-Macintosh while the historical background of their scenarios are entirely different.

The process of gaming in "Nobunaga's Ambition" has been characterized by the computerized seasonal change such as spring, summer, autumn and winter. Sometimes, natural or social disasters such as a typhoon or an epidemic, once they happen, will become the obstacles to fighting capability of each warlords. Revolution, *coup d'état*, people's uprising or the occupation of their own land by other warlords would be the serious source of their power devaluation. It could be partly compensated by the use of money in the form of loans from feudal age merchants or by the armed force if they win. Commands which players could choose are movement of force; war decision; change of rice tax; development of rice field; marriage relations among warlord's family; financial support by merchants; employment of soldiers and spies; training of soldiers; information gathering of other warlords; construction of towns and castles; betrayal by people; change of ratio of soldiers within one warlord system; rest of soldiers for a short period etc. Various maps which restricts the space of fighting, decisions by dependent warlords and waiting time for decisions are also significant factors for the process of gaming/composition of each battalion of soldiers could also be decided via keyboard of the computer. Composition of battalions among a gun-equipped part, riders and foot-soldiers show a quite different fighting capability depending on the opponent's situation and the environment. Geographical factors such as low mountains, high mountains, rivers, lakes, oceans, towns, and castles are also significant for the result of each battle. Decisions on military movements, on active attack or on defensive posture, betrayal, surrender and subsequent observation of the opponent are also significant for the short-time future development of gaming scenarios.

"The Nobunaga's Ambition" was the most successful game software available for NEC-PC-9800 series of personal computers. Nothing is compared with this software as the numbers of copies already sold in the market of Japan. KO-EI also produced a similar type of game-software such as "The Romance of the Three Kingdoms" which is based on the scenario of ancient Chinese History and "White Wolves and Blue Danube" which is based on the scenario of the formation of the Mongol Empire. However the idea to develop the next generation peace game is already in full swing by the company's planning team. KO-EI is already successful in producing an entirely new game, "Storm of the Meiji Restoration". This is the most interesting historical peace game ever produced in the world. The game scenario is such that about 500 nationalist leaders compete against each other by way of peaceful persuasion and by other violent methods to unite feudal Japan of Tokugawa Shogunate. If peaceful persuasion would be totally successful, Meiji Restoration could emerge. In the theoretical framework of N-generation models of person-computer simulation, as originally defined by Professor Harold Guetzkow, "Storm of the Meiji Restoration" should be called an entirely new generation game software different from previous types of war games or balance of power games. It is also different even from a revised form of the neo-realist type of B-P (Balance of Power) game model.

Tron Concept of Computer System and Human Centred Global Simulation

The idea of the "Storm of the Meiji Restoration" could be easily applied to the new idea of developing "Storm of the Global Politics" or "Storm of the Global Transformation" For the past 25 years, Northwestern's origin of international simulation step-by-step moved to the direction of all-computer global simulation symbolized by the "GLOBUS". The new direction is surely based on the contemporary technological level of computer system. In this framework of global simulation, validity could not be improved to over 50%. This is formally mentioned by Professor Harold Guetzkow in his memorial lecture at Ritsumeikan University in 1987. Then, why is the remaining 50% factor difficult to be tested? Remaining factors are generally considered as human factors which are difficult to be measured, and sometimes to be marginal development factors. In the present global developmental situation, reform of universities and think tanks or development of global telecommunication networks and global leadership ideas would be such marginal factors of global modelling. If we could be successful in taking up these marginal factors step-by-step, remaining marginal factors will decrease in the order at $1/2 + (1/2)(1/2) + (1/2)(1/2)(1/2) + \ldots$ They could be stepwise validated.

Perspectives for global simulation now faces a new turning point in which the fifth generation computer or TRON concept of computer system would be applied to the development of person-computer simulation system. The concept of TRON (The Real-Time Operation System Nucleus) is recently proposed by Professor Ken Sakamura, a young computer scientist specialized in the so-called fifth generation computer. According to Professor Sakamura, TRON is the effort to reorganize computer system as human centred system by which human beings need not adapt to the computer logic irrespective of human character. Moreover, it could also overcome each person's brain capacity by the network formation among many brains. The total network in which TRON takes the role of nucleus could be called the sixth generation computer, one level higher than the level of the fifth generation computer, that is, a model of individual brain. In the middle of the 1960s, Professor Harold Guetzkow tried to construct a person-computer communication network station with which I myself co-operated. The design was such that gaming type simulation and computer type simulation could be technically integrated through many terminals of the big scale central computer CONTROL DATA (CDC 1604) in the Coordinated Science Laboratory at University of Illinois-through time-sharing terminals by this network formation and simultaneously total system indicators could be displayed by immediate calculation of the result of interaction based on the operational procedure programmed in the large computer CDC-1604. (PLATO system).

In 1980s, the person-computer interface system has been greatly improved by the development of personal computers. Why could such research project not be revived? Such a question is now easily stated as many personal computers are far superior than CONTROL DATA in the middle of 1960s with respect to costs of computer use for simulation and gaming as well as capability of software particularly for input and output of any language via any terminal. At present, gaming type of simulations are generally conducted in classroom courses of international relations. Sometimes it is big scale gaming in which over 300 participants play the roles of various global actors. Pre-gaming training, exercise of gaming, and post-gaming phase of analysis of the results are extremely time consuming and sometimes nearly impossible because of the complicated character of the pattern of communication and the mutual language content. This system, once completed, is like the sixth generation computer through which the given situation of the gaming world is simultaneously displayed by various forms of variables and indicators processed in the computers.

References

Bremer, S. A. (ed.) 1987. *The Globus Model, Computer Simulation of Worldwide Political and Economic Developments.* Frankfurt & Boulder: Campus/Westview.
Chadwick, R. W. 1966. *Development in a partial theory of international behavior: A test and extension of Inter-National Simulation Theory.* Ph. D. diss., Northwestern University.
Chadwick, R. W. 1972. Theory development through simulation: A comparison and analysis of associations among variables in an international system and an Inter-Nation Simulation. *International Studies Quarterly* **16**:1.
Guetzkow, H., Chadwick. F. A., Brody, R. A., Noel, R. C. and Snyder, R. C., 1963. *Simulation in International Relations.* Englewood Cliffs: Prentice Hall.
Guetzkow, H. and Valadez, J. J. (eds.) 1981. *Simulated International Processes: Theories and Research in Global Modelling.* Beverly Hills: SAGE.
Guetzkow, H. 1987. Sekai heiwa to simulation kenkyu (Global peace and simulation studies). *Ritsumeikan Kokusai Kenkyu* (The Ritsumeikan Journal of International Studies) 1:1.
Ricci, D. M. 1984. *The Tragedy of Political Science, Politics, Scholarship, and Democracy.* New Haven & London: Yale University Press.
Rummel, R. J. 1972. *The Dimensions of Nations.* Beverly Hills: SAGE.
Sakamura, K. 1987. *TRON karano Koso* (Constructive Idea through TRON). Tokyo: Iwanami Book Co.
Seki, H. 1966. *The Use of PLATO in Inter-Nation Simulation.* Evanston: Northwestern University.
Seki, H. 1967. *Towards an N-generation model of international process simulation theory.* Evanston: Northwestern University.
Seki, H. 1969. *Kokusai Taikei Ron no Kiso* (Foundations of International Systems Theory). Tokyo: University of Tokyo Press.
Seki, H. 1971. *Kokusai Kincho Kanwa no Simulation* (Simulation for Inter-Nation Tension Reduction). Annual of Japanese Political Science Association 1970. Tokyo: Iwanami Book Co.

Strategic decision making and simulation in Shell

D. J. A. Kalff
Shell Nederland B. V.

ABSTRACT: This keynote address describes an experimental programme in Shell designed to involve management teams in the building of simulation models of the strategic decisions they face. First the organizational context is sketched in which the experiments are being conducted. Second some characteristics are outlined of the strategic problem large institutions in general and Shell in particular face. From this outline conclusions are drawn as to the new requirements strategic decision-making processes have to meet. Subsequently the role model building and simulation can play to improve these processes is explored by means of the description of a representative experiment. Concluding remarks are devoted to some of the barriers that have been encountered and to ways to overcome those in future projects.
KEYWORDS: Strategic problems; requirements for strategic decision making; simulation; Shell, system dynamics; management teams.
ADDRESS: Dr. D. J. A Kalff, Manager, Corporate Planning, Shell Nederland B.V. Dept. PL, Hoflein 20, 3032 AC Rotterdam.

The Context of Strategic Decision Making in Shell

As no doubt some of you are aware, Shell is a heavily decentralized organization in which traditionally the country organization, in our jargon the operating company, fulfils a pivotal role. In the recent past we have gone further down this path by the introduction of the business unit concept. It is strongly felt that only by organizing ourselves along those lines can Shell take full advantage of opportunities different markets have to offer, in culturally, politically and economically very different countries. This article of faith is not a grandiose statement for the benefit of the public in general and conferences of learned societies in particular. It is very much reflected in the way strategic decisions are being taken.

The initiative to build a chemical installation, to acquire oil reserves, to revise a marketing policy or to start a new line of business rests with the operating company. Specific investment proposals are checked by our service companies for technical soundness before they are submitted to the shareholders or their representative to acquire support and eventually approval.

Such proposals are part and parcel of the companies strategic plan or, put more dynamically, constitute the outcome of the companies strategic planning process. Without going into any detail you can well imagine that on the basis of some form of business segmentation and definition both demand and the competition are analysed. This gives rise to a number of options either to defend or to enhance the position of the company or business unit. These options are tested, as systematically as possible, against a variety of possible future commercial, technical and political developments. Finally, in case of investments, financial evaluation methods are applied to differentiate between options quantitatively as part of the development of a proposal for the shareholders. Of course such a process is not linear at all, options are being moulded all the time, the business outlook is influenced by tentative financial evaluations, new data regarding markets and technical developments are being injected continuously. Also the process tends to be a blend of a formal planning cycle, strategy development studies involving formal steering and working groups, and a wealth of informal, partially overlapping, circuits.

In line with our decentralized character these blends have different flavours in different operating companies. However, many companies have planners with the mandate and the obligation to enhance the quality of strategic planning. This can take many different forms: planners conduct studies on the competition, organize workshops to generate options, design planning procedures, and are responsible for our annual Group wide 5 year forward business planning cycle. You will appreciate that all these activities gyrate towards senior management and it is indeed obvious that in the operating company the quality of the relationship between the planner and the general manager and his management team is of crucial importance.

Another observation worth making is that we strive to make planning and decision making synonymous. In other words the general manager and his team are the true planners. Consistent with this approach the planner becomes the custodian of the planning system, the guardian of the quality of decision making, and the facilitator of actual decision-making processes.

The Character of Strategic Problems

Building chemical installations, acquiring oil reserves and revising marketing policies commits the company irreversably and has consequences which will be felt for anything between 10 and 30 years. The scale of these commitments is going up continuously both in absolute terms and in terms of proportion of corporate financial resources. The classical example in offshore oil exploration and production is the exponential increase in capital expenditure with waterdepth. Another example is investing in oil refining. A simple crude distiller calls for a capital expenditure of approximately 10 million guilders per 1.000 MT throughput per day. The comparable figure for a more advanced installation, say a catcracker, is Dfl. 80 million, whilst the most advanced installation, the illustrious hydro conversion installation presently under construction in our Pernis refinery requires capital expenditure to the tune of 150 million guilders per 1.000 MT throughput per day.

Strategic decision making 53

At the same time multinational oil companies like any other institution no longer live in a linear world in which the uncertainty in demand poses the relatively simple problem of the correct timing of capacity adjustment. We find great merit in the views of Ackoff (1974), Mason and Mitroff (1981), Rittel (1972) and others who characterize today's problems as wicked. Not necessarily wicked in the perverse sense of being evil but in the sense of resistance to conventional analysis and problem-solving techniques.

Some of the most important characteristics of these problems are:

- *First*. They lack a definite expression. The process of formulation and reformulation will never come to a conclusion.

- Second. The formulation of the problems and the options to solve them are inseparable. Reformulation leads to different options. Equally, further work on options leads to shifts in the problem formulation. An interesting example is the design of a refinery where the basic configuration, the response to the original problem, continuously changes as a result of more detailed studies of different possibilities within the starting configuration.

- *Third*. They have no closure. For ever evolving organization structures are an example in cases, where the implementation of a revised structure already sows the seeds for the next round of restructering.

- *Fourth*. They are interconnected with other problems and opportunities. Shell and other multinational oil companies got involved in mining and trading of coal, partially as a response to the oil supply crisis of the early seventies and partially in anticipation of government policies to reduce dependence on oil. The very growth of a worldwide coal business including its transport and distribution system has forever changed the economics of electricity generation and the trade in high sulphur fuel oil as the main competitive fuel for this industry. The ripple effects go further: high sulphur fuel oil prices are linked to low sulphur fuel oil prices which in turn are found in the price formulae used in many natural gas contracts. The wide availability of coal and the impact on fuel prices is one of the many forces that determine crude oil prices etc., etc.

- *Fifth*. Problems are essentially unique, history provides no guidance. The reason is at least threefold. The type of commitments I discuss are prompted by structural change, are inspired by the anticipation of further change and, almost by definition cause structural change.

- *Sixth*. Problems are moulded by personal and organizational characteristics, loyalties and interests. Clearly Exxon's problem formulation must have been completely different from Shell's in the aftermath of the oil price collapse. After the collapse Exxon embarked on a very radical reorganization of their central offices and world wide organization which led to a very

considerable reduction in manpower. Shell's reaction was both more measured and differentiated. It was considered illogical and in violation of established policy to reduce support for sectors like oil marketing and chemicals that had only to gain from an oil price collapse.
- Finally problems are dynamic in nature, each strategic commitment triggers action by competitors, customers and governments which makes the original problem formulation rapidly obsolete.

The Implications for Strategic Decision Making

First and foremost the quality of the problem formulation becomes of paramount importance. The core of a problem formulation in a corporate setting is a gap perceived by management. A problem does not exist on its own. It is a management creation or construct. In business a gap is frequently expressed in financial terms. Net income after tax, cashflow costs are the most straightforward indicators amongst a bewildering array. If these are considered measures of performance different gaps could be perceived whilst looking at the strategic position of a company or business unit. This is generally expressed in terms such as "market share" and "fixed costs relative to the competition". Obviously gaps are differences and the problem formulation should therefore contain a basis of comparison. Frequently used comparisons are, past and present performance, present and potential performance, and the corporate strategic position in comparison to the competition. A gap could be perceived here and now and could be seen emerging.

Proper formulation of problems is not at all trivial. A literature survey that I conducted yielded anywhere between 20 and 30 quality criteria for problem formulation. Some of those criteria will appear rather unobtrusively in my description of the archetypical experiment.

A second implication is that it becomes crucial to place problems in a context of other problems and opportunities.

The building of an installation, an acquisition or a change in commercial policy designed to resolve a specific problem, say the projected loss of marketshare, could compound other problems or could compromise opportunities. Ideally, such options should also be evaluated in terms of the creation or destruction of other options. To give a specific example, the decision to build an installation generally implies a choice for a particular technology. Given limited human and other resources such a commitment also destroys options to pursue rival technologies with potential to solve other problems. In general terms strategic management can be conceived as the management of strategic options. Options can be created; for example the decision to build a high technology pilot plant will, if all goes well, create the option to build one or more commercially sized plants. Other options can be bought outright; take the purchase of a licence. Equally options can be sold.

Refineries sell the right to process minimum and maximum quantities of crude oil to traders. Moreover options can be kept open albeit most of the time at a price (as in the extension of drilling rights) and options can be left to expire by not putting in bids for exploration acreage. Pushing the analogy to its limits contracts are often little else but an exchange of options, with any imbalance expressed in money.

Following Mason and Mitroff (1981), the third implication for the strategic decision-making processes of the need to increase the size of commitments in an inherently unpredictable world is that the number of people involved in decision making has to be broadened. This becomes unavoidable as a consequence of the identified interdependencies and the need to satisfy, or at least not to alienate, a range of legitimate interests. It is also inescapable as a means to build joint commitment once an option has been selected and to co-ordinate implementation. Another aspect is that this broadening is likely to cut through organizational boundaries. Some cynic has labelled existing organizational divisions of tasks and responsibilities as monuments erected to commemorate the solution to past problems. The requirement to broaden has in turn consequences for the design of the decision-making process. Without special measures broadening implies delays, extension of veto rights, communication problems and dilution, or even abdication, of responsibility.

A fourth and at least in the context of this presentation final implication is the need to base decision making on a wider spectrum of information. Each manager has powerful notions as to what is, what will and what ought to be. They also have strong views on the factors that will determine the success of certain options as well as profound insights in the dynamics of a situation. This calls for the collection, storage and retrieval of data that are very different from data collected to solve non-strategic problems. Moreover, managerial insights and preferences need to be complemented with information generated by corporate staff, and from external sources. Again this has considerable process implications.

It is quite apparent that each of the four implications pose a considerable challenge. In combination the prospects appear to be very bleak indeed.

The requirement is to design a new process involving a larger than usual number of managers and experts with different organizational loyalties who can commit themselves to irreversible decisions on the basis of a rich and joint perception of a problem and its context.

The central thesis of my presentation is that processes designed to enable management teams to build their own simulation models meet many of these requirements and have shown considerable promise in experimental settings.

Facilitation of Strategic Decision Making

Let me now turn to a general description of such an experiment. I hope that you will understand that for reasons of confidentiality I cannot provide you with a real life case. I like to think that it is a measure of our success that I have to be as cautious as I am.

For demonstration purposes I use a general problem in a fictitious project.

The business area that I use in my example is oil-refining and the general problem I have selected is the emergence of so-called source refineries. These are newly built refineries in oil-producing countries, such as Saudi Arabia, that are highly competitive due to the application of the latest technology and the use of cheap gas as a refinery fuel.

The first phase consists of a gradual focusing on a problem or opportunity. During informal discussions it turns out that at least a few managers believe that a strategic commitment of some form or other might be required either to fend off a threat or to capitalize on an opportunity. Generally a few options are already under consideration but there is no pretence that all possibilities have been exhausted. Moreover there is a great deal of uncertainty regarding the consequences in case any of these options would be exercised.

These discussions produce a tentative problem formulation including at least one perceived existing or emerging gap and a few problem boundaries. Inevitably, the question whether to focus on the Pernis refinery, on Shell Nederland or on all Shell companies in NW Europe has to be addressed.

Such a tentative problem formulation is essential to select managers whose units might be effected. More importantly, the group of managers should in combination carry the responsibility to resolve the problem and have the power to act. The selection should be broad enough to explore a range of interdependencies. A second function of the tentative problem formulation is guidance in broadening the information base. In practice this requirement translates into the selection of indispensable staff and external experts both with access to potentially relevant sources.

The project starts in earnest with one and a half hour interviews with managers, staff and experts. Questions are geared to the identification of the full range of existing and/or emerging gaps.

Put in the simplest of terms the interviews are designed to unearth perceptions of what is, what will, and what ought. Some managers might focus on the existing capacity in the European refining industry add the existing Middle East capacity and the new capacity under construction, deduct local demand both in Europe and in the Middle East and calculate a new total overcapacity. Other managers might reason that such an addition of high quality overcapacity will force the early exit of large numbers of weaker players possibly reducing overall capacity in Europe.

The interviewees are subsequently invited to speculate about causes and effects of the identified gaps. Certain interviewees might consider the addition of Middle East refinery capacity as incidental and only prompted by the enormous cashflows available immediately after the oil price increases in 1973. Others might consider that oil-refining is one of the few available

Strategic decision making 57

routes towards industrialization in capital rich countries with small populations and that therefore more investments should be expected. Analysis of effects centre on the impact of the overcapacity on the so-called refiners margin, the difference between the value of the total product package produced by the refinery and the crude oil price. This leads naturally into speculation about the exit of certain competitors. It appears that interviewees lured into this type analysis quickly turn to options available to the company, commercial steps or investments with a potential to close the gap. The interviewees are deliberately discouraged to arrive at conclusions of any kind. Confidentiality is stressed to stimulate the free flow of thinking.

In summary, the interviews are designed to help individual managers and experts to make their insights and preferences explicit. Of each interview a protocol is made for future reference. We use in-house developed software to categorize and store interview statements for easy reference.

The next stage in the process is all important and determines largely the overall success of the effort. The question is whether the generally rich but equally highly personal problem perceptions can be integrated and can be turned into a view shared by all relevant managers. In order to achieve this our process design calls for a workshop away from the office followed by series of meetings. The workshop is devoted to feedback and validation of the interview results and further exploration of the identified gaps. Subsequent meetings serve the building of a shared perception of the problem in its setting. The key instrument that we use is the influence diagram, a graphical account or map of the perception and reasoning of individuals or teams. Our experience is that in a first joint session a very simple, highly transparent influence diagram should be presented, which is inevitably a hopelessly inadequate description of the situation, but is likely to be perceived as a good starting point as long as each participant sees his interview at least to some degree reflected.

Depending on the problem different mapping conventions have been used. One set of conventions that we have applied with considerable success are the system dynamics symbols (Morecroft, 1988). As most of you will be aware there are only 5 or 6 (see Figure 1).

The box on the right is a level that accumulates action. In the centre is a composite symbol that represents an action flow (shown as an arrow) and a flow regulator (shown as a T) which controls the size or volume of the flow. Finally, on the left is a source (shown as an irregular "blob") which supplies the action flow.

Total capacity installed for example which is the accumulation of all past investments in capacity.

The investment rate is controlled by the flow regulator. The investments originate from a "pool" of possible projects.

The last but most important symbol is the decision function or converter. The decision function "receives" information shown as dotted lines,

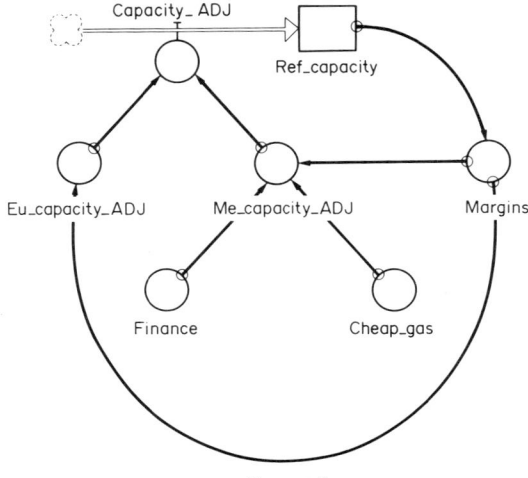

FIGURE 1

processes it and generates an output in the form of action or more information. There is always a decision function or converter attached to the flow regulator. In our example the rate of investment is a composite of investments in the Middle East and disinvestments in Europe in turn determined by the variables identified in the original problem formulation. Of course you have already noted the feedback loop in this very simple map. It was assumed in the tentative problem formulation that margins were largely a function of refinery capacity. It was also assumed that margins caused capacity adjustments in EU. These very few symbols provide a flexible toolbox to represent a large number of problems, whilst imposing a considerable discipline on management discussions.

Starting from the simple map standard workshop techniques are used to alter and expand. In a neutral setting, with an unusual task at hand, following rules set by the process-consultant, hierarchy and organizational affiliation can be played down considerably. The crux of the matter is that the joint effort to build a representation of a corporate problem forces the participating managers to express preferences and to make their assumptions and insights explicit. This makes it, at least in principle, possible to establish their relevance and to test their validity. Much of this takes place during the workshop when the collective experience of the team is brought to bear on the suggested critical parameters and presumptions. In addition the iterative character of the effort means that staff work and research can be done between sessions either to settle differences of opinion amongst the participating managers or to address questions raised by the group as a whole.

During the workshop and meetings the original interviews fulfil two functions in addition to the source of material for the creation of the starting influence diagram. One is that the interviews enhance awareness of preferences and insights considerably and participating managers tend to draw heavily on that encounter. The second is that the workshop facilitators are aware of the views of all participants and can ensure that all strongly held views are being explored.

An important additional advantage of relying on system dynamics symbols is the small step from what we call the mapping phase to the modelling phase. For this third phase in our view fairly advanced software is now available. The software we use trades under the name STELLA, runs on an Apple-Macintosh and is specifically designed to conduct policy experiments. With apologies to those who are intimately familiar with STELLA I like to stress the following features that we consider of utmost importance:

- A series of icons representing the system dynamics symbols can be manipulated with a mouse to draft and redraft a map on a high resolution screen.
- The map steers the mathematical model building. Clicking on a converter or a stock opens up a screen on which all the variables that logically should be part of the equation (because they are linked up in the map) have already been specified.
- The drafting of nonlinear development paths of parameters.
- Flexibility in the variables that are displayed during the simulation runs.

Most importantly, the system allows for large numbers of runs to explore the impact over time of possible strategic commitments under a variety of exogenuous developments and it greatly facilitates the analyses of the dynamics that gave rise to the results. These features, in general the emphasis on visual representation and the user friendliness of the software, make it feasible to involve management actively in the model building.

This involvement can take a variety of forms. There have been instances whereby managers built parts of models themselves. In other projects the facilitator has presented small submodels based on discussions in the previous session. In all instances managers were able to verify whether their assumptions and insights were correctly reflected in the equations. This process of testing within a team of managers eventually produces a model which enjoys a high degree of ownership and confidence. The latter not in the conventional way of being able to replicate the past satisfactorily but in the sense of being a fair representation of their insights and assumptions. This way the credibility problem that every model builder haunts can be circumvented and a model can play an appropriate role in the decision-making process.

The process of model building has a profound impact on the original problem formulation. Inevitably, the loosely identified gaps have been defined precisely, the tentative problem boundaries have been redrawn, the first approximations of causes and effects of gaps have been specified.

Finally and most importantly, the exploration of options to close the gaps had to be first broadened and then narrowed down to the most promising options for detailed analysis.

Barriers in Management Team Model Building

In addition to a considerable number of petty problems three fairly fundamental questions need to be addressed.

First. Management time in Shell, as elsewhere, is a scarce resource and in absolute terms the total claims being made by a model building project are quite considerable.

It is extremely difficult to gauge how much time would be involved in addressing the problem within the framework of normal management practice and it is even more difficult to demonstrate conclusively that the quality of problem formulation will be higher. This pushes future applications of the approach towards mega projects where the high risks involved help shape the management agenda.

Second. Initially and for very good reasons the process has to be separated from the normal management processes. However it is inevitable, unless the process leads to an organizational restructuring that the results have to be absorbed by the existing organizational and management structure.

In the future we hope to control the inevitable risks this entails better by having the final model building sessions take place during regular business meetings.

Third. We have to face the fact that many strategic problems cannot be modelled mathematically. Particularly a company like Shell operates in politically heavily charged environment and is subjected to very real but unquantifiable forces with considerable impact on the shape of our strategic problems. Conversely, many of the options Shell considers have potential consequences for governments, special interest groups and other stakeholders which go beyond the quantifiable.

We know that the process I sketched is still relevant if only an influence diagram can be produced. We do intend to go one step further by applying an experimental semi-customized software package with very considerable graphics capabilities to see whether if we cannot quantify we can at least codify. If we can create a shared language suitable to describe the dynamics of the problem we will definitely attempt to evaluate strategic options by tracing consequences through such an advanced influence diagram.

In Conclusion

Let me conclude by summarizing my presentation in the form of three theses:
 The characteristics of simulation model building by management teams meet many of the new requirements for strategic decision making.

Active management team involvement in building simulation models of strategic problems is feasible.

The combined pressure of the need for ever growing commitments and a world posing wicked problems will prove to be a fertile ground for further experiments.

References

Ackoff, R. L. 1974. *Redesigning the Future.* New York: Wiley.

Mason, R. O. and Mitroff, J. J. 1981. *Challenging Strategic Planning Assumptions.* New York: Wiley.

Morecroft, J. D. W. 1988. System Dynamics and Microworlds for Policymakers. *European Journal of Operational Research* **35**.

Rittel, H. 1972. On the Planning Crisis: Systems Analysis of the "First and Second Generations". *Bedriftsokonomen* **8**.

Section Two

Organizational Change

Editor: **Cees A. Th. Takkenberg**

Organizational change, implementation, social simulation, cultural change and leadership, model switching

Cees A. Th. Takkenberg
Utrecht University, The Netherlands

There is an old parable that has made the rounds about the grasshopper who decided to consult the hoary consultant of the animal kingdom, the owl, about a personal problem. The problem concerned the fact that the grasshopper suffered each winter from severe pains due to the savage temperature. After a number of these painful winters, in which all of the grasshopper's known remedies were of no avail, he presented his case to the venerable and wise owl. The owl, after patiently listening to the grasshopper's misery, so the story goes, prescribed a simple solution. "Simply turn yourself into a cricket and hibernate during the winter." The grasshopper jumped joyously away, profusely thanking the owl for his wise advice. Later, however, after discovering that this important knowledge could not be transformed into action, the grasshopper returned to the owl and asked him how he could perform the metamorphosis. The own replied rather curtly, "Look I gave you the principle. It is up to you to work out the details!"

Nowadays, consultants have been forced not only to formulate wise principles, but also the guidelines for implementation: organizational change. Whenever the idea of change is brought up, it is almost automatic to think about organizations. The spread of religions or ideologies, wars on poverty or ignorance, and fights against crime or pollution are all organizationally based. The success of these efforts at large-scale change depends on the successful organization of resources and people. Small-scale changes are also organizationally dependent.

In this section gaming and simulation are seen as effective instruments for organizational change.

Alan Coote and Maarten van Mens organized a workshop "on implementations of changes" and organized it as a metagame and reviewed four games after the conference: SWITCHER, illustrating that unforeseen problems call for organizational and/or personal switches, GHOSTS IN THE

MACHINE, demonstrating that you need symmetry between strategy, culture, structure and action, A SOCIAL SIMULATION OF ORGANIZING AND ORGANIZATIONAL CHANGE, showing the need to combine personnel, experience, skill and structure harmoniously to realize aims, and PACT, to explore and design new organizational concepts and gain acceptance by involving people in their design.

They use a metaphor of a garden to classify the games and also followed the notions of Jan Berting, presented in the opening address of the conference.

Bart van Linder presented SWITCHER, a game to give people more insight in their sometimes dominant model(s) of organization. A user-oriented language for game design is presented and several questions will be answered: what are models?, are there different kinds of models?, can we switch from one model to another? is switching related to competence?. Also SWITCHER is discussed as a frame game and as a computer network.

Frans-Bauke van der Meer and Ton Roodink presented a work shop session with the title "Social simulation of organizing and organizational change". In social simulation the core processes are generated by human participants, engaged in social interaction, constrained by initial and boundary conditions, but not by role-prescriptions. By appropriate tuning of these conditions the simulation can be induced to be a valid reflection of real life organizations on the structural and process levels. With respect to improving competence in dealing with problems of organizational change and managing implementation processes the method can be used in several modes: as a research setting, as a method and setting for training and education, and as a tool in shaping implementation strategies. The method is especially fit to help find ways to deal with unintended or unexpected consequences of change in advance and to investigate the processes at hand systematically. One of the major conclusions in the review of the workshop is related to estimating the simulation time needed to realize a socially and psychologically valid organization pattern.

Maarten van Mens presented PACT. This game was used by the NMB Bank to explore the issues of a less hierarchic and more decentralized organizational concept guiding the process of organizational change and to support a change of culture. The aim is becoming a more professional, more productive, competitive and innovative organization. This is supported by a parallel development of new infrastructures. PACT is designed as a chess game, a shell that can be translated to demonstrate and stimulate discussions about all sorts of organizational concepts. The NMB bank used it to illustrate and explore consequences of a new automation and communication infrastructure, at the same time introducing a modern logistic approach to banking.

The editors also put the game MIDAS into this section. It is discussed by Dolf Dekker. MIDAS is an awareness game on innovation management and is a powerful tool to motivate managers and their staff in taking the decision to initiate an innovation programme. The name MIDAS suggests an analogy: using it turns everything into gold with a simple touch!

Objectives of the game are: first to give managers of small and medium-sized companies the feeling that innovation can be managed if they use specific knowledge and skills and second to start a structured discussion on innovation. The diffusion techniques in innovation consultancy can be improved considerably by introducing games. During this conference the game was applied to a case concerning a carpenter's shop.

Alan Coote and Clive Loveluck discussed the development and use of GHOSTS IN THE MACHINE, a flexible, interactive computer-aided simulation game designed as an aid for management development. It enables participants to make judgements about their capacity to estimate compatibility between strategies, cultures and tactical responses. Experience with this flexible and adaptable game gave the designers a basis for making some important statements about cultures, subcultures and the influence of leadership.

Organizational change: workshop review

Alan Coote and Maarten van Mens
Polytechnic of Wales and NMB Bank, Netherlands

ABSTRACT: The sessions "on the implementation of changes" were organized as a metagame. We ran four demonstrations and reserved time afterwards for a structured discussion, about the gaming experience, the gaming-concept, structure and possible applications. By comparing the implementation of changes in organizations to a process of gardening, we tried to evaluate the games within the constraints of a common structure. Four games were presented; SWITCHER, illustrating that unforeseen problems call for organizational and/or personal switches; GHOSTS IN THE MACHINE, demonstrating that you need symmetry between strategy, culture, structure and action; A SOCIAL SIMULATION OF ORGANIZING AND ORGANIZATIONAL CHANGE, showing the need to combine personnel, experience, skill and (organizational) structure harmoniously to realize your aims; and PACT, to explore and design new organizational concepts and gain acceptance by involving people in their design.
KEYWORDS: GHOSTS IN THE MACHINE, PACT, A SOCIAL SIMULATION, SWITCHER, game design, organizational application of games, implementation of change, organizational competence and culture.
ADDRESS: *Coote* Alan, Polytechnic of Wales, Pontypridd, Mid Glamorgan CF37 1DL, Wales.
van Mens Maarten Ernst, Kastanjelaan 24, 1214 LH Hilversum, The Netherlands.

Introduction

At the ISAGA 88 conference in Utrecht we convened a workshop series on "the implementation of changes". We had a wonderful audience of about 16 regular participants, that kept up interest until the very last minute. The topic was selected for a number of reasons:

1. There is a lot of interest in literature on policy formulation and strategy development. However, we seldom read about the experiences of putting these ideas/concepts into practice.
2. Gaming looks like an excellent management tool not only for gaining personal competence in an educational setting, but as decision support as well (to implement new concepts) and as an explorative tool (to develop and test new concepts), but it is seldom used by managers.

Both convenors not only shared these opinions but had some practical experience in using games for implementing organizational change as well. We decided to make use of the opportunity ISAGA offers to demonstrate and discuss gaming concepts and structures that illustrated our point. The workshop ran like a metagame; the problem setting was what design you need for games that support an increase in organizational competence. A garden metaphor was developed to structure the workshop sessions, with four games as process elements. As in our opinion the main elements in gaming are *structure* to create the necessary setting and *process* to deliver the message, we thus devised a simple meta-game.

The Metaphor

Gardens and organizations both are artificial creations; using living matter. Organizations – like plants in a garden – grow, compete for space, co-operate, change over time, are influenced by outside pests and blights, showers and storms, sunny spells and winter. A manager like a gardener needs an eye for the upkeep, regular attention for individual plants, timely intervention to keep a balance and different approaches for different phases in the garden life/cycle. Using this metaphor a number of conditions can be described:

1. Deciding on the architecture, i.e. a process of organizing and structuring to create a proper culture and a proper clientele.
2. Short-term upkeep i.e. keeping competition at bay, trimming, pruning, mowing etc. In short, day-to-day management.
3. Long-term upkeep, rooting out old plants, introducing new plants, making a beautiful around the year growth scheme, keeping harmony and structure in the organization.
4. The art of gardening i.e. rules and conventions that are helpful to strike a good balance between personal capacity, energy and income and the work that has to be done.
5. Technical aspects of gardening, useful approaches, tools and machinery including the specific techniques to use/introduce them.
6. Supportive garden elements like ponds, stone slabs etc. that can be used for changing design, adaptation to new views on upkeep and art.

This list could certainly be longer. Suffice to say that we had four games (and comparatively little time) to illustrate a number of these aspects. GHOSTS, for example, fits in category number 4, PACT in number 6, SWITCHER in number 3 and A SOCIAL SIMULATION in number 1. The garden metaphor structured our thinking and talking about games. The metaphor only played a superficial role in our evaluation sessions.

Preparations

Our aim was to discuss game concepts, their applicability and content, their structure and outlook. For organizational learning, group processes are essential. Organizational games often take up a lot of time. Creating a group needs time and the concept is often complex. This type of a game often needs several trial runs before playing. Time for evaluation is often extensive as well. Settings for organizational games are:

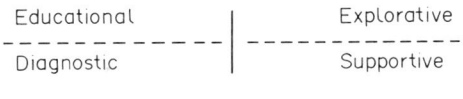

Educational	Explorative
Diagnostic	Supportive

FIGURE 1

In organizations, games can be put to a normative use (for educational and diagnostic purposes) or for development (explorative/supportive use), for a new (education/exploration) or given (diagnosis/support) environment. Each of these aims need different approaches, plots and settings. Aside from this, either the structure or the process of a game can carry the message. Thus there were many elements for discussion and evaluation.

The Start

We started with a brief presentation and an introduction of the games. Four theses were presented:

1. Strategy needs transformation to company reality before implementation.
2. Strategy, culture, structure, skills and knowledge must be in harmony.
3. Both 1 and 2 must be supported with coherent and consistent action.
4. Simulation/gaming supports these aspects cognitively and interactively.

In our opinion all four games supported these theses.

For SWITCHER, the structure was essential. It had to be rigidly kept to realize its educational aim i.e. making people experience the ineffectivity of a hierarchy. In an educational and diagnostic setting the outcomes must be controlled. With a more flexible structure (e.g. having players decide on the setting) the game could be put to an explorative or supportive use. Discussion afterwards centred around its construction; free-form versus deterministic, imposed versus spontaneous action, collective versus personal experience, game culture versus company culture. The conclusion was that a game should preferably be self-explanatory, self-learning and self-steering.

GHOSTS was designed to explore the need for symmetry between strategy, culture, structure by relating these aspects to incidents and responses. Using a (any) scoring system and a pre-determined structure was controversial. As in most educational games, the structure had to be preset to support its outcome. Used more interactively (having two groups compare each others scores) and in a freeform fashion (having people produce their own cards) the concept could be used to reconstruct company reality (diagnostically). Using stacks of cards (strategy and culture statements) and a preparatory construction cycle, GHOSTS could be used exploratively.

A SOCIAL SIMULATION had people explore the interdependency between culture – a collective human product – and structure (as a set of tools, materials, tasks and goals). The concept centres on the process. Participants can structure their own interactions and communications and can explore alternatives, train different management styles and explore different organizational options. For these they need much more time than to experience a predetermined structure. Process oriented games seem to be more generic, supportive and experimental and less experiential and individually oriented than structure oriented games (like SWITCHER and GHOSTS). They are closely akin to decision support and developmental systems (information engineering and knowledge engineering).

PACT had people explore a common language to develop and plan a process (making a traditional Dutch meal for 20 persons) using logistic principles (as soon as possible). The game is a way to collectively define and explore an organizational concept. The war room approach, a formalized step by step evaluation of alternatives is promising for conceptualizing ideas and gaining acceptance. The game is useful if it gets a high reality content. After discussion we discerned two basic families one *constructive*, process oriented and one *descriptive*, structure oriented family.

Conclusion

Looking back at the discussions in our workshop we eventually followed Professor Berting's notions as presented in his opening address on the conference. In our wrap-up session we presented this in the following order:

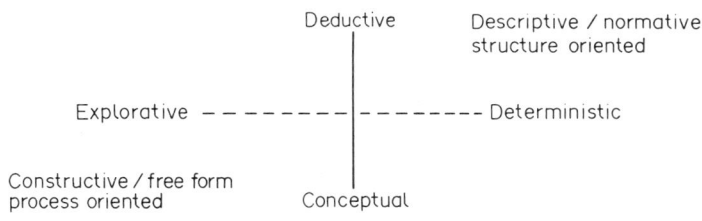

FIGURE 2

Process oriented games support a collective approach, they try to mobilize different skills and actors. Structure oriented games have more individual aims. Gaming applied to an organizational setting is very promising indeed. It will probably generate new gaming families with a more experimental and constructive orientation. Looking at the conference, this line of thought certainly creates a completely new audience for ISAGA and opens up opportunities to interactively relate theory development to practical experience.

SWITCHER: an organization support system for improving reflective competence

Bart van Linder
University of Amsterdam

ABSTRACT: This paper provides a short introduction of the game SWITCHER and discusses its basic concepts. The game is designed to give people more insight in their sometimes dominant model(s) of organizing; the problems these models can cause and how to cope with them. After the introduction we show how the concept of a user-language can be used for game design. Thereafter we give answers to questions like: What are models? Are there different kinds of models? Can we switch from one model to another? Are there different kinds of switches and/or different levels of modelling? Is switching related with competence? In the last paragraphs we discuss SWITCHER as a frame-game and as a computer communication network. We finish with a review of the game session we ran at the conference.
KEYWORDS: Serial switches; parallel switches, monopoly-models; dialogue-models; reflective competence; user-language.
ADDRESS: University of Amsterdam, Research Project OOC, Grote Bickersstraat 72, 1013 KS Amsterdam, The Netherlands.

Introduction to the Game

An organization has arranged daily meetings within a two week period to cope with a problem. However, despite this extensive information exchange there still is a lot of ambiguity between the participants. A special kind of meeting is set up to tackle this very situation. Contrary to expectations things are getting worse.

In this example the model of meetings is perhaps not the most suitable instrument for improvement. However, the people within the organization are not aware of this problem of problem-modelling. They blame the problem instead on their own model of the problem.

SWITCHER is built to cope with this particular kind of problem. The game divides about sixteen players into four groups. Each group gets a different colour: red, green, black and blue. The aim of all players is to solve as much tangram-problems as possible (Fig. 3 shows a tangram). However each group has a different type of tangram-puzzle, for example a rectangle or an egg, and each player has two tangram-figures which can be solved with this group-puzzle only. According to their rank players are rewarded with stars every 10 minutes, for instance one player gets four stars and another just one.

Reds have a special role in the game. They are highest in rank and responsible for the pay-off. Only reds are allowed to walk around whereas the other players have to stay at their workplaces. The reds also have to check the puzzle-solutions of players. If a player has solved a puzzle (s)he has to mention this as soon as possible to a red player. The reds keep records of the level of productivity of players and groups. The most important game rule is that players with a higher rank can impose new rules on players with a lower rank. All other rules are organizational rules which can be changed if necessary.

At the beginning the organization is running smoothly. But as the number and the complexity of problems are gradually increasing participants run against the boundaries of the models they utilize. Additional problems start transcending each individual's ability. Time-pressure is increasing and the budget is getting low. In order to survive players have to switch to other perspectives of organizing. They have to negotiate, to invent new rules, to cooperate, to make decisions and to puzzle in order to succeed in switching to a more competent organizational model.

In short SWITCHER tries to generate a specific organizational environment, that can be characterized as a machine-bureaucracy. The game first seduces players to adopt this particular dominant organizational model and then it confronts them with the boundaries of that model.

This process should teach players to switch between other models and model usages. It serves as a model-generator and not as a model-(de-)terminator because participants can experiment with different models and different ways of organizing, for example by organizing a meeting between all players that needed more than an hour without having one (tangram) problem solved they were challenged to reflect on their organizational performance and to develop better models. The idea is that by switching from perspectives one can discover the blind spots of the present sometimes dominant image of organization.

Concepts behind SWITCHER

Since we are still in the process of prototyping, we will not describe the current game in great detail. Instead we will offer an overview of SWITCHER at the conceptual level. Our impression is that this level deserves special attention since it is the foundation the game is built on.

Gaming-simulation: a User-language

For the design of SWITCHER we used the notion of gaming-simulation as a language (Duke, 1974). We needed specific elements, and relations between elements that had to make the support system a real one. Therefore criteria were formulated for the quality of a SWITCHER. They can be compared with the minimal grammar that is necessary to understand a language. However SWITCHER is not a natural language, nor is it an interaction-language in the sense that it regulates the use of (social) systems, i.e. like a meeting does. It is a language that can be used by the users themselves to improve interaction-languages or as something that can set limits to (social) systems. In these terms we view SWITCHER as a *user-language* (De Zeeuw, 1987).

The language, SWITCHER, is considered a token-system. We therefore distinguish between three elements: a. the token, b. that what the token is referring to, c. the person who is using the token.

Corresponding to this distinction we have three important criteria for the construction of our game (Klabbers, 1987), (Ledermann, 1984):

1. By playing SWITCHER and studying the relations between the elements within the token-system, we concentrate ourselves on the *syntax* of the game. The reliability of the syntax is important. We try to discover whether the same structure repeats itself in another context or over time. We can make a further distinction in product-reliability and process-reliability. In the first case one studies whether the outcome is being repeated (predictable) and in the second one whether the processes are convergent (predictable). For SWITCHER to be reliable we are looking for syntax that produces a reliable learning process. We are not interested in behaviour, interactions or gains and losses *per se*.

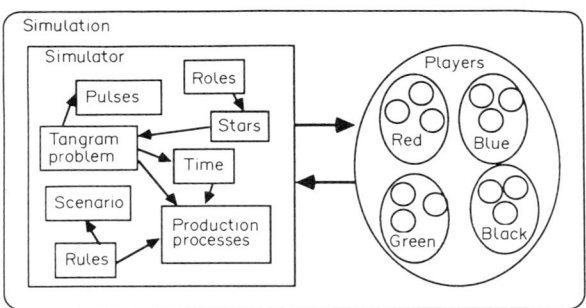

FIGURE 1   The ingredients of the game with a possible syntax

2. When we look at the way reality is represented and to the conditions which influence the correspondence between play and reality, we study the *semantics* of a game. We then focus on validity. We distinguish construct validity (realism) and face validity (verisimilitude). Face validity means that the correspondence between game and reality does not have to be isomorphous. However a certain verisimilitude is needed, a representation that resembles reality without being a perfect copy. Like a sketch of a few lines can say more about a person than a photograph or an actor who imitates a

character with a few simple movements. We decided to make a sketch of the crucial things needed for the learning process.
3. When we study subsequently the language-user we examine the *pragmatics* of the game. We then test the level of utility of the game. The users draw up a balance of the gains and losses in terms of time, money, energy and emotions. The pragmatic criteria support the game operator in structuring the debriefing.

The levels of validity, reliability and utility are often used as criteria for evaluating games. Our experience was seeing a game as a user-language, can support the formulation of a consistent conceptual framework that facilitates the design process as well.

Models

What are models?

Why do we use social process like meetings to solve problems? Our assumption is that models are used to cope with reality in this case the model of meetings, to cope with organizations. We could describe a model as a scheme of reality, or as a cognitive map (Axelrod, 1976). But a model is more than just a scheme. We select a specific model because we have particular objectives. Furthermore the selection of the model itself is one of the determinants of model quality. We prefer to describe a model as a metaphor (Morgan, 1986) or more generally as a *distinction-system* (Heylighen, 1988). "A distinction can be defined as the process or its results of discriminating between a class of phenomena and the complement of that class (i.e. all the phenomena which do not fit into the class" Heylighen, 1988).

Models and competence

By coping with reality with the help of models we construct and change reality. Our level of competence depends for a great deal on the models we use. Competence tell us something about our abilities in handling affairs.

What kind of competence are we talking about here? We can make a distinction between individual and collective competence. *Individual competence* is a specific ability that not necessarily leads to higher competence of others. In the game we use tangram-problems. When one person solves such a puzzle this does not imply that the other can solve their respective puzzles.

Certain kinds of competences are sometimes less bounded to one particular context. They have a transcending character. For instance when a certain individual competence increases the competence of others may increase too. As a consequence a network of shared competences emerges which is called *collective competence*. SWITCHER stimulates people to reflect on the quality of their models and their individual and collective competence. So we try to improve *reflective competence* (Schön, 1984). Reflection is a kind of internal dialogue (Nelissen, 1987). It is this kind of competence that can produce keys to open closed and dominant models.

Monopoly-models and dialogue-models
We can make a distinction between two types of models: *monopoly-models* and *dialogue-models* (Bråten, 1986). Important is the way in which models are used. Usage is a part of the quality of the model. This is important because the use of models, of a specific type can also cause specific problems.

We can select excellent models, like meetings and procedures to cope with organizational problems, when we use them in a wrong way the results are counterproductive. One dangerous way of using models is as follows: The improvement of the model is often indicated by the level of truth. One tries to construct a model that is correct, that is the best model. As there is only one best model in a certain context, this model or type of models is thereafter used to cope with reality. There may be other models but they are less correct, for it makes no sense to use another model when you already have the best model. Such a model mode can have a character that it leaves no space for other models, it swallows them. We call that a monopoly-model usage. It is very convenient to use only one model. Perhaps that is why we see so many monopoly-models in science, economics, religion, organization and even in gaming-simulation.

Sometimes the very use of monopoly-models does cause problems. This happens when a certain monopoly-model used by one individual dominates in such a way that it diminishes the collective competence. We then use a model that reduces our variety. For instance in daily meetings the problem solving capacity, the collective competence, is reduced by those meetings instead of enlarged because of the use of a monopoly-model. There is no time left for other perspectives. In such a situation the model or the model use as such is never the main subject of discussion. One stays within the model as if it were a cage.

Another way of using a model is dialogue-model usage. Characteristic for this kind of usage is that more than one (type of) model is being used. Whether these models are true, false or less correct is not the most important criterion. They may be suitable just as long as they maintain or enlarge the competence. Many great discoveries are the result of having a wrong model but using it in a special way. We call this dialogue-model use. The model is used for coping with reality instead of using it as reality. There is a conversation or dialogue between several models.

If we use models as monopolies there is no attention for other models. One fixates on solutions within the model. However one could think about solutions outside the model, including the use of the model Instead of changes within the system of meetings or procedures in an organization we

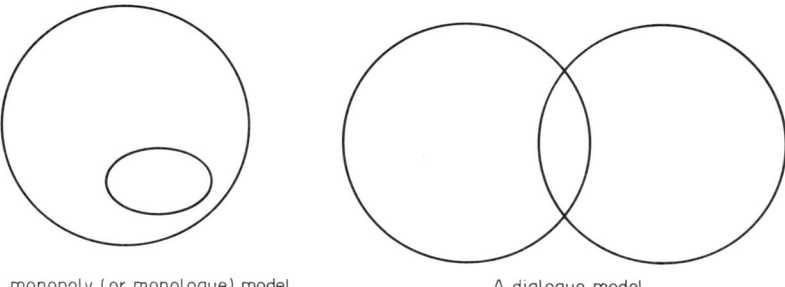

A monopoly (or monologue) model   A dialogue model

FIGURE 2

step outside the system. We then switch between models and also between usages. This may cause a dialogue dependent on the kind of switch that is made. This does not mean that a dialogue-model mode is the ideal use of a model. Sometimes a monopoly-model mode is more adequate. It is just that there has to be the possibility to change modes; a key to open the cage. This key should be standard in every model.

How this relates with the game? In the first part of SWITCHER we try to create a context that stimulates players to use an organizational model in a monopoly mode. We do this with a specific scenario and rules. In fact for a certain period of time such a model works perfect. However after a while problems multiply and players run against the limits of their model and its usage. So something has to change.

Switches

Kinds of switches

We describe a *switch* as the process that takes place as one steps or jumps from one (type) of model to another. A distinction can be made between serial switches and arbitrary switches.

A *serial switch* is a switch of the kind that Kuhn describes (Kuhn, 1962) as he speaks about scientific revolutions. Such a switch supposes that the old model has to be rejected because there is another, better model. It is a unique switch because within this model mode one does not switch to a model that is false or worse. One dominant model is replaced by another dominant model. In organizations we see serial switches i.e. Japanese style of management, bureaucratic management etcetera.

An *arbitrary switch* or *parallel switch* is not necessarily a switch to a better model. One uses another model mode and because of this a switch or switches between different models are possible. These kind of switches can evoke a dialogue between different types of models. It is possible to switch there and back. In order to be able to make these switches, to shift from thinking within one model to a border position or to another model, we need reflective competence.

A tangram-puzzle is an elegant example of a game that stimulates reflective competence although not on the organizational level as is the aim of our game.

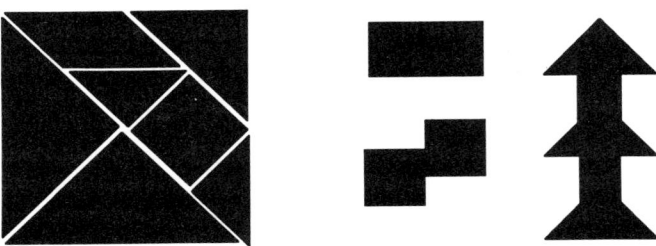

FIGURE 3

SWITCHER should be a support system that gives organizations the opportunity to reflect on the existing organizational model and its use and if necessary to (self-) improve it. It should stimulate users to make switches themselves on the basis of their expertise and the variety of models they use. They result in different problem definitions. In this way the players focus also on the different relations between problems and context as seen from different perspectives. That is different from the hiring of an expert in problem solving which leads often to a serial switch; the old dominant model is replaced by a new dominant model.

By creating a dialogue and consequently a switching perspective, a better problem orientation and a better competence might be developed. The quality of a SWITCHER therefore has to be such that a real dialogue and not a pseudo-dialogue becomes possible.

Switch levels

We distinguish four levels of switching:

- an individual level,
- an organizational level,
- a level of the relations between individuals,
- a level of the relations between organizations.

Suppose, we want to switch from monopoly-model usage to a dialogue-model usage and vice versa. In that case we could create a switch on the individual level. The question is whether this does increase the collective competence. We doubt this.

Starting to switch on the individual level can lead to frustration or even resignation, because the conditions of such a switch might involve high risk. SWITCHER shows immediately what happens if players only change their individual perspectives. It causes chaos and frustration. It makes perhaps more sense to stimulate switching on the level of the relations between individuals. Then we focus on the process of organizing. By switching on this level we can receive insight in the pattern that (re-) produces the organization and in the specific models it uses.

Potential Users and Possibilities

In the first place the game is made for machine types of organizations which deal with relatively many rules, which have to cope with high levels of complexity and a fast changing environment whereas the use of the system of rules often leads to many negative side-effects. We think of Social Services, like "doles" and other local, regional and national government. Especially in these organizations dominance of monopoly models (system driven) has raised a debate. Therefore these organizations need support to improve reflection on their models and for the development of new models.

Secondly we think that SWITCHER might also be appropriate for other types of organization which use a monopoly-model. For obvious reasons we started with a prototype with only one scenario to start the game with a bureaucratic scenario. However, we intend to design it in such a way that other starting scenario's can be used. We would like to construct a game where the players could develop their own scenario or select a scenario out of a scenario-base stored in a computer. Organizations can add new scenario's or exchange scenario's with other organizations. SWITCHER then gets the character of a frame-game which can be loaded with different scenarios.

We are also planning a computer-version for a local area network giving the game-operator the opportunity to control the degree of interaction between players. In strongly hierarchical organizations for example it can be an advantage to stay anonymous for a while. Secondly people can stay at their work places, which may be more convenient for them. It is like playing chess by post. Of course not all organizations have computer networks. In fact such a set up will generate a completely different game, even if keeping the same goals.

Session review

It was the fifth time we played SWITCHER at the conference. So we knew that the game had still all kinds of sharp edges which could be irritating and reduce its potential force. We therefore used the ISAGA-conference as a try out and as a kind of super-game-polish. The game was played briefly in order to get familiar with its dynamics. Very soon already the participants triggered a debate on the level of rigidity the game should contain, which is one of its major design issues. Many suggestions were made. The result was that we got a lot of suggestions that could improve the game. Players suggested for example to alter the timing of the input of tangram-problems in the game, to build in review sessions for reflection, to give the starting-scenario more backbone etc.

One of the most important observations made during the conference session was that a lot of switches were made. However those switches could not transcend the individual level while in previous try outs more switches on the collective level had occurred. A possible explanation is that the participants were experienced gamers who as such were primarily interested in the game characteristics and shared less interest in interaction-processes.

The main conclusion of the participants was that the game concept of SWITCHER proved to be appropriate and promising. We realize however that many improvements are still ahead of us. The feedback we received gave us a turbo-boost to play on into the chosen direction.

References

Bono, E. de. 1967. *The Use of Lateral Thinking*, Middlesex: Penguin Books.
Braten, S. 1986. The third position: beyond artificial and autopoietic reduction. *In* Geyer (1986).
Coote, A. 1985. Managers and Reorganization: Can Simulation/Gaming help old dogs learn new tricks? *Perspectives on Gaming and Simulation,* **10**.
Crookall, D. *et al.* (eds.) 1987. *Simulation-Gaming in the Late 1980s*. Oxford: Pergamon Press.
Duke, R. D. 1981. Managing Complexity: Gaming- A Future's Language. *In* Greenblat *et al.* (1981).
Geyer, F. and Van der Zouwen, J. (eds.). 1986. *Sociocybernetic Paradoxes*. London: Sage.
Greenblat, C. S. and Duke, R. D. 1981. *Principles and Practices of Gaming-Simulation*. London: Sage.
Heylighen, F. 1988. Formulating the Problem of Problem-Formulation. *In* Trappl (1988).
Klabbers, J. H. G. 1987. A user-oriented taxonomy of games and simulations. *In* Crookall (1987).
Ledermann, L. C. 1984. Debriefing, A Critical Reexamination of the Postexperience Analytical Process with Implications for Its Effective Use. *Simulation and Games,* **15**:4.
Morgan, G. 1986. *Images of Organization*. London: Sage.
Nelissen, J. M. C. 1987. *Kinderen leren wiskunde. Een studie over constructie en reflectie in het basisonderwijs*. (proefschrift). Gorinchem: De Ruiter.
Schön, D. A. 1983. *The Reflective Practitioner*. New York: Basic Books.
Trappl, R. (ed.). 1988. *Cybernetics and Systems '88*. Deventer: Kluwer
Zeeuw, G. de. 1985. Problems of Increasing Competence. *Systems Research* **2**.

Simulation References

DESTRUCTURE, Coote, A. 1985. Managers and Reorganization: Can Simulation/Gaming help old dogs learn new tricks? *Perspectives on Gaming and Simulation,* **10**.

Social simulation of organizing and organizational change

Frans-Bauke van der Meer and Ton Roodink

University of Twente, The Netherlands

ABSTRACT: In social simulation the core processes are generated by human participants, engaged in social interaction, constrained by initial and boundary conditions, but not by role-prescriptions. By appropriate tuning of these conditions the simulation can be induced to be a valid reflection of real life organizations on the structural and process levels. With respect to improving competence in dealing with problems of organizational change and managing implementation processes the method can be used in different modes: as a research setting, as a method and setting for training and education, and as a tool in shaping implementation strategies. The method is especially fit to identify unintended or unexpected consequences of change projects in advance, to help finding ways to deal with these, both for management and other organization members, and to investigate the processes at hand systematically.
KEYWORDS: Social simulation; organizational change; implementation strategies; ex ante evaluation; learning by experience.
ADDRESS: "De Boerderij", University of Twente,
P.O Box 217, 7500 AE Enschede, The Netherlands.

Introduction

The social nature of organizations is increasingly recognized. This is not to say that the notion that organizations consist of people and patterned social interactions between them is new. But practitioners and researchers in the field of Organizational Development have gradually developed an image of organizations in which the implications of its social nature are more fully integrated. One aspect of this image is that the dynamics of organizational processes is mediated by the meaning participants attach to the course of events. Generally these meanings stabilize in ongoing interaction to constitute a pattern which enables the participants to make sense of the situation, to decide what to do, to interpret the behaviour of other participants, and hence to function and to make the organization function (Weick, 1979; Pfeffer, 1982; Morgan, 1986).

From this view a first set of problems for organizational change can be deducted. Organizational change implies a change in patterns of organizational behaviours. Neither initiators nor other participants in the organization know in advance precisely to which new patterns certain measures or interventions will eventually amount, because these will be the result of ongoing interaction which will only stabilize during or after the implementation process. Therefore, organizational change creates uncertainty and ambiguity, and is difficult to predict or direct.

A second set of problems stems from the fact that different actors or groups within (and around) an organization hold different views or interpretations of what is going on. We found (Van der Meer, 1983) that these differences are strongly related to the positions actors occupy within the organization. Furthermore actors do not only use the existing organizational patterns to make sense, but also draw from their cultural backgrounds (Swidler, 1986) and from other organizations or groups they belong to. This implies that different actors utilize different repertoires to decide what is going on and which behaviours should be selected. In case of organizational change this enhances the unpredictability of its outcomes.

This brief analysis is of course only a raw sketch and somewhat abstract. Here it suffices however to clarify the background of the introduction of *social* simulation as a tool for (implementing) organizational change and increasing competence in dealing with it. Both organization theorists and practitioners in the field of organizational change increasingly are aware that in general unforeseen and unintended effects come about during implementation processes and afterwards. Therefore there is a clear need for methods and instruments that enhance insight in potential future developments in a given situation, characterized by specific structural and cultural factors. In many respects social simulation provides such a tool.

Characteristics and Methodology of Social Simulation

Simulations are processes that model processes in some real life reference situation (Schultz and Sullivan, 1972). Simulations of social processes need not be *social* processes themselves, but *social* simulations are. To put it differently, in a social simulation real social processes are invoked under artificial conditions in order to reflect relevant real life social processes. This is done by including human participants in the simulation in such a way that the complex network of their interactions constitute the core of the simulation.

It is vital that no role prescriptions, or goals to be attained, be given to the participants. It will be illustrated below that constraints induced by the situation and generated by the participants themselves in ongoing interaction during the simulation suffice to provide a valid reflection of social processes in the reference situation. By leaving it at that, the simulation can be used, both by participants and observers, to gain insight in these processes of organizing and organizational change. It is not necessary to *know* these processes beforehand – which is difficult, as argued in the introduction. Social simulation can be used to *study* them in a simplified and controllable setting.

Example

In one of the simulations we developed, the participants are provided with a description of the initial situation in a house building company (see below). After a general introduction the participants are randomly assigned to the different positions in the organization as it is structured in this initial situation. The participants are responsible for what is going to happen in and with the simulated company. They are free to choose their own line of action, strategies, etc. Starting from the predefined initial situation, everything may be changed: the production process, product design, the organization structure, decision procedures, division of work and responsibilities, wages, who holds which position, etc. Participants are explicitly asked to try to realize during the simulation labour relations and organizational structures as they themselves think desirable. There is only a limited set of boundary conditions, intended to represent almost physical limitations in real life organizations. A simulation lasts about two days, giving change and feed back processes ample time to show.

Initial situation. For a typical run of the simulation (25-30 participants) the initial positions and their hierarchical relations are mirrored in Fig. 1.

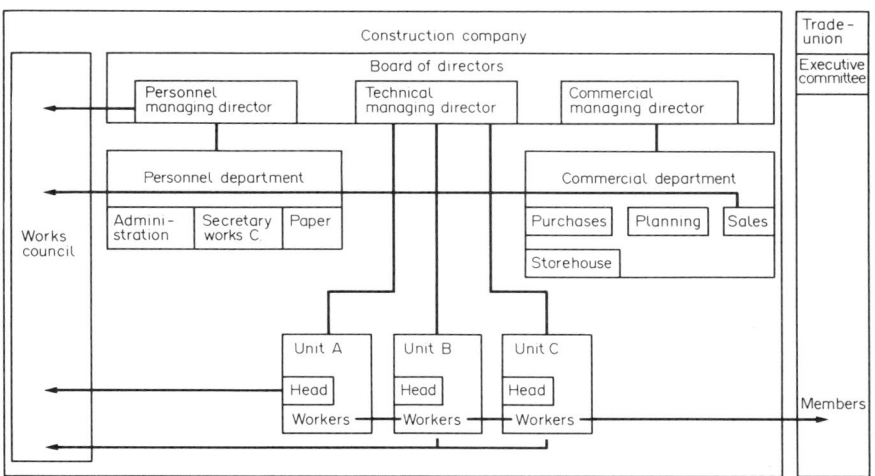

FIGURE 1   Organization chart initial situation

In addition to the general introduction participants receive, depending on their assigned positions, specific information. This is intended to represent "educational background" and "experience". In the initial situation there is a real production process. "Small" and "large" houses are made of raw materials (like chipboard, cardboard and glue), using capital goods (like saws, scissors and knives). These goods have to be bought by the commercial department (CD) and be distributed among the production units, B and C making parts, and A assembling them. End products are delivered to the CD again, which sells them on the external market. The CD has also tasks with respect to (financial) planning. The personnel department administers the wages and supports the personnel managing director. The three managing directors together form the board of the company and decide about general policy. Within the organization there are, in the initial situation, a number of consultative bodies, a.o. the advisory Works Council. Furthermore there is a trade union organization with a professional Executive Committee.

Boundary conditions. The developments during the simulation are subject to the following boundary conditions:

- Ongoing time. There are cycles of fixed duration (usually 80 minutes), called "year". By the end of each year financial and economic results are calculated and fed back as input information for the next year, which begins immediately after the preceding.
- External markets. The behaviour of the markets on which the company has to buy its raw materials and capital goods, and to sell its products, is preprogrammed, the company not having complete knowledge about it. Prices depend on quantities ordered or offered and on the quality of the products.
- Communication rules. In order to represent physical restrictions to communication in organizations a number of communication rules are introduced.
- Livelihood. During the simulation, consumptions must be paid from the salaries the participants earn, depending on the position they occupy and the wage policy of the company.

Validity. One of the interesting phenomena of this type of simulation is that under the relatively few and simple conditions indicated above, a social reality emerges, within about 2 hours, which can be considered a valid dynamic model of real life production organizations at structural and process levels.

When a simulation starts, participants typically are somewhat uncertain. The information given, the forms and materials around etc. of course give some clues about what it is all about, but there is not yet a stabilized pattern which makes obvious what should be done. Therefore the initial interaction usually is playful. The participants play at their assigned positions, utilizing the available clues, in the meantime communicating that they are playing. After some time however the interaction becomes more "serious". The participants find themselves in a situation they have created themselves, in which they are related to each other in terms of the house building company.

So, the initially playful behaviours develop to a pattern that is utilized to make sense of the situation and to decide what to do. After, say, 2 hours, hardly any references are made to the fact that the simulation is only a game. Real conflicts, frustration or satisfaction about the functioning of the company and labour relations can be observed. Also results of quantified measurements show that characteristic organization patterns are actually reproduced in the simulation (Van der Meer, 1983, 1986).

Applications

Social simulations as defined above can be used in a number of ways to improve designing, implementing and evaluating organizational change.

Research

The method provides a setting for systematic research into the complexities and dynamics of "organizing" as analyzed in the beginning of this paper. The simulated situation appears complex and "social" enough to be considered a valid reflection of real life organizations on the structural and process levels. This implies that on these levels the simulation can be approached as a field setting and be investigated accordingly. Important advantages over real field research are:

- the possibility of replication, which makes a clear distinction between systematic and incidental effects possible;
- the controllability: conditions can be kept fixed or varied systematically within or between replications;
- the simplified character, which enables almost complete monitoring of developments;
- the relatively short cycles, which makes the study of dynamic processes manageable within a relatively short time span;
- a less complex relationship between researcher and subjects.

Therefore, social simulation can very well be used to study the dynamics of change and to gain insight in types of unforeseen/unintended problems that may arise, e.g. as a consequence of incongruencies between an implementation strategy adopted and the existing culture of the organization. "Culture" could be controlled by drawing participants from an existing organization or by selecting them on relevant attitudes. "Implementation strategies" could be introduced by including confederates of the experimenter among the participants with the instruction to follow a predefined mode of acting, reflecting a strategic conception.

Actual use: The social simulation described above was used in a research mode in a series of 24 replications in which a structural interactionist theory about the shaping of labour relations, related to structural and cultural conditions was tested and to a large extent supported. In these simulations it also appeared possible to realize different strategies of change by means of confederates in such a way that different strategies are perceived differently by participants in the expected direction on a large number of indicators. This means that the strategies were not only executed correctly, but were social meaningful interventions as well (Van der Meer, 1983 and 1986). Currently the dynamics of these trade union strategies is being investigated. In another series of simulations two strategies of science policy in an inter-organization network were studied (Wilhelm and Bolk, 1986).

Education and training

A second option is to use social simulation as an element in an implementation process. Organizational change expresses itself not only in changes in isolated tasks or functions but also in transformations of labour relations, shifts in flows of information etc. As said before, such changes cannot always be foreseen. Such unanticipated changes or unanticipated reactions of organization members might impede an optimal implementation process. Social simulation can provide a sensible complement of job-specific training etc. by offering participants a setting in which they can acquire experience with new structures, procedures, or with their implementation, and can become aware of its social and organizational implications. If such a simulation is evaluated carefully with the participants it may contribute to their going about with the changes on the one hand, and/or to necessary adaptations in the implementation strategy on the other.

With a view to improving the change process, it could be advisable to let organization members participate in a social simulation in a position which differs from their own, in order to get an impression of how things look from other positions and hence to develop more adequate reactions.

In any case evaluative sessions are necessary if optimal learning effect are to be attained. This is not only so because explicitation enhances the strength and consolidation of learning experiences, but also because the full learning potential of social simulation is only tapped if participants exchange their experiences. Then it becomes clear how – and in part why – perceptions and problem definitions of groups and individuals differ. This will add significantly to the understanding of their behaviours and reactions with respect to each other. It is, so to speak, possible to look back to the processes in the simulation from different positions simultaneously, which implies "collective learning".

An interesting possibility is to alternate a number of simulation cycles with evaluative sessions. In intermediary evaluations, analyses can be made of current organizational problems and plans for improvement can be discussed, whereupon the ideas generated can be brought into practice in subsequent simulation cycles. Perhaps, then new types of unexpected problems can be discovered, which constitutes input for a next evaluation round.

Actual use: The simulation of a company has been used numerous times in courses and classes on management problems, organizational change and labour relations. It appears to generate valuable experiences for discussing organization theories or strategies of change. Theoretical notions are complemented and confronted with observation and own experience.

A simulation of a governmental organization was played a number of times by members of governmental departments in order to gain insight in the complexities of transforming their own organization to a more decentralized mode of decision making and responsibility and to get an image of how the new structure might look, which consequences it might have for their own functioning, etc.

Shaping strategy

It will be clear from the sections above that both research and educative applications of simulation give information and insight that may be useful in designing and executing implementation strategies. But the method can also be used in a more direct mode to support the shaping of such strategies.

First, it may be used as a heuristic device. Performing several runs of a social simulation with different groups and/or under different conditions can stimulate the creative generation of ideas about how implementation can be structured, which types of problems might arise, how these can be tackled, which conditions seem to be vital for success or failure, or how success/failure may be defined by different groups or individuals.

Second, ideas about interventions or detailed implementation plans may be tested in a simulated setting. In this way unexpected results or problems may be detected and their causes be traced, before actual implementation takes place.

Third, the dynamics of new projected organizational arrangements, after the implementation phase, may be demonstrated in a simulation. Here, however, there is a serious drawback. For, in reality there will never be a new organizational arrangement without a preceding implicit or explicit implementation process, which shapes the eventual pattern. Therefore if this implementation phase is left out of the simulation, the result may not be relevant, or even misleading with respect to the real organization to be changed. Nevertheless this option may be useful to discover inherent problems of organizational designs.

Fourth, the method can be used when in the real implementation process unexpected problems manifest themselves. In a heuristic as well as in a testing mode simulation can give valuable information on potential effects of marginal or substantial adaptations in the strategy.

Fifth, social simulation is a suitable instrument to support ongoing implementation and monitoring processes. In fact, the image of organizing with which this paper started, enforces a new conception of strategy. Organizational processes cannot be controlled from one position and will never develop exactly as intended. Therefore continuous planning, ongoing adaptation of strategies, and participatory shaping is increasingly considered imperative. Social simulation provides a setting for developing, testing and improving such continuous and dynamic strategies. It may help to prevent major failures in real organizational change, or to react to it adequately.

Projected use: Currently we are building a social simulation of administrative automation in a service organization. It is designed both for research applications and for supporting automation processes in real (service) organizations in the ways suggested above.

Conclusions from Actual Use

Social simulation of the type presented in this paper appears to be a strong instrument, both for researchers and for participants, to study processes of change in organizations and to support their implementation. It also seems to be a valuable tool in actually shaping implementation strategies, especially if these are seen – as they increasingly are – as continuous processes in which in principle all groups within the organization play a part. The multiple nature of organizational reality (as defined by the participants) can be validly mirrored in a social simulation, which makes the method fit to experiment with preliminary ideas, or to identify unintended or unexpected consequences of more elaborated organizational change plans in advance.

Session Review and Concluding Remarks

In the workshop session the initial phase of a somewhat reduced and simplified version of the simulation was demonstrated (15 participants; duration 1.5 hours). Remarkably soon, a high level of involvement of the participants was attained, probably in part reflecting the "culture" of the group.

The evaluative discussion afterwards was largely devoted to the usefulness of this type of simulation for implementation of organizational change. The following conclusions can be presented as a summary.

- One or two hours of simulation suffice to realize a socially and psychologically valid organization pattern. Also the participants are, depending on the position they occupy, confronted with a number of organizational

problems, within this limited period. However, it will take far more time to analyse these problems, and to design and implement strategies to deal with them. Two days for a typical simulation can therefore be considered sensible. Intermediary evaluations to enhance analyses and invention of strategies seem very desirable if educational purposes are in order.
- The complexity of this type of simulation, which is in fact largely generated by the interactions of the participants, makes the method especially fit to study processes of change in which the multiple nature of the organization plays a major part, i.e. comprehensive changes whose fate depends on the complex interplay of different organizational groups with their own frames of reference.
- Application with existing groups or organizations might be sensitive, but, under appropriate conditions, also the most effective mode to contribute directly to actual change processes.

References

Becker, H. A. and Porter, A. L. (eds.) 1986. *Impact Assessment Today*. Utrecht: Jan van Arkel.
Guetzkow, H. Kotler, P. and Schultz R. L. (eds.) 1972. *Simulations in Social and Administrative Science*. Englewood Cliffs (NJ).
Meer, F. B. van der. 1983. [*The game of Organizing: social simulation as method in organizing research*], (Dutch). Enschede: University of Twente.
Meer, F. B. van der. 1986. Social Simulation: a research methodology and learning strategy. *In* Becker and Porter (1986).
Morgan, G. 1986. *Images of Organizations*. London: Sage.
Pfeffer, J. 1982. *Organizations and Organization Theory*. Boston: Pitman.
Schultz R. L. 1972. Developments in Simulation in Social and Administrative Science. *In* Guetzkow *et al.* (1972).
Swidler, A. 1986. Culture in Action: Symbols and Strategies. *American Sociological Review* **51**: April.
Weick, K. E. 1979. *The Social Psychology of Organizing*. Reading (Mass.): Addison-Wesley.
Wilhelm P. W. and Bolk, H. 1986. Strategy and the Art of Structure Maintenance: on the interplay of strategy and structure in science policy. *In* Becker and Porter (1986).

PACT

Maarten van Mens

NMB Bank, The Netherlands

ABSTRACT: In a process of organizational change NMB Bank decided to use a gaming approach to explore the issues of a less hierarchic more decentralized organizational concept and to support a change of culture. The aim is to become a more professional, more productive, competitive and innovative organization. This is supported by a parallel development of new infrastructures. The game PACT was developed as a simple language to describe working processes and exchange information; a carrier for the new organizational concept; a tool to redesign working processes; and a way to mobilize knowledge and skills within our organization.

PACT is designed as a chessgame. NMB uses it to illustrate and explore the consequences of a new (automation and communications) infrastructure, at the same time it is introducing the possibilities and consequences of an industrial, logistic approach to banking. It is a *generic* freeform game; a shell that can be translated to demonstrate and stimulate discussions about all sorts of organizational concepts. PACT needs a well defined organizational concept translated into simple rules for application, good process descriptions and a mixed audience of skilled and experienced company personnel to play.

KEYWORDS: PACT, war room gaming, end user involvement, logistical management, process orientation, organizational conditioning and design, industrial production concepts.

ADDRESS: *van Mens*, Maarten Ernst, Kastanjelaan 24, 1214 LH Hilversum, The Netherlands.

Introduction

NMB is one of the larger Dutch Banks, it has a staff of about 12000, 450 branch offices in the Netherlands and about 120 in other countries all over the world. A few years ago the Board of Directors concluded that NMB developed the wrong way. We produced services, but couldn't pinpoint exactly how they were made – or to exactly what purpose they were used by our clients. We had large productive processes, for example for transferring money but we knew little about the cost, or the work that was involved. Hierarchy was important, control predominant, Head office a "waterhead". Cost growth (9-10%) had begun to exceed profit growth (8%).

Within the Bank the administration process is predominant - i.e. a *supportive*, non-productive process. The result; our services are good, but inflexible, clients have to use *our* (administrative) standards i.e. when they transfer money. In a way we behave like Henry Ford I, inviting clients to buy a car any colour as long as it's black. Most Banks are like that. In this setting adaptation to new developments is slow, costly and will often meet an awful lot of resistance. At the other hand, competition is becoming fierce, Europe 1992 heralds an international integration, the situation has to change, become more innovative and client oriented, to keep competitive edge.

Example of industry. The idea was to take the example of industry as being professional, highly efficient and market oriented and translate their concepts to banking. There are a few examples, Citibank is one, of banks taking comparable decisions. In 1987 our research group tried to find out if these concepts were really applicable to a banking organization, and if so, what scenarios could be developed to introduce this major organizational change. We found out that industrial concepts are applicable and even highly interesting, but they needed translation to the specific setting of a bank, moreover; that a modern *industrial* approach is quite contrary to existing culture within a bank.

We concluded that we had four needs: 1. to chart out and reconnoitre existing processes; 2. an idea of how an "industrial" approach in banking would look; 3. an idea of the organizational consequences of our new infrastructure; 4. a programme to illustrate, talk through and gain general acceptance and understanding for the proposed changes.

Background

Banks today don't transport money, they handle information about money. Thus their main activity is information handling and their primary technology today is informatics (information + automation). In an industrial approach you concentrate on handling your core technology extremely well. The question was; if *we* did, the answer was no. Our business had a very large number of separate computer applications, if few people knew what their neighbours did, neither did our computer programs. The result of course is an enormous loss of information, inefficient data handling, a necessity to enter data superfluously (they often are already available somewhere else) etc.

An industrial approach puts people at the workbench back in control and supplies them with all necessary concepts, tools, means and skills to do it right. So in industry, the meaning of "market orientation" also is tranferred to an *intra*-organization concept – processes have to become user (worker) oriented as well.

Our first conclusion was that we had to start an internal communication process about work, have different corporate entities discuss about their respective roles in production and have them analyse their situation according to new rules. Besides hierarchic, our organization is both vertical product oriented and specialist, so there are a lot of different corporate entities with partial responsibility for specialized aspects.

In industry, production is tangible; you can say that production capacity is offered to the public via the intermediary of a product. Stocks can be used as a temporary receptacle of surplus production capacity.

Services are not tangible. Production capacity is offered to clients without the intermediary of a product and you can't temporarily stock overcapacity in for example half-products to relieve peak production periods. Nowadays however, banks do stock a lot of capital goods (computers/software). They also stock a large labour capacity and have some very large bulk production processes, especially for order entry. Therefore the concept of logistic management, taking care that the right things, are in the right place, at the right moment, with the right instructions (van Rens and de Leijer 1986), was applicable.

A look at some very successful applications in transportation demonstrated that something akin to an "industrial" approach, i.e. integrated process management, a logistic approach, modular concepts etc., was afoot in that sector. We derived some interesting notions from it.

Change of culture. Our concept had to support a change of culture. One of the main aspects of an industrial approach is decentralized responsibility. Industry has found out that this highly improves reaction time and flexibility. Russian "glasnost" demonstrates that even the Soviet state has perceived the value of this notion, but that it meets a lot of resistance within their hierarchy. We decided we would certainly meet a comparable resistance. A practical solution would be to involve *all* organizational levels (and partly disregard hierarchy) and our clients, building up internal as well as external pressures for change. We also needed markets; if you want to convert a Soviet economy into an open market economy, you need to create markets i.e. an open exchange of ideas and recompenses. We decided on a differentiated bottom-up approach of our organization.

The NMB Approach

We would first describe reality in co-operation with the executory level. Their recompense would be emancipatory; more grip on the work at hand, more responsibilities and a more interesting job. Then we would confront lower and upper management with their perceptions, introduce them to our conceptual approach and give them some ideas about how to cope with the new situation. Their recompense would be better control and more competence as a manager. This approach has been thoroughly described in the thesis of J. van Mens-Verhulst (Utrecht 1988).

We tackled our action as a companywide learning process. As the company had decided to redesign the information infrastructure, we had a good motive for this approach "acquaint yourself with the new infrastructures". It also provided a strong carrier for our message "a change of structure asks for new competences" to support this aim. Thus PACT is introduced as a training programme, part of a process to rebuild our infrastructure with the support of our personnel; to mobilize skills, knowledge and experience at all organizational levels, to realize quick acceptance and develop practical user friendly applications.

Industrial approach. Industrial concepts like for example; Flexible Production Automation, Just In Time, Total Quality, CAD/CAM. MRPII and many other industrial applications, offer a rich variety of concepts to support integrated, flexible, productive and very expedient industrial processes. In a way they represent a number of conflicting aims and solutions.

(a) *FLEXIBILITY*: is the ability to cope with a variety of specific client wishes. This is often solved by modularity. Modularity is costly however, you need a high degree of standardization, and rather complex management- and planning processes.

(b) *ADAPTABILITY*: implies being able to quickly start up new product lines, introduce new techniques etc. It is costly as well; you need a highly adaptable (= highly trained) staff; processes designed for easy replacement, easy adaptability to new techniques and "systems independent applications".

(c) *EXPEDIENCE*: i.e. quick responses. You need expedience to quickly comply to clients wishes, without keeping large stocks and to quickly react to market development. Responsiveness is expensive as it means being able to cope with a varying demand even at peak levels.

(d) *QUALITY*: Production and products will have to meet all internal and client specification 100%, no material wasted, do it 100% right the first time, no need for corrections. This type of quality costs a lot of money, but is essential.

These purposes are comparable to a training programme for top sportsmen. The aim is fitness to compete, our task, that of a trainer trying to build up condition for the Olympic games. As in life, the athlete (our company) has to win the medals himself. He has to develop his condition and pay for it with his energy. So line-management has to be primarily responsible. The interesting aspect is that industry tries to win these medals with the least possible exertion. The investment in external variety (flexibility), in reaction speed and quality, is compensated by a parallel reduction in internal variety; by integrating tasks, by substituting costly tasks, by reducing the number of components in products, by using robots, computers etc. extensively. This goal is only realized with a high degree of standardization. But standards (and automation) tend to reduce flexibility. The solution is standardization by component; not only product components, but information, organization, tasks and skills etc. The effect is impressive. The larger a collection of components is, the greater the variety of combinations that can be made. However such a solution is immanageable without a computer.

A Model. To find out what the effect of these notions would be on a banking operation, we had to develop a model. Thus we could test the applicability of these notions, find out what aspects exactly were important to manage the process and what conceptual differentiation would be necessary. Essentially, a banker manages two things, a flow of money and labour capacity for order processing. The basic process actually is quite simple, you need order collection, order evaluation, data-entry, information processing data transport and clearing.

Banks use computers for a dual purpose, both for management support and production (order processing). Our main product, MONEY, is also used productively. So, in our model we had to differentiate between four different flows and study their logistic aspects separately. The end result had to fit both our organization and our information concept.

Our research ended up with the following conclusions: We had to transfer a clear understanding of informational and organizational possibilities. We had to instir a discussion about their applicability throughout the bank. We needed a comprehensive translation of some very complex organizational and informational notions (not what they were but what they *implied*). These should be expressed as a simple set of rules for designing processes. We concluded that a gaming approach would be profitable.

A standard means for expressing and exchanging thoughts, a simple symbolic language would do. With this language, "virtual worlds" could be described (Klabbers, 1988). A workshop that follows certain rules had to support this action. We ended up with a gaming concept that looks like a game of chess (a war room game). There is one difference with a chessgame, the main aim is *collective organizational* competence. So there are no winners or losers. We developed a "toolkit", containing everything necessary to describe and analyse a process, a game and a programme to further organizational competence.

The Game

Ideally PACT is played in several rounds. It confronts people with theory and periods of practical application in between. Each theoretical input is first applied in practice. Workshops are held to compare experiences, exchange ideas, evaluate the value of the concepts and get some new ones. PACT follows a learning cycle (Kolb, 1971); *first*, to acquaint everybody with the symbols, the concept of logistic management and the procedures for standardly describing a process; *second*, to stimulate an exchange of ideas, skills and knowledge between fellow workers, to get a good model of reality (perception model); *third*, to evaluate the perception model on logistic, decision making and managerial aspects (like timing, specialization, synchronization, parallelization etc.); *fourth*, to introduce and assess our new infrastructure (e.g. a translation in potential effects i.e. integration of tasks, automation of routine jobs, interactive computer communications etc) to make an anticipation model (Niemeijer, 1983); *fifth*, a round to evaluate the anticipation model on its practical applicability and translate the results in practice.

This approach has several snags. An organization accepting it, must accept to do away, at least temporarily, with hierarchy. If management doesn't want to demotivate their personnel, they will have to do something with the outcome of this process as well.

The organization never will be the same afterwards. Involving people in developing a change perspective, giving them a grip on their own job will deeply influence existing culture. It's an irreversible experiment.

Playing PACT

After a brief introduction to the aims, symbols and conventions of the game, people are asked to design the process of making Nassi Goreng for twenty persons (a simple traditional recipe originating from the Dutch East-Indies) using our "toolkit". The aim is to do this in the least possible time. The group is divided into several sub-groups. Care is taken that there is a more or less equal repartition of skills and knowledge. Moreover, everybody can freely exchange information, ideas, suggestions etc. The result must be a process design, that demonstrates the elements time, capacity, quantity, productive steps and process organization.

Each group is then invited to cook the meal in practice, following their design *exactly*! Results are compared and tasted, the group making the tastiest meal within the shortest time gets a commendation. After this groups can reconvene, compare results and redesign their process. Then a first workshop is held with the aim to integrate individual results into one process design in collective agreement.

We took the example of cooking, because the essence of cooking is logistic. The advantage of cooking is that everybody is familiar with it, but few people (excepting professional cooks) consciously plan their *activity* aforehand and few men cook meals in practice. The resulting experience will give a basic understanding in logistics and the use of symbols.

Application. We then invite people first to describe and to analyse their own job and confront each other with the results in another workshop. This will instir a vivid communication between participants, that must be well led and documented; an elicitation of group knowledge. The result is a perception model. The group becomes conscious of interdependency, major problems, essential productive steps etc.

Process language. In this phase of the game, we start giving a language lesson. A process can be characterized by a string of verbs (a sentence) each singular activity (process component) is a verb. Designing a process can be compared to applying a grammar. A good use of grammar will yield an elegant sentence. Bankers have a preference to use complicated language, often making themselves difficult to understand. As we found out in practice, this complexity is often mirrored in their process design. In a process language, you need different grammars for different purposes. For example, when organizational concentration is deemed to be important, you will see many verbs like sending, batching, checking etcetera in process design. If decentralization is an important goal, these typical *auxiliary* verbs will occur much less frequently, but there will be a much more repetitious occurrence of certain primary verbs.

The string of verbs of a good process description gives away a lot about its designers in the past. At the ISAGA 88 conference I gave a short course on banking archeology to illustrate this point.

The main reason for this, is that process upkeep is often fragmented. As new organizational concepts are added on top of old ones, vestiges of all past concepts remain. Old processes thus contain quite a few inconsistent steps.

Cross Referencing. We already indicated that communications between the different departments involved in a working process, is often stagnating. For this, we organize specialized, interdepartmental workshops. We invite them to present an analysis of their own part of the job to each other, to study the process as a whole, to challenge the necessity of certain steps, to state their problems etc.

Such a session typically starts with a discussion if the situation presented on the table could be considered as true to life, optimal or not. Each department will point out what measures are taken to cope with problems, how they *manage* their part of the process and what steps they take to ensure good quality.

In the following session, the departmental groups are mixed, they are instructed to analyse a random part of the total process, to pinpoint the verbs that are essential/characteristic for it, auxiliary and potentially superfluous verbs. They can look for possibilities to change the order of verbs in their part of the process. This season very often leads to shorter process sentences already.

Participants then present their results. If there is general consent, they may change the order of steps. The aim is to leave as little symbols on the table as possible, but to preserve the essential aspects and performance of the existing process. This is achieved by asking the groups to edit logical i.e. coherent and consistent subcollections of verbs.

The new infrastructure. When a process has been described, cross-referenced, analysed and restructured, we introduce NMB's new infrastructure. Participants are invited to apply their knowledge and skills and evaluate the consequences of applying a decentralized computer network. We translated our network concept into a number of practical organizational possibilities. This round is combined with simple prototyping tools, using NMB's standard PC computer programs. In this network people can define their own process-steps, link up to each other and see the consequences for themselves. In most cases these consequences are;

- doing away with nearly all remaining routine jobs (administrative work often can be produced as a data-entry by-product).
- a better and quicker information and production/planning support if a good *distribution of information* is organized.

We actually want to demonstrate that applying new organization concepts has a much greater impact than introducing a new computer infrastructure. The effect is that people find out that "informatics" are a supportive tool giving them a far larger variety of action and far greater chances for "enactment" than before (Weick, 1979).

Conclusion

In banking, the major investment is in labour. Our aim is to flexibilize labour. It means integrating tasks, using computers and giving different tasks to the same person. Above all, it means using sound judgement. Logistical management, taking care that the right things, are in the right place, at the right moment, is important for this development.

The effect of logistical management, e.g. stressing the supportive managerial aspects and leaving the productive aspects to the executory level, is very fundamental. It puts good old handiwork (in a new coat), back into the front seat. PACT has been designed to support a companywide process to redefine and reorganize work, on the light of new organizational options. The PACT concept has proved itself a good and very flexible carrier to present this message to our company.

References

Klabbers, J. H. G. 1989. Representation of virtual worlds through gaming-simulation. *Proceedings of Simultec 1987*. Geneva, Mondavio: TE.COM.
Kolb, D. 1971. Learning and problem solving. *In* Kolb *et al.* (1971).
Kolb, D. *et al.* 1971. *Organizational Psychology, an experiential approach*. Englewood Cliffs (N.J.): Prentice Hall.
Mens-Verhulst, J. van. 1988. *Modelontwikkeling voor vrouw-en-hulpverlening*. Thesis with English summary. Utrecht: University Press.
Niemeyer, K. 1983. A contribution to the topology of games. *In* Stahl (1983).
Rens, J. van, and Leijer, I. de, 1987. *Thematrans, transporttechnologie en organisatorische ontwikkelingen*. NVI.
Stahl, I. (ed.) 1983. *Operational gaming, an international approach*. Oxford: Pergamon.
Weick, K. E. 1979. *The social psychology of organizing*. Reading, Mass.: Addison Wesley.

MIDAS: an awareness game on innovation management

Dolf Dekker

Dekker & Hekman bv, Management Consultants, The Netherlands

ABSTRACT: Innovation is currently one of the main issues in management practice. Managers are reluctant however to start an innovation programme. They feel uncomfortable in taking such a high risk decision because they don't know how to manage innovation processes. Therefore, a powerful tool is needed to motivate managers and their staff in taking the decision to initiate an innovation programme.

In this paper such a tool, an awareness game on innovation management, is presented. Its objective is to give managers the *feeling that innovation can be managed*. This is done by:

1. making managers discover that the way in which they normally manage the new product process will not meet future goals of the organization
2. decreasing the fears and anxieties about innovation by giving them an experiential learning exercise in the steps and techniques used in innovation processes
3. making managers aware of the factors which influence the innovation process and allowing them to experiment with decisions which foster (or block) the innovative climate in the organization
4. starting a structured discussion whether or not to innovate and establishing the requirements for an effective programme.

KEYWORDS: Innovation; awareness; management; game; consultancy;

ADDRESS: Dekker & Hekman, management consultants, Schoutenkampweg 28, 3768 AE Soest, The Netherlands.

GHOSTS IN THE MACHINE:
a computer-aided simulation/game to explore the relationships between strategic policies, tactical action and organizational cultures

Alan Coote and Clive Loveluck

The Polytechnic of Wales, UK

ABSTRACT: This paper discusses the development and use of GHOSTS IN THE MACHINE – a flexible, interactive computer-aided simulation/game designed as an aid to management development. The authors discuss the rationale behind the game before outlining how it was developed and suggesting different ways in which it may be used and adapted to suit a variety of user needs.

KEYWORDS: Computer-aided; simulation; game; organizational culture; strategy; tactics; organizational change.

ADDRESS: Alan Coote and Clive Loveluck, Department of Management and Legal Studies, The Polytechnic of Wales. Pontypridd, Mid Glamorgan CF37 1DL, Wales, UK.

Section Three

Business Simulation

Editor: **Willem J. Scheper**

Business simulation: an introduction

Willem J. Scheper
Utrecht University, The Netherlands

Papers in this section are dealing with business simulations. Themes like uncertainty reduction, corporate planning, organizational behaviour, organizational structure and -culture and leadership styles are addressed. However, additional to these general themes is a discussion of topics that far exceeds the realm of business simulation. These topics have general significance for all those who are working in the field of gaming and simulation. Examples are: group dynamics, methodological issues such as the student-instructor distinction and the use of gaming-simulation as a research tool, the effectiveness of gaming and simulation, gaming and simulation and ethics.

Schulein and Borawitz both discuss methodological aspects of gaming simulation when they are using simulations as a research tool. *Schulein* uses gaming simulation in the realm of crisis management. Crisis being a situation defined by extreme uncertainty. Players of a crisis game have the opportunity to define the management structure of the organization they are part of, allowing the comparison of different structures. Other aspects that are dealt with in his paper having significance for designers of business simulation are the modelling of organizations and the handling of unreliable data. *Borawitz* discusses wargaming. The SOLTAU wargame has been used as a research tool, investigating the effectiveness of different army tactics. From this discussion, researchers interested in corporate planning might gain some valuable insights concerning the use of gaming simulation as a research tool in the process of strategy-formulation and -implementation. Borawitz also discusses the assistance of computers in *gaming-simulation*, the advantages of graphics in simulations and the organizational embedding of (war)games.

In his contribution *Freeman* contemplates on the use of gaming-simulation as a research tool. He is interested in the relation between gaming-simulation and the corporate planning process. WHOLETRAIN is a simulation designed for training managers in goal-setting skills. These skills are being viewed as necessary for long-term viability of the organization. A study on the effectiveness of WHOLETRAIN has been conducted, showing that it has theoretical as well as practical value. Theoretical value because it can be used as a tool for advancing theoretical understanding in the area of work motivation, practical value because it can be implemented as a training tool in goal setting skills.

Whiteley and Faria question the pedagogical value of simulations. They conclude that the evidence from past research as to whether business simulations are a more effective teaching tool than other instructional approaches is inconclusive. The authors set up their own experiment using the simulation LAPTOP. The results of this experiment show that "simulation games are an effective means by which to improve quantitative skills but not an effective means by which to improve the acquisition of applied or theoretical knowledge." However, they do not discuss the theoretical basis of LAPTOP.

Lobuts introduces the notion of "perturbatory management", being a style of management that perturbs people and causes anxiety, frustration and dysfunctionalism and thereby affecting organizational operations. In the long term, viability of an organization characterized by this management style is at risk. Perturbatory management is caused by a one-sidedness of perspective, an attitude of "winning at any cost". However, in the end this type of attitude has destructive effects. Lobuts designed the ESCALATION GAME in order to prove this thesis. Including ethics and morality in the organizational life of every day is in his view a necessary condition for overcoming the problems associated with perturbatory management.

The importance of ethics in business and business education is also recognized by *Bates, Moore and Christopher* in their paper on the Australian Rehearsal Technique (ART). Simulation games are perceived as a form of drama. In ART several roles are distinguished: actors, audience/commentators, drama director and game overall director. The authors describe a game called LADY JANE HOSPITAL in which the participants are faced with the ethical implications of decision making. During debriefing it is the audience that dominates the conversation, producing more variety in their comments as compared to the more traditional role-plays.

In their contributions (included as abstracts) both Teach and Christopher are dealing with differences in ethical standards, attitudes and opinions, as they are experienced by companies taking over other companies or by companies doing business in other countries with different cultures (like e.g. multinationals). In order to make managers more aware of the problems resulting from intercultural differences *Christopher* designed TALKING HEADS. Its purpose being to make players more aware of a range of cultural factors that affect the conduct of a business meeting involving people of different cultures. *Teach* directs the attention to the designing of an intercultural business simulation. His objective is twofold. Firstly, an intercultural business simulation should improve the ability of recognizing cultural differences between people, and secondly, it should increase the understanding of the effects cultural differences have on business. Teach proposes a game structure that in his opinion might succeed in accomplishing these goals.

Also included as abstracts are the contributions of Brech, Thavilkulvat and Chang and Crookall and Saunders. In his contribution *Brech* focuses on the methodological consequences of multiple realities for designing business simulations. In his opinion, computer simulations can be used to facilitate learning by doing. In this process computer simulations help students to recognize and correct mistakes they may have made while running the simulation. Students are in the position of experimenting with different policies and they may develop a certain feeling for errors to occur as a result of particular (non-) decisions. Brech argues that the role of the instructor has to change in order to match the students' needs. Given multiple realities there is no ultimate truth. This implies that the instructor/tutor does not give the answers but asks the pertinent questions. Subsequently, students try to find possible answers.

Thavilkulvat and Chang are dealing with the differences between students and instructors. There main objective is to increase the user friendliness of a simulation for both students and instructors. In their paper the simulation MANAGEMENT 500 is presented as an example of a simulation designed to meet user-friendliness criteria.

Crookall and Saunders discuss wargames. They study possible ways in which a wargame can be adapted for other issues. Game materials used during the workshop are presented to enable others to pursue ideas.

As the reader of this introduction will have noticed, many topics related to gaming-simulation in general and business simulation in particular are addressed. Despite the many differences that can be listed between the simulation-games described in this section, there is one property they all have in common: they all are designed for the improvement of competence of those who participate in one of these simulation-games.

Crisis gaming for research and training

Peter Schulein
FEL-TNO, The Hague, Holland

ABSTRACT: In business, management games have been played for a long time, although most applications have been in the financial or economic area. Within FEL-TNO, management gaming is confined to controlling an organization in times of crises. Crises gaming consists of handling complex decisions in a short time period, based on incomplete and/or unreliable data.
Players of a crisis game do not play against an opponent, but try to survive in a supervisor-controlled (mostly hostile) environment. An extra dimension is added by allowing the players to define the management structure they will play in themselves. This allows the comparison of different structures. Besides research into the design problems of such a game, the following aspects are being studied:

- how organizations can be described and modelled formally;
- in what way crises can be studied and defined;
- how management can be defined and modelled;
- in what way unreliable data should be handled;
- how leadership can be defined, and how people should be prepared for it.

An application of the research effort is the development of a crisis game for the Royal Netherlands Air Force Staff College. Our experience in designing and developing this game will be discussed.
KEYWORDS: Management game; crisis game; process model; management model
ADDRESS: FEL-TNO, P.O. Box 96864, 2509 JG The Hague, The Netherlands.

Introduction: What is a Crisis Game?

This paper is about organizations. An organization is defined as a set of activities bound together by a common goal. These activities can be thought of as processes. The actions of the organization that are directly related to the achievement of the common goal are called the primary processes. Unfortunately most organizations are not isolated, but are placed in an environment. This environment can influence the organization both in beneficial and detrimental ways. A detrimental influence that cannot be handled within the primary processes of an organization is called a crisis.

This paper is about crises. Assuming that a detrimental effect is recognized within an organization, two possible solutions exist to end a crisis: either the crisis destroys the organization (or a part thereof) or the organization includes a set of processes specifically meant to counter the threat and uses them to survive. These processes are called the Corrective Processes.

A crises game is a simulation that allows one or more players to control the use of the Corrective Processes to minimize the detrimental effects of one or more crises on a given organization.

In the context of this text the environment and its influences are supervisor controlled, thus allowing direct control of the problem scenarios the player is confronted with.

Modelling an Organization

If an organization is to be the subject of a simulation, any kind of simulation, a model of the organization must be build. To gain the knowledge needed to build such a model, a description method is needed that will suit both the research analyst and the members of that organization. This means the method should be straightforward, easy to learn and preferably graphical. The method that is currently used in the development of games is based on 6 symbols (see Fig. 1).

An activity in the organization is modelled by a process (symbol: circle). Processors are related to objects (symbol: rectangle). Objects which cannot be produced within the organization are called resources (symbol: rectangle with double sides).

Processes can consume, produce, refer to, or use objects (resources can only be consumed, referred to, or used). This is indicated by a relation (symbol: arrow). Objects that are used are returned after the process has been active. This is indicated by two parallel arrows. Objects that are referred to, are considered not used but copied and consumed. This is indicated by a dashed arrow.

One important feature of this method is that the definition of an object as an input for a process, does not necessarily mean that this object is available at a given time. Only if the object is available, a process can become active, otherwise it will stay passive. The process description is time independent. If a description of the organization *at a given time* is needed the description must be augmented with a list of available objects and process stati.

With these symbols a variety of organizations can be modelled. See the example in Fig. 2.

This method is further elaborated and illustrated in (Schulein, 1988).

Defining Management

If a model of the operations of an organization has been defined, it is possible to build a simulation of the organization according to that model. Using this model in a management game however requires a management model as well.

The first problem is: how to define management? A figure such as Fig. 3 is common in the literature but has one large drawback: the interaction between the management and the organization is not defined clearly. For modelling and gaming purposes it would be most advisable to describe an organization and its management using the same symbols.

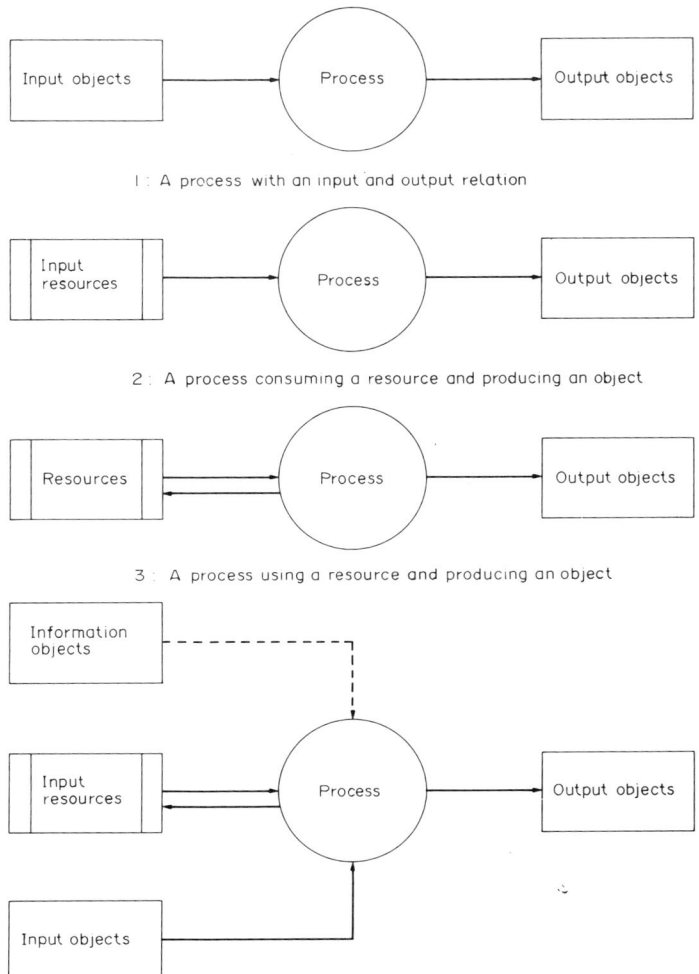

FIGURE 1 Symbols for organization modelling

Therefore an attempt has been made to integrate the management and the organization in one scheme. This scheme is presented in Fig. 4.

In this scheme the players of a management game function as the management process. Their task consists of three parts:

1. to get information about the elements of the scheme,

2. to assess if corrective action is required, e.g. to assess if a crisis has developed,

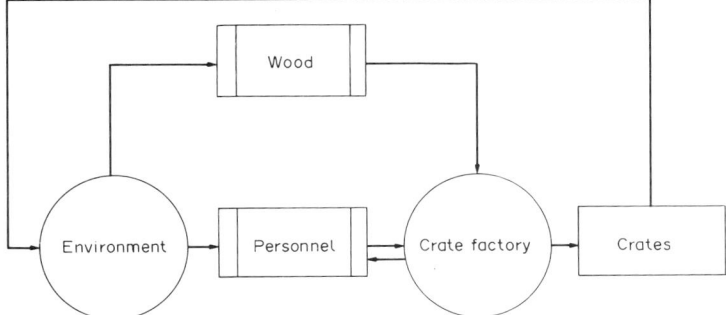

Process scheme "crate production"

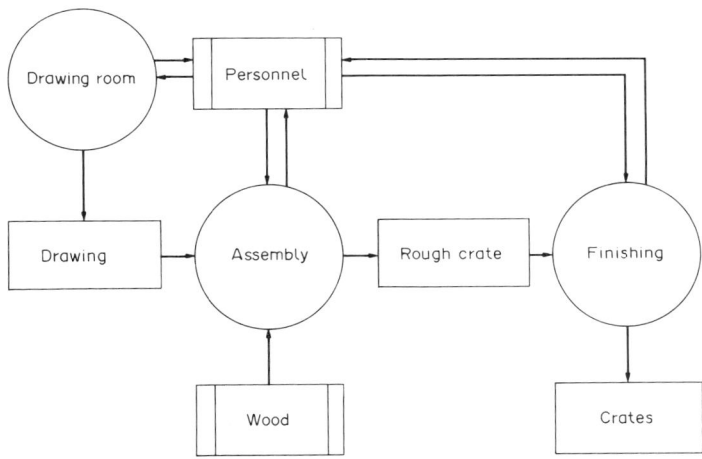

Process scheme "crate factory"

FIGURE 2 An example of a description: "Crate Production"

3. to activate the corrective processes that are available to counter the crisis by reallocating the resources of the organization to those processes. This is also elaborated further in (Schulein, 1988).

Defining the Player

What is a player within the management game definition that is presented above? The player or players represent the management of the organization. In model- or game-terms the player is related to the game by the *information he can extract* from those parts of the organization he has access to and the *commands he can give* to the parts of the organization under his authority.

The sum of all information the players can extract and the sum of all the

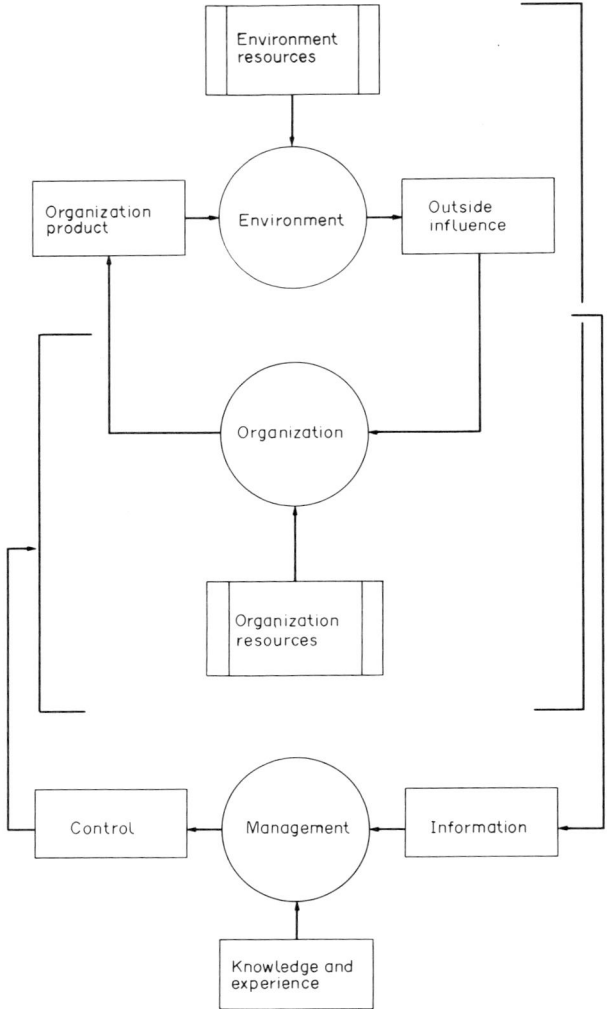

FIGURE 3 Management and its organization

orders the players can give define the total set of information the model can generate and the total set of commands the model can respond to. Any management structure can be defined by redistributing the information flow and order acceptance over a variable number of players. This is illustrated in Fig. 5.

Applications of Crises Gaming

The applications of crises gaming are threefold:

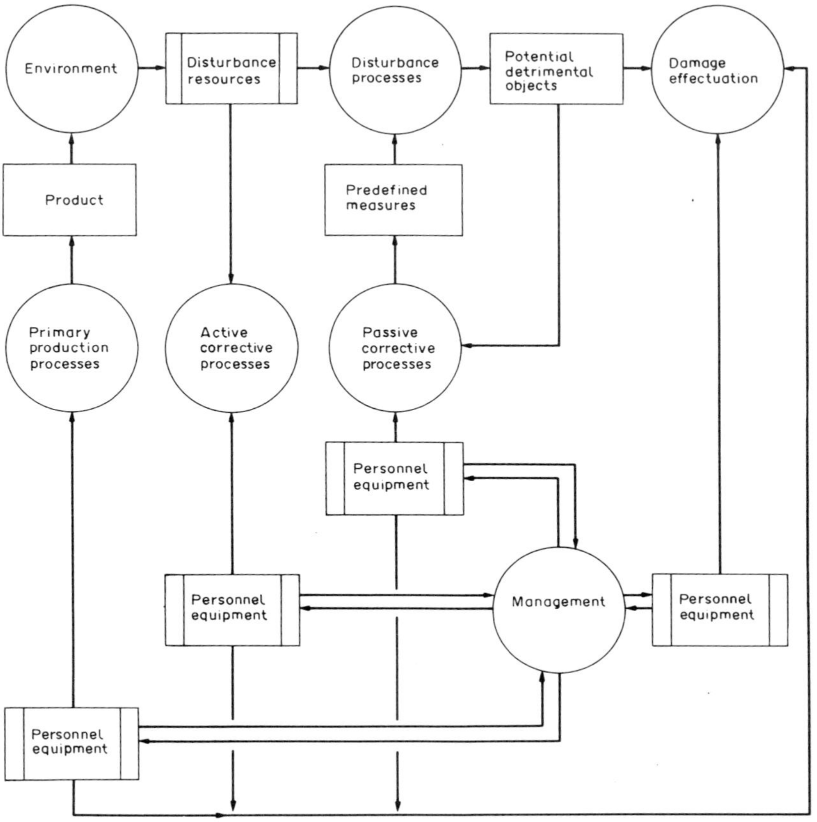

Crisis gaming for research and training

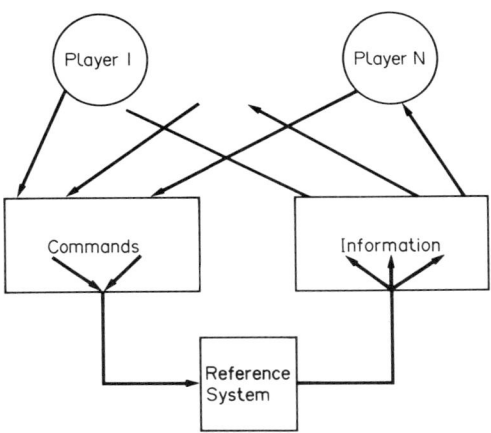

FIGURE 4 Integrated Organization Scheme

FIGURE 5 Defining a management structure

1. Since a crisis game is based on an underlying simulation of the organization, it can be used to train players in *handling the organization*. The game transfers the facts and empirical knowledge about the organization that were collected to design the game in the first place. With these facts the player can gain experience in handling the organization himself.

2. After the players have acquired the necessary knowledge about the organization, the game can be used to create situations in which players are trained in *handling crises*. The training staff will present the players with incomplete, untimely or wrong information without the players knowing it. Emphasis lies on learning management skills, not getting to know the organization.

3. Because of the special way in which a player is defined it is possible to redefine the number of players or their associated relations. As the player structure represents the management structure of an organization it is possible to use the game as a research tool to design an adequate management structure for the organization. There is no difference between a management game and research tool.

Experiences So Far

In preparation for the use of the management game under construction now, a number of scenarios have been tried in case-study formats. Based on those try-outs and our experiences in the design and the construction phase a number of interesting issues can be generated:

The use of computers in gaming is indispensable if the effects in time and the interrelationships of simultaneously occurring problems and their solutions, have to be simulated. Without the support of computer-assisted games it is too difficult to keep the large dataset consistent throughout the gaming session.

On the other hand it appeared that a paper case study is more suitable for problems that demand a single elaborate solution without the need for knowing detailed consequences in time.

The modelling of the organization can be fairly abstract and as such relatively small in terms of programming. An important factor is to recognize which details are important for the simulation and which details are only needed to give the players the "feeling of the real thing".

A difficult design problem is the contradiction of designing a game that will fascinate the player by simulating the organization as realistic as possible on the one hand, and the necessity to analyse the problems in the scenario from a distance to be able to generate the most promising solutions on the other hand. This could be the underlying difference between a research game and a player game. A player needs to be enthusiastic about the game and needs to engulf himself in it. An analyst should try to think as open and from a distance as far as possible. Thus if a game is used in both applications, it is advisable to separate the simulation essentials from the game details.

The implementation of a game based on a process model description method as introduced in this paper is fairly straightforward. An implementation method based on lists and an event simulation mechanism has been developed in the Pascal language. Extended use of standard procedures makes expansion easy to accomplish. It proved also possible and practical to build a meaningful game on microcomputers such as the IBM-PS/2.

One problem that has not been solved yet is the use of computers by the players. We raise the question: should players play by computer while they don't have access to computers in the real world? On the one hand the use of computers enables the design and construction of very user-friendly and flexible games; on the other hand they require an introduction to use, tend to create a "beat the computer" atmosphere and tend to introduce "computer glamour" that often distracts the attention from the real game essentials.

It is difficult to rate the players work. The problems given to a player in a gaming session typically have no best solution. There exists no ultimate answer to a given situation. Thus the players cannot be judged. So in order to make playing the game a worthwhile learning experience, the design of scenarios in which possible and probable decision moments are incorporated is the most important part of the preparation of a session. Evaluation of the players decisions by real world experienced personnel is almost a necessity.

A last point of interest is the cost/benefit analysis of a game. The costs of the development of a game are significant. The benefits are twofold: the game can be used as a flexible training tool and while the game is under construction a lot of information about the target organization is generated (in easy transferable form such as the schemes presented earlier). Still, a game is only a good training device alternative if there are no other means to create the same experience, because of the difficulties in designing and the cost of constructing a good game.

The game that is being built now should be ready in prototype format by end of 1988. It is expected that the game will be in use for the training of the students of the Royal Netherlands Airforce Staff College by the second half of 1989.

Session Overview

During the session it became apparent that the problems presented are problems that bother or have bothered us all. An important point to note was that some of the underlying thoughts, in the design and construction of a *crises* game, are equally applicable to a pure *business* simulation in the classical sense of the word.

Leadership skills and uncertainty handling are prime focal points in crises gaming. These skills are mandatory for the management of any organization and are a logical follow-up on more traditional business handling skills.

The quality of shared responsibility was placed in the context of fast decision making that often is required in crises situations. On the other hand the need to slow down the actions of a decision maker until enough information is gathered to analyse the whole situation, was explained by some examples. The discussion about accepting a small loss now for a larger gain later is very difficult and probably an inhuman discussion if human life is threatened.

References

Schulein, P. 1988. A process description method for organization modelling. *TNO-IR 1988-19* (in Dutch). The Hague: TNO-FEL.

Wargaming*

W. C. Borawitz

Physics and Electronics Laboratory TNO, The Netherlands

ABSTRACT: In this paper the command post exercises are described that have been supported by the SOLTAU wargame. A number of positive and negative aspects as a result of the experience with the aforementioned exercises are discussed.

As it is well possible to use wargaming techniques for training we look ahead to planned developments within the Netherlands regarding the computer-assisted Command Post Exercises.

KEYWORDS: Command post exercises, wargaming, training, computer assistance, research games.

ADDRESS: Ir. W. C. Borawitz, Physics and Electronics Laboratory TNO, Oude Waalsdorperweg 63, P.O. Box 96864, 2509 JG The Hague, The Netherlands.

Introduction

Development of combat-simulation modelling was started in the 1960s at the former Physics Laboratory[1] of the Netherlands organization for applied scientific research (TNO).

The first running model was EINFALL (1968), originally developed by Scientific Control Systems Ltd for the German Forces (Vink and Van Veen, 1975) and later adjusted to the Netherlands situation.

In 1973 EINFALL was used in an experiment (Swinkels, 1973) for real time wargaming.

However the long computer response times and the relative user unfriendly input/output restricted the further applications of the model only to research simulations.

Since the 1960s the tactical exercise SOLTAU has been conducted annually at the Army Staff College, for the advanced "education" of senior staff officers. The exercise was manually controlled and the request for computer assistance was frequently made. In 1981 the former Physics Laboratory was tasked to provide computer assistance for the SOLTAU exercise, hence the name for the current model: SOLTAU-wargame. The Army Staff College defined the following important requirements for this development:

- the total number of players and controllers should be no more than approximately 25 officers:

---

*"The paper is a revised version of an earlier paper, published in the proceedings of the NATO AC/243(Panel VIII) Symposium on Wargaming, February 1987".

- the model must perform in real time with minimal technical assistance with four available I/O stations at the Army Staff College connected to the Physics Laboratory CDC-CYBER computer system.

This means that the model has to be simple and fully user-oriented (Wierink, 1982 and Van der Velden, 1983). The SOLTAU wargame therefore contains only the most important aspects of the G2 (intelligence) and G3 (operations) chains: combat units, artillery units, terrain, obstacles and some aspects of engineer activities.

Computer-assisted Command Post Exercises

Basic organization

As depicted in Fig. 1 the wargame is the beating heart of the lively battle of the player staff or staffs.

Only exercise control and the player staffs are not directly connected to this centre. The exercise control unit is managing the exercise through direct contact with all of the functions and is directing the enemy players and the control units to reach the exercise goals.

The player staffs are linked to their own operational commanders and command their own units (lower-control). Depending on the exercise set up, the player staffs can be in the field in their own tactical deployed command posts or in another location. When not operating from tactical deployed command posts, communications should be as real as possible.

Higher control commands the player staff and must perform functions such as engineer or artillery in support of the player directly into the wargame. Higher control does not control the enemy players, but does control the tactical actions of the player and as such is responsible for a large part of the exercise goal(s).

Flank control fills the information gaps left by higher control to give both enemy players and player staff(s) a complete battle picture.

The enemy players are traditionally fighting the battle according to some basic settings made prior to the exercise to make the exercise successful.

The monitor control is introduced to sort out or redirect situations due to wargame shortcomings and control unit errors. One important task is to inform both higher and exercise control of tactical situations prior to their "discovery" by lower control, player staff(s) and higher control. This task if neglected will make exercise control impossible due to the large number of imperfect interpretations and long delay times from lower control through player staff(s) to higher control.

Lower control is a crucial function to be performed in the computer-assisted CPX. Lower control has a number of difficult tasks: to order his units in the wargame, to ensure correct results (both tactical and technical), to interpret wargame messages, translate these to tactical information and then send this to his commander in the appropriate way. The purpose is to make his commander believe he is in actual battle (and not playing a game with the computer).

The technical assistance is tasked to run the wargame on a set of computer systems, to provide the control-units access to the wargame (input and output) and to assist where necessary to sort out technical or wargame problems.

The basic organization for large-scale exercises does not differ much from small exercises.

The larger scale spreads the functions over a wider area and communications will accordingly be over longer distances.

As far as computer assistance is concerned, the number of computer connections will be larger and more attention must be paid to the system reliability (back-up procedures and systems) and security regulations.

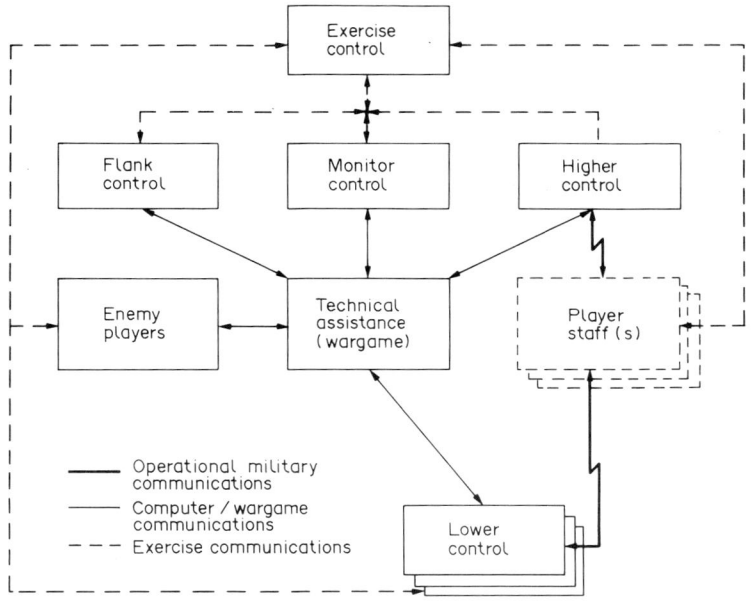

FIGURE 1   Computer assisted CPX, basic organization

1 (NL) Corps Battalion CPX's

Since the introduction of the SOLTAU wargame at Army Staff College (1983) six exercises in total were successfully conducted by the Army Staff College.

In 1984 a request was made by the G3 of 1 (NL) Corps to investigate the possibility of using the SOLTAU wargame for battalion CPX's.

The questions to be addressed were: is the terrain digitization used in SOLTAU (1 × 1 km square grid) accurate enough and do all implemented algorithms perform correctly for platoon level units instead of company level units.

In the field exercise with the 59 Armoured Battalion staff (June 1985), the wargame proved to be a valid tool to assist in a battalion CPX. The battalion CPX is used to train a mobilizable battalion staff once every 2 years.

The battalion staff is located in its tactically deployed command post at the training grounds nearby.

Communications with the control organization and technical assistance cell is established through the normal military communication systems. The exercises are on 24 hour basis for 2 to 3 days. Currently (February 1987) six exercises have been computer-assisted and the general opinion is favourable. The exercises contain more realism and contribute to the better understanding of all participants of the consequences of combat actions and the related reaction times.

In the next 2 years at least 10 battalion CPX's will be computer-assisted. (More are expected when brigade CPX's incorporating standing battalion staffs are computer assisted.)

Research games

Since 1984 four research games have been carried out with the SOLTAU wargame to investigate army tactics for implementing scatterable mines and attack helicopters.

These research games were carefully prepared by project teams studying the introduction of these weapon systems and were executed at the wargame facility of the Physics and Electronics Laboratory under close control of the project teams.

The research games have contributed to the confidence in the research results and experience of the project teams and have pointed out the basic problems and assumptions.

A small number of other research games is anticipated during the coming years, but is dependent on the availability of 30–50 officers/specialists.

Exercise "First Impression"

In June 1985 (after the successful computer-assisted CPX with the 59 Armoured Battalion) the Commander of 1(NL) Corps decided to organize his first computer-assisted Corps CPX. The goal of the exercise was to acquaint the participants with the computer and to learn from the playing of five different Corps defence scenarios. Three pre-exercises were carried out with a restricted number of participants: one to test the data-connections and I/O system performance (August 1985), one to test the control set-up (October 1985) and one to instruct the lower controls a week before the real exercise. The Command Post Exercise, First Impression, was conducted in December 1985 at Twente Airbase and lasted 1 week.

In total 800 persons participated from the brigade staffs, Corps staff, artillery staff, and the staffs of 4 divisions (1, 4 and 5 Division of 1(NL) Corps and 3 Division of 3 (GE) Corps[2] acting as "players". The players (the division staffs) were not tactically deployed but were located in buildings at Twente Airbase, a short distance from the Corps HQ and the control organization.

The technical assistance cell was centralized in one room. All I/O from and to lower control was distributed by runners. The computer system used in the CPX was the CYBER in The Hague at ca 200 km distance, connected by two scrambled parallel communication links (one civil data link, 16 channels wide and one military data link, 16 channels wide). When one of the two link types was inoperable, reconfiguration along the remaining link type was realized within 5 minutes.

Two micro-VAX's were on line performing the graphics assistance and also were used as back-up capable of taking over the wargame as a whole. Alas even two parallel data links and 2 back-up computer systems can be out of order at the same instant. Fortunately this only happened during the last pre-exercise and lasted less than 2 hours. This finalized the decision to work toward the full implementation of the SOLTAU wargame on local computer capacity with internal back-ups.

The influence of the computer assistance on the exercise was considerable:
(a) The degree of realism was excellent, for example:
   – deployments and activities were played real time; thus movements of reserves took a realistic time to achieve.
   – There could be no argument over the results of engagements since the computer is neutral.
   – Lower control, players and the enemy were confronted with the results of their decisions, or of their failure to make timely decisions.
(b) constant foward planning and anticipation of events was required by lower control and players to pursue the exercise in a realistic time frame.
(c) proper co-ordination of various staff functions at the various levels of command is measurable. For example: if fire and movement were not properly co-ordinated, the results from the computer would clearly reflect the error.
(d) Lower controls also received important training since they were, in a sense, also players and therefore neutral toward the computer. However lower control must learn not to fight the computer but to use its output and react to it tactically.

Exercise "Second Impression"

For 1986 the new Corps commander decided not to have a full computer-assisted Corps CPX but a smaller "research-like" Corps staff exercise.

This research-like exercise was to concentrate on tactics and the consequences of major assumptions. To investigate at least three main scenarios, a full two-weeks exercise (12 hours of real-time battle per day and 2 hours of debriefing) was planned. The exercise was performed in September 1986 at the Physics and Electronics Laboratory wargame facility.

As the SOLTAU wargame only represents the basic G2 and G3 chains, a number of battle support functions were not played such as logistics, air defence and air support. In order to minimize the number of participants the "division player staffs" were permitted to interact directly with the wargame, making lower control obsolete.

The overall number of "players" was 50 officers and NCO's representing the G2-G3 chains down to brigade level. A number of important insights were gained by the participants in the planning and execution of corps and division operations.

Exercise "Kastanje Dubbel"

The computer-assisted division CPX "KASTANJE DUBBEL" was conducted in February 1987.

In this exercise two player brigade staffs operated from their tactically deployed Command Posts at locations some 30 kilometres away from the control location at Zelhem (NL).

Problems and Advantages

Mobile versus stationary wargame facilities

Much discussion is ongoing in the Netherlands about how wargaming "facilities" are to be integrated into the organization. There are two options: mobile or stationary facilities.

The advantages of stationary facilities are obvious. One can optimize the lay-out of the facilities eliminating frequent movement and installation of vulnerable computer equipment.

The advantage of mobile facilities is that one can go to a unit's location. In fact one can go to the actual terrain where the exercise is played both in the field (terrain reconnaissance) and in the wargame. In the Netherlands this will probably lead to one stationary set-up with a computer system and (lower) control housed in a central part of the country to satisfy the needs of larger exercises and one or two mobile systems for the smaller exercises and for special purposes.

Wargame model assumptions

In the SOLTAU wargame as within every wargame a large number of assumptions is incorporated.

It is crucial for the control organization to know *all* these assumptions as well as their estimated impact on the battle evolution. One cannot accept the wargame as an instrument without knowing its "ins" and "outs", especially if the wargame is used for training and teaching.

It is very important to keep remembering that if the computer is fed with "wrong" rules, data or assumptions, one will train/teach wrong ideas and tactics! (Garbage in equals garbage out.)

It should be clear that a wargame directed to support a command post exercise will not reflect all aspects of reality. The purpose of a command post exercise is to train a staff in decision making, planning, procedures, operational control, and operation in a simulated combat environment. To realize these goals it is necessary to simulate the battles and operations at an accelerated pace to ensure enough battle within a limited training time (several days).

For the participants it should therefore be clear that the results of a computer-assisted exercise are not to be taken literally. However wargames are suitable for the testing of battle plans of own forces against enemy force battle options. Wargames for education or in support of a command post exercise are often simple to adjust for a particular purpose and in that case give an important contribution to the testing of existing or newly developed tactical doctrines and logistic procedures.

SOLTAU Wargame model shortcomings

The SOLTAU wargame by its design has the following shortcomings:

- the terrain, manoeuvre and artillery unit resolution will only credibly support brigade and higher level CPX's.
- the lack of battle support functions give too simplified a picture of how the battle is organized and
- these shortcomings are impossible to resolve within the current model implementation without great effort.

A better representation of the simulated battle is important, but entering more and more detail can have three basic counter effects:

1. elimination of real-time performance or requirements for enormous additional investment in software,
2. increase in controller personnel to feed in actions at the desired detail and
3. "troops" could be burdened with so much detail that battle preparation will take days and battle execution will be extremely slow (as in reality?).

The new wargame development therefore will be followed by extensive user-oriented evaluations and it is more or less accepted that a new wargame will tend to change with time for at least a number of years (prototyping). Imperative is a careful, "easy to maintain and update", design.

The advantages of computer-assisted CPX's

The preparation of a computer-assisted CPX takes less time than a traditional CPX. One has only to define initial forces and dispositions with a global description of the scenario to be played. The computer assistance will handle details during the exercise, so detailed and time consuming preparation of "scenario-books" is unnecessary and an "open-ended" game can develop.

A computer-assisted CPX is a two party exercise where the opponent is played by a group of specialists adhering to the doctrine and order of battle of a possible opponent.

Experience thus far shows that in this way the players and lower control acquire considerable knowledge about the organization and doctrine of the opposing party.

The most important operational enhancements compared to a traditional CPX are:

- the consistent representation of time and space factors for the players as well as for lower control,
- consistent representation of the battle evolution,
- measurable influence of good planning and timely reaction on the actual and anticipated situation,
- measurable influence of artillery support on manoeuvre and
- intense involvement and good learning effect for lower control and (somewhat less) for higher control, meaning that in fact three levels are trained.

All these facts indicate that computer-assisted exercises are more realistic and enhance functional skills of all participants (players and control organization).

Another advantage is the possibility to re-enter previous combat situations for instruction, learning or just exercise purposes. Also it is possible to re-enter a previous combat situation, change this and start again to get a more interesting battle-play.

Since all combat-situations can be stored during an exercise, the exercise can be replayed for analysis, instruction or demonstration.

The advantage of graphics

During the development of the computer assistance for the command post exercise, more and more attention was paid to the graphics within the wargame concept. For a number of reasons the graphics are the most direct way for the control organization to look at situations within the wargame.

Speed and detail of the graphics displays give ample opportunity for monitor control (and also higher control) to see how the battle evolves and to make corrections or re-adjustments with regard to the exercise goals.

Graphics are also very useful for instruction and evaluation purposes during or after the exercise, i.e. one can show *exactly* how the situation in the battle developed compared to the interpretation of the player staff.

Future Developments

Replacing the SOLTAU wargame

In June 1986 the project was started to replace the current SOLTAU wargame by developing and implementing a new wargame as specified by Kievit (1987), designed for battalion and brigade level.

In this new design the terrain resolution will be 100 × 100 square grid, possibly using Digital Land Mass System[3] data combined with off-the-road traffic ability.

Units will be specified as groups of assets (weapons, personnel, engineering equipment and stocks), usually platoon or company-sized with terrain or road lay-out, making it possible to introduce single weapon system problems (line of sight, traffic jams).

The new design will allow weapon system characteristics to be exploited although the units losses will still be calculated on the basis of an adjusted weighted unit value system.

Separate development of a higher level wargame out of this concept will start after the initial laboratory version is accepted.

Introduction of logistics/engineer/air defence/air support functions

All important stock items (ammunition and POL) together with personnel numbers, and (repair) status of the weapon systems are represented in the datastructure.

Logistics functions are represented through logistics units and "automatic" repair cycles within the units.

Logistics installations and a road network are introduced, leaving the possibility to enhance the logistics function with more detail.

Engineer functions are introduced to represent fully the bridge laying and river crossing operations, the defensive preparations (minefields, ditches, demolitions) and the attack functions (minefield breaching). The air defence and air support functions will be implemented, introducing the 3rd dimension within the wargame.

Upgrading hard/software performance

The introduction of more detail and more functions will require more computer capacity than for the SOLTAU-wargame. Therefore the project team will start an early implementation process to deliver a model version to be tested on these aspects. Regarding the currently available computer systems, no problem is foreseen in finding an appropriate system set-up for the new wargame to perform well under real time.

However it would be desirable if the developed wargame could perform well within real time in small exercises on easily transportable computer systems such as the micro-VAX. The SOLTAU wargame demonstrates that large parts of the wargame system can be subdivided and run in parallel on a network of computer systems. This technique will be exploited in the new development. The implementation language is ADA and a graphics package GKS (Graphics Kernel System) is used for the graphics functions. A software performance analyser will be used to evaluate and optimize critical algorithms with regard to CPU-performance.

Organizational embedding of a wargaming facility/unit within the Netherlands Army

Considerable experience was acquired during recent years about how to organize the wargame support group to support the computer-assisted command post exercise. Many aspects are still in development and there is still no agreement about the configuration of such a wargame support group.

At this moment the basic support group is detached to the Physics and Electronics Laboratory and consists of four conscripts (three of whom are computer engineers) based on the experiences and workload of 1986 (12 exercises and pre-exercises averaging 1 week). This group is headed by one senior officer, responsible for the overall technical support (heading the monitor cell). The support group is responsible for the equipment installation and running the wargame system on a 24 hour basis.

As the workload increases the support group will need to be enlarged or changed to give more flexibility and continuity.

Plans exist to enlarge the staff of 1 (NL) Corps with a section "computer-assisted exercises" consisting of:

- one senior officer heading the co-ordination,
- one officer heading lower control,
- one officer and one NCO as basic team within the opponent players cell, and
- four conscripts (as mentioned before).

Computer-assisted CPX policy of 1 (NL) Corps

Based on the positive results with the computer assistance of the command post exercises thus far the following policy items are agreed for the next years:

- in principal all command post exercises from battalion level up to corps level will be conducted using computer assistance.
- whenever possible command post exercises are conducted in such a way that all player-subcommanders under higher control are participating, so no exercises for one battalion or one brigade player staff only,
- whenever possible mobilizable staffs will participate to the aforementioned exercises of standing units

The tactical training of platoon and company commanders will be realized as much as possible with the aid of simulation techniques based on wargaming.

Conclusions

Not only is it possible to use wargaming techniques for training and tactics, it is an excellent means to conduct command post exercises more efficiently and more instructfully.

Computer-assisted CPX's are not a new way of training. The computer assistance is a supporting means to be used with an already existing method of training. The organizational form and size of computer assisted CPX's are like the traditional CPX's but the quality of training as compared to traditional CPX's is far better even with the simple SOLTAU wargame. Due to the open ended games, not only the player staffs are trained, but also large parts of the control organization (especially lower control and parts of higher control and the enemy player staff).

Wargame techniques can be used to gain insights in tactical doctrines. One should however refrain from taking wargame outcomes absolutely. Wargaming will remain an attempt to reflect reality and will never be an exact copy of that reality. Experience shows that the best method to introduce wargaming techniques in the army is to introduce them gradually and "experimentally" trying to improve the computer assistance every following exercise. Well-supported practical use will show best (to exercise control, control organization and players) how and to what degree wargaming is useful and what the effects on the exercises are. With this knowledge and experience the sort and size of wargaming for training can gradually be optimized.

For the coming years increasing numbers of command post exercises will be computer-assisted.

Session Review

The SOLTAU wargame asks for a certain minimum level of: players, computer equipment and "military knowledge" to operate, which was outside the scope of the conference.

Therefore the SOLTAU game and our experience using it was introduced by a presentation.

After the presentation a lively discussion was centred around the subject of validity.

Notes

1. In December 1984 the Physics Laboratory and the Laboratory for Electronics Developments merged to form the Physics and Electronics Laboratory.
2. All I/O was also generated in German.
3. Internationally agreed data format, provided by the Defense Mapping Agency in the United States.

References

EINFALL MARK 2 1969. *A landbattle simulation* Scientific Control Systems Ltd.
Kievit, J. 1987. *Een wargameconcept voor computerondersteunde stafoefeningen.* Den Haag: FEL-TNO.
Swinkels, W. J. A. M. 1973. *De doorbraak bij SOLTAU, een spel met het HKS/TNO wargame.* Den Haag: PHL-TNO.
Van der Velden, A. M. 1983. *Een wargame ten behoeve van de HKS-oefening SOLTAU.* Den Haag: PHL-TNO.
Vink, J. H. en Van Veen, A. J. J. 1975. *Beschrijving van een interactieve versie van de wargame "EINFALL".* Den Haag: PHL-TNO.
Wierink, J. M. H. 1982. *Een voorstel tot een wargame ten behoeve van de HKS-oefening SOLTAU.* Den Haag: PHL-TNO.

Goal-setting and business gaming

James M. Freeman
University of Manchester Institute of Science and Technology, England

ABSTRACT: Goal-setting is a vital stage in the corporate planning process – *goals* being the means by which an organization's long-term *objectives* are operationalized. A rigorous goal-setting discipline is necessary for effective implementation of the strategic plan – the resulting network of goals acting as a model of the organization's strategy over the planning period.
 The study, described, confirms the value of business gaming as a medium for improving goal-setting competence. Analytical findings support the case from work motivation theory, of a positive linear relationship existing between level of goal difficulty and corresponding task performance.
KEYWORDS: Goals; Corporate Planning; Business Gaming; Motivation; Logistic Regression.
ADDRESS: School of Management, UMIST, P.O. Box 88, Manchester, M60 1QD, England.

Introduction

Goal-setting theory has implications for corporate planners and organizational psychologists alike.

Corporate planning

Corporate planning is a total planning process encompassing strategic, operating and project plans (Denning, 1971). While the *strategic plan* defines the long-term objectives of the business and the means by which those objectives are to be attained, the *operating plan* – derived from the strategic plan – concern "the forward planning of existing operations in existing markets with existing customers and facilities"; *project plans*, in contrast, cover particular capital investments, marketing or other operations.
 Corporate planning is comprehensive, embracing all types of activity and all time horizons within the company.
 Typically, corporate planning is defined (see Hussey, 1974), only in relation to its strategic and operational components e.g. for Irving (1970) corporate planning is:

1. The formal process of developing objectives for the corporation and its component parts, evolving alternative strategies to achieve these objectives, and doing this against a background of a systematic appraisal of internal strengths and weaknesses and external environmental changes.

2. The process of translating into detailed operational plans and seeing that these plans are carried out.

Planners have approached the subject from a variety of different perspectives e.g. Drucker (1955), whose stance is essentially that of the man-manager, sees corporate planning as involving

> the setting of objectives, organizing the work, people and system to enable those objectives to be attained, motivating through the planning process and through the plans, measuring performance and so controlling progress of the plan, and developing people through better decision making, clearer objectives, more involvement and awareness of progress.

The corporate plan, itself, is sometimes considered as a system of interlinking strategic and *tactical plans* (a tactical plan, in this context, representing a "shorter term action plan which leads to the implementation of actions proposed in the strategic plan" (Hussey, 1974)). Each of these plans needs to be organized by responsibility, time and information, in order that senior management can effectively direct and control the future of the organization.

Organizational control is normally exercised by monitoring the organization's performance, relative to a network of goals, defined after its strategy has been determined. (Note that for corporate planning purposes, "goals" are time-assigned targets, derived *from* strategy – unlike "objectives" which are understood to be desired outcomes, set in advance of strategy.) Goals need to be expressed in quantitative terms since they are designed as a standard of management. Goals must be purposeful: there is no value in defining targets which the organization either cannot or does not intend to measure. An advantage with goals derived from strategy is that they should be compatible with each other. There is no reason why goals cannot be adjusted in the light of experience e.g. if stock levels are found to be too high. Adjustments are entirely permissible so long as revised goals are followed up by plans for achieving those revised goals.

It is preferable that the goals adopted are those that can be measured as a matter of routine. Where goals require the collection of non-routine measurements, the costs of obtaining data may well outweigh anticipated benefits.

Goals commonly defined in business (Hussey, 1974) relate to:

- Percentage market share (by product and/or geographical area)
- A ratio, such as return on sales
- An absolute figure for sales
- A minimum figure for customer complaints
- A maximum figure for hours lost in industrial disputes
- A labour productivity ratio
- Total number of employees
- A maximum employee "wastage" rate
- A standard cost

- A cost reduction target
- A date by which a particular event must take place e.g. a new product launch

Work motivation

Of all the major theories for *work motivation*, goal-setting theory is recognized as the theory having greatest scientific validity (Pinder, 1984).

Claims for the goal-setting approach are impressive. In particular, recent reviews of the goal-setting literature – see e.g. Locke *et al.* (1981), (Latham and Lee 1986) – agree a remarkably high proportion (90% and 97% respectively) of goal-setting studies showing a beneficial effect of goal-setting on performance.

Early laboratory-based results for goal-setting theory have since been substantiated in the field. Key research findings to date – paraphrasing Locke and Henne (1986) – are summarized below:

1. Difficult goals lead to higher task performance than easy goals (with moderate goals in between)

2. Specific and difficult goals lead to higher performance than no goals or vague goals such as "do your best"

3. Goals affect task performance by directing attention and action; by mobilizing effort; by increasing persistence and motivating the search for appropriate performance strategies.

4. Feedback appears necessary for goal-setting to work. In fact feedback and goals have been found to work together far more effectively than either one alone.

5. Goal commitment is necessary for goals to affect performance.

6. Goal commitment appears to be generally unaffected by participation but is influenced by, for example, the value placed on success.

7. Money may stimulate goal-setting to take place where otherwise there would be none. Alternatively it may lead to higher goals being set or goal commitment being increased.

8. Goal-setting effectiveness varies considerably between individuals: in general, it does not relate well with either their demographic or psychometric characteristics.

9. Goal acceptance and choice depends on many different factors includ-

ing: ability, competition, difficulty, direct instructions, expectancy, previous success and failure and self-efficacy.

The first finding in this list viz. the linear relationship between goal difficulty and task performance was evidently received with some scepticism when it was first discovered. At the time, an inverted U-shape relationship seemed much more plausible and had the advantage, also, of being consistent with expectancy theory (which essentially aims to explain work motivation in terms of the outcome expected by an individual for a given amount of effort). The two theories are now reconciled following the discovery lately, that while goal-setting theory holds *between* groups, expectancy theory is applicable *within* groups.

The Study

The study focuses on the application of a computer-based business game, WHOLETRAIN (Freeman and Badcock, 1986) for training managers in goal-setting skills. The game, developed at UMIST for the major Electrical Wholesaler, GTE Sylvania, was used recently to support practical sessions at a four-day seminar, the EWF*/Sylvania Distributor Management Programme (Keenlyside, 1988). The package had been originally launched at the 1986 "Young Wholesalers Forum" (Freeman and Armstrong, 1988) but modified subsequently to meet demands from EWF for greater "realism". As with the Forum, the Management Programme was aimed at branch managers from the Electrical Wholesale Industry.

However, whereas the former event was directed specifically at improving managers' business planning skills, the Management Programme adopted a broader "business practice" theme.

The first session of the Programme was the WHOLETRAIN briefing session. At this, the nineteen delegates were provided with comprehensive printed materials on the game and assigned to their respective teams (there were three in all). After being quickly led through the rules and objectives for the exercise by the game facilitator, teams were given the opportunity to compete in a trial WHOLETRAIN round. Apart from allowing a preliminary exposure to the game model, the round was also helpful in kindling initial rapport between team members (Freeman, 1987).

Following on, teams met in the separate "boardrooms", they had been allocated, to decide on their goals, for the four rounds the game was due to run. There was no limit on the number of goals, teams could set themselves but, for each of their chosen goals, they had to assess on a scale of 1-7 the degree of difficulty they expected to experience in trying to meet that goal. A special proforma sheet (see Appendix 1) was circulated for this purpose. Teams retained their goal sheets for the duration of the Programme. A copy

*Electrical Wholesale Federation

of each team's goal sheet was also held by the game facilitator. At the end of the fourth WHOLETRAIN round, teams were provided with a second proforma, on to which their goals had been transcribed from the previous circular, and asked to score, using a scale of 1-5 (see Appendix 2), the extent of their achievement against each goal. Some of the "achievement scores" obtained at this stage, were plainly unconvincing, in the light of actual output generated by the WHOLETRAIN model, and were therefore moderated by the facilitator (with group support) at the debriefing session.

Statistical Analysis

A statistical analysis of data arising from the latter business game operation has been undertaken, taking into account, for each of the team goals specified, the corresponding team code, H (Hardness) Score and A (Achievement) Score.

Table 1 provides a statistical summary of these details by team. Clearly

TABLE 1 *Statistical summary of team performance*

Team	Number of Goals	Average H Score	Average A Score	Average A/H Ratio
1	5	4,800	2.8	.6200
2	6	4.333	2.0	.4556
3	6	4.000	3.5	.8917

there were marked differences in team performance – particularly in relation to the average A/H ratios. However, whereas an analysis of variance of A/H ratios by team is found to be significant – see Table 2 – that for all other variables in Table 1, is not.

The statistical significance observed for Table 2 is due to team 3's average

TABLE 2 *Analysis of Variance: A/H Ratio by Team*

Source	df	ANOVA SS	MS	F
Teams	2	0.5807	0.2904	4.28*
Error	14	0.9493	0.0678	
Total	16	1.5301		

*($P(F(2,14) > 3.74) = .05$)

A/H ratio being quite distinct from (and very much higher than) that for team 2. As corroboration, the plot of A score against corresponding H score, for each of the team goals, in Fig. 1, reveals that, with one exception, team 3 demonstrated a higher achievement outcome for a given hardness level than

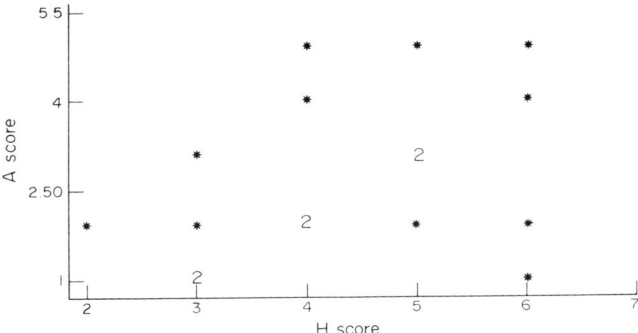

FIGURE 1   Achievement (A) Score by Hardness (H) Score for each Goal

did team 2. The reason for this unlikely result lies with team 2's performance in the third round of WHOLETRAIN. In an effort to maximize its earnings relative to those of its competitors (it was claimed), team 2 set the gross margin %s of all of its eight product groups to the maximum value permitted by the programme. Unfortunately, this ploy had disastrous consequences for most other aspects of its simulation performance (e.g. sales volume and market share slumped while stock levels rocketed) and the overall effect was to badly distort corresponding A scores.

Allowing for the impact of this distortion on Fig. 1, it appears A and H scores are positively related. Further, there is the suggestion that the relationship between the two variables is that of a growth curve rather than a straight line. To investigate this possibility more fully, a logistic regression analysis (Dobson, 1983) has been performed for the data, based on the formulation:

$$\text{logit}(A/5) = \beta_0 + \beta_1 x \qquad (*)$$

where logit (A/5) represents the natural logarithm of the odds $A/5/(1 - A/5)$ and x, alternative expressions based on the H score variable.

(Logistic regression analysis is a particular example of "generalized linear modelling" which allows conventional regression analysis to be extended to situations where the dependent variable is a binomial proportion, for example.)

In all, four variants of the basic model (*) have been tested. Relevant details are given in Table 3.

Of the four models described here, model 4 can be seen to have the best technical characteristics. (Models 2 and 3, though statistically significant, clearly represent the data less effectively. Model 1 is, in fact, just short of significance – most likely because of the problems with team 2's game input, commented on earlier.)

For model 4, the H scores were adjusted by the corresponding mean H and

TABLE 3 *Results of Logistic Regression Modelling*

MODEL 1: Logit $(A/5) = -1.282 + .357 H$
                        $(.824)\ \ (.184)$
         $D = 24.28$ Tarone-Gart statistic $= 1.96$

MODEL 2: Logit $(A/5) = .275 + .419 (H - H_k)$ $(k = 1,2,3)$
                        $(.226)\ (.192)$
         $D = 23.27$ Tarone-Gart statistic $= 2.20$

MODEL 3: Logit $(A/5) = -1.571 + 1.847 (H/H_k)$ $(k = 1,2,3)$
                        $(.854)\ \ (.831)$
         $D = 23.05$ Tarone-Gart statistic $= 2.25$

MODEL 4: Logit $(A/5) = -1.926 + .814 H (A_k/H_k)$ $(k = 1,2,3)$
                        $(.728)\ \ (.264)$
         $D = 16.95$ Tarone-Gart statistic $= 3.26$

A scores in an effort to offset the significant difference in A/H ratios between teams, discovered earlier. With this formulation, the A score for an individual team goal depends on both the average A score for the team as well as the relative hardness of the goal.

Note: 1. The figures in brackets, here, are the standard errors of corresponding estimated regression coefficients.
2. H_k and A_k are the mean hardness and achievement scores recorded by the k'th team (k = 1, 2, 3).
3. The log likelihood ratio statistic, D, measures the "goodness of fit" between observed and predicted A values. (Hence, the lower the value of D, the better the fit and vice versa.)
4. Under $H_0: ß_1 = 0$ the distribution of the Tarone-Gart statistic has been found to be approximately standard normal (Tarone and Gart, 1980).

(expressed by the quotient H/H_k (k = 1, 2, 3)).
A comparison of observed A scores with those predicted by the model, is shown in Table 4.

Summary

Results from the study confirm the value of business gaming as a research tool for advancing theoretical understanding in the important area of work motivation – also as a practical approach to training in goal-setting skills for business.

Gaming lends itself naturally to the goal-setting application, drawing, as it does, on the strengths of both laboratory and field approaches. As a relevant experimental medium, business gaming has much to recommend it as an aid for future research in motivation.

TABLE 4 *A comparison of observed A scores and those predicted using Model 4 (see Table 3)*

		Initial Estimate	First	Second Approximation	Final
	β_0	0	−1.590	−1.857	−1.874
	β_1	0	.653	.768	.776

Team	Observed A Scores		Fitted Values		
1	3	2.5	3.035	3.142	3.149
	3	2.5	3.492	3.657	3.668
	3	2.5	2.036	1.974	1.970
	4	2.5	3.492	3.657	3.668
	2	2.5	2.537	2.561	2.563
2	1	2.5	1.662	1.543	1.536
	2	2.5	2.372	2.367	2.367
	1	2.5	1.662	1.543	1.536
	5	2.5	2.372	2.367	2.367
	2	2.5	2.007	1.939	1.935
	1	2.5	2.743	2.803	2.808
3	2	2.5	1.976	1.904	1.900
	3	2.5	3.947	4.138	4.150
	2	2.5	2.696	2.748	2.752
	5	2.5	4.352	4.525	4.535
	4	2.5	3.384	3.538	3.548
	5	2.5	3.384	3.538	3.548

Session Review

Evidently, the session was received with some interest by the twelve or so delegates who attended – feedback to my 20 minute input lasting for nearly a full hour!

A number of issues arose in the discussion: the fact that goals had been treated in the analysis as being of equal importance was rightly judged to have been unrealistic. (This is a problem that I hope to address in the near future.)

My formal presentation of key research findings for goal-setting theory was clearly too terse for many of those present. I was questioned at length afterwards about different aspects of motivation theory that had not been covered e.g. the nature of the laboratory studies that had been carried out, whether findings were different for goals that were self-selected or imposed etc.

Throughout, the discussion was lively and good-humoured – the ISAGA President-Elect (an East German) and I arguing at one point on our different understandings of the term "democratic" (which I had used in response to a question on gaming conditions)! It was a memorable moment.

A number of the points raised I found particularly helpful and I hope to build these into research, planned for later in the year. I thank all those who attended my session and sincerely hope they benefited from it as much as I did.

Appendix 1
Team Goals: Pre-game Questionnaire

At the beginning of the business game event, team goals for the game should be as summarized below. Remember that goals should be measurable.

Against each goal, teams should indicate the difficulty they expect to experience in trying to meet their stated objectives – using the self-assessment scores shown to the right-hand side of the goal descriptions.

GOAL DESCRIPTION	Impossible goal to achieve						No effort needed to achieve goal
I	7	6	5	4	3	2	1
II	7	6	5	4	3	2	1
III	7	6	5	4	3	2	1
IV	7	6	5	4	3	2	1
V	7	6	5	4	3	2	1
VI	7	6	5	4	3	2	1
VII	7	6	5	4	3	2	1
VIII	7	6	5	4	3	2	1
IX	7	6	5	4	3	2	1
X	7	6	5	4	3	2	1

Team:

Appendix 2
Team Goals: Post-game Questionnaire

At the end of the game event (and not before) teams should assess their performance against stated objectives – duplicated below – using the self-assessment scores shown to the right-hand side of their goal descriptions.

References

Ackoff, R. C. 1970. *A Concept of Corporate Planning*. New York: John Wiley and Sons, Inc.
Cooper, C. L. and Robertson, I. (eds.) 1986. *International Review of Industrial and Organisational Psychology*. New York: John Wiley and Sons, Inc.
Denning, B. W. 1971. *An Introduction to Statistical Concepts*. New York: MacGraw-Hill.
Dobson, A. 1983. *An Introduction to Statistical Modelling*. New York: Chapman and Hall.
Drucker, P. F. 1955. *The Policy of Management*. New York: Heinemann.
Freeman, J. and Badcock, D. 1986. A computer-based training aid for wholesale management. *Electrical Wholesaler*. 25:2.

GOAL DESCRIPTION	Goal attained with complete success						No success whatever in attaining goal
I	7	6	5	4	3	2	1
II	7	6	5	4	3	2	1
III	7	6	5	4	3	2	1
IV	7	6	5	4	3	2	1
V	7	6	5	4	3	2	1
VI	7	6	5	4	3	2	1
VII	7	6	5	4	3	2	1
VIII	7	6	5	4	3	2	1
IX	7	6	5	4	3	2	1
X	7	6	5	4	3	2	1

Team:

Freeman, J. 1987. Computer-based Simulation Games and Induction Training. *Training Officer*. 23:1.
Freeman, J. A. and Armstrong, A. J. 1988. Training in Business Planning and Budgetary Control: A Business Game Application for the Electrical Wholesale Industry. *Simulation/Games for Learning*. 18:1.
Hussey, D. E. 1974. *Corporate Planning, Theory and Practice*. Oxford: Pergamon Press.
Irving, P. 1970. Corporate planning in practice: a study of the planning in major UK companies. *M.Sc. Dissertation*. Bradford: University of Bradford Press.
Keenlyside, N. March 1988. *Electrical Wholesaler* 25.
Latham, G. P. and Lee, T. W. 1986. Goal Setting. *In* Locke (1986).
Locke, E. 1986. *Generalising from Laboratory to Field Settings*. Lexington, Mass.: Lexington Books.
Locke, E. A. and Henne, D. 1986. Work Motivation Theories. *In* Cooper and Robertson (1986).
Locke, E. A., Shaw, K. N., Saari, L. M. and Latham, G. P. 1981. Goal setting and task performance: 1969-1980. *Psychological Bulletin*. 90:1
Pinder, C. C. 1984. *Work Motivation*. Glenview, Ill.: Scott Foresman.
Tarone, R. E. and Gart, J. J. 1980. On the robustness of combined tests in proportions. *Journal of the American Statistical Association*. 75: 369.

Simulation References

WHOLETRAIN. Freeman and Badcock. 1986.

A study of the relationship between student final exam performance and simulation game participation

T. Richard Whiteley and Anthony J. Faria
University of Windsor, Ontario, Canada

ABSTRACT: Despite the proliferation and widespread usage of simulation games in the field of business education, the pedagogical value of this instructional aid remains unclear. The present study, using a controlled setting, sought to determine whether incorporating a business simulation game in a principles of marketing course improves the acquisition of marketing knowledge. The results suggest that simulation games are an effective means by which to improve quantitative skills but are not an effective means by which to improve the acquisition of applied or theoretical knowledge.
KEYWORDS: LAPTOP; marketing; exam performance; participation; simulation game; quantitative knowledge; theoretical knowledge; applied knowledge.
ADDRESS: Faculty of Business Administration, University of Windsor, Windsor, Ontario, Canada. N9B 3P4.

Introduction

It has now been over 30 years since the first business simulation game was used in a college class. Since that time, the number and variety of business games has grown enormously. Interest among academics in the teaching and learning possibilities of business games has grown as well. At present, over 200 business games are being used by approximately 8,500 teachers at over 1,700 colleges (Faria, 1987). Empirical research in the area has also been extensive. Comprehensive reviews can be found in Greenlaw and Wyman (1973), Keys (1976), Wolfe (1985), and Miles *et al.* (1986).

Despite the proliferation and widespread usage of business simulation games, a review of the literature reveals that the pedagogical value of such games still remains unclear. The present study sought to determine whether incorporating a business simulation game in a principles of marketing course improves the acquisition of marketing knowledge. Other potential benefits of game playing were not considered.

Past research

Accompanying the development of business games and their increased usage has been an active research track. A great deal of research has investigated (1) the factors thought to affect the simulation learning environment and simulation performance (e.g. Brenenstuhl and Badgett, 1977, Walker, 1979, Gentry, 1980, Hsu, 1984, Faria, 1986), (2) the learning aspects of the business game approach to instruction (e.g. Biggs 1975, Biggs and Greenlaw, 1976, Chisholm, 1979, Edwards, 1987, Hall, 1987), and (3) the relative educational benefits of simulation games versus other teaching methods (e.g. Wolfe, 1985, Miles *et al.* 1986).

The majority of the research in the third area identified has compared games to cases, since both are experiential teaching tools and the case method has long been accepted by business teachers. Four major review articles exist which summarize the published comparative studies in this area. After reviewing 22 studies published between 1961 and 1972, Greenlaw and Wyman (1973) concluded that there existed little clear evidence to indicate what was learned from business games or whether business games were a superior, or even adequate, method of instruction. Keys (1976) reviewed fifteen studies that compared simulation game sections of a class with sections using some other form of instructional approach and came to a similar conclusion. Wolfe (1985) updated the Greenlaw and Wyman (1973) study by reviewing 39 studies published between 1973 and 1983. Because of the wide variety of study conditions utilized in these studies (e.g. simple versus complex games, variable number of decision periods played, different methods of end-of-course evaluations employed), Wolfe concluded that no statement about gaming effectiveness could be made. Finally, Miles *et al.* (1986) reviewed sixteen studies that used student self-judgement of skill acquisition through cases and simulation games as the dependent variable. They also came to the conclusion that the mixed results uncovered were difficult to interpret and compare because of the wide variety of study environments used.

As these four review articles show, the evidence as to whether business games are a more effective teaching tool than other approaches is inconclusive. As well, because of the highly dissimilar study designs employed in past research, it is difficult to compare the findings of one study with those of another. Beyond this, the previous studies undertaken suffer from several drawbacks. Principal among these are: (1) where studies compare performance of students in two separate sections of a course, there has been a lack of control across sections as to similarity of students and instructor teaching; (2) most studies have involved small numbers of students, such as 20 or 25 per section or treatment, possibly making the findings unstable; and (3) where self-judgement is used, it is very questionable as to whether students, having just completed a course, are a good judge of what they have learned.

Purpose and Design of the Present Study

The present study was designed to overcome the drawbacks found in previous research and to determine whether student learning can be improved through the addition of a simulation competition to the normal course requirements in a principles of marketing class. The drawbacks of differing student populations and instruction as found in previous research were overcome by using treatment groups from the same large principles of marketing section. Thus, all of the students were exposed to the same instructor, lectures, films, readings, exams, and exam times. The similarity of treatment groups was measured, and assured, through the use of a pre-test, the course mid-term exam. End of course performance/learning was measured through the use of an objective final exam, not by means of subjectively graded cases or student perceptions. Finally, the testing was undertaken with an audience of 189 students.

Methodology

One hundred and ninety students from the same section of an undergraduate principles of marketing course served as the respondent base for the study. Prior to the mid-term exam, the students in the class were given the opportunity to sign up to play a simulation game entitled LAPTOP. This is a simple marketing game designed for use in an introductory marketing course.

Participation in the game gave each student the chance to earn up to seven bonus grade points. These points, which were eventually added to the student's final grade in the course, were based on team and individual results. Offsetting the potential for bonus points was the additional time required by the student to read and understand the simulation game, to make decisions, and to analyse the period-by-period results.

Sixty-nine students signed up to play the simulation while 121 students chose not to participate. The data for one of the students in the latter group was randomly selected for elimination in order to have proportional cell sizes during the analysis stage of the study.

Design

A $2 \times 3 \times 3$ factorial design was used to analyse the data. The first two variables are between subjects variables while the third variable is a within subjects variable. The final exam grade served as the dependent variable.

Between subjects variables. The first of these variables, Game Status (GS), is the most important variable in the study. A student's game status was either "Played" or "Did Not Play". As previously indicated, 69 students played the game and 120 students remained in the "did not play" condition.

The second between subjects variable, Mid-Term Exam Performance Level (MTL), was used as a blocking variable in order to reduce the level of experimental error (see Neter and Wasserman, 1974).

The mid-term exam contained 30 multiple-choice questions. Thirteen of these questions were classified as applied while 17 were classified as theoretical. In order to give equal weighting to each question type, the student's percentage grade for each question type was computed and averaged to determine the student's overall mid-term exam grade.

On the basis of the above mid-term grade, an equal number of students were assigned to each of the high, moderate, and low MTL categories. The cell size for each MTL level was 23 for the "Played" game-status group and 40 for the "Did Not Play" group.

Within subjects variable. Question type (QT) on the final exam served as the three-level within subjects variable. The three question types on the final exam were applied (21 questions), theoretical (26 questions), and quantitative (20 questions).

Question classification

In order to determine the question-type classification of each of the multiple-choice questions on the mid-term and final exams, five faculty members, all of whom have taught principles of marketing, were asked to classify each question. The faculty members were asked to classify the question as *quantitative* if they felt that it requires a knowledge of or use of a computational approach to arrive at a correct answer, to classify the question as *applied* if they felt that marketing knowledge is required to understand the scenario (or situation) described, or to classify the question as *theoretical* if they felt that it focuses on one's knowledge of a particular theory or concept (but excludes computations and applied scenarios).

Dependent variable. The grade on the final exam served as the dependent variable. More specifically, the student's percentage grade for each question type was used instead of the raw score in order to give equal weighting to each question type.

Results

The data were analysed using the ANOVR analysis of variance programme developed by Games *et al.* (1979). When significant results were uncovered in the analysis of variance, each pairwise contrast was investigated using the FOLUP programme developed by Yancey *et al.* (1979).

Data transformations and the assumptions of the analysis of variance model

As the initial analysis of the final exam scores indicated that the data violated the assumptions of the analysis of variance model pertaining to the within subjects variable, the percentage grade values were converted to Z-scores on a variable-by-variable basis (i.e. by question type). This approach to data transformation resulted in equal means and standard deviations across question type, thereby removing any within subjects QT main effect which might have existed. The transformed data met all of the assumptions of the analysis of variance model.

Analysis of variance of final exam scores

The analysis of variance of the final exam scores reveals only three significant results: the MTL main effect [$F(2, 183) = 26.42, p < .001$], the GS × QT first-order interaction [$F(2, 366) = 4.05, p < .05$], and the MTL × QT first-order interaction [$F(4, 366) = 7.62, p < .001$]. The following effects were not significant at the .05 level: GS [$F(1, 183) = 1.54$], GS × MTL [$F(2, 183) = 1.27$], QT [$F(2, 366) = 0.00$], and GS × MTL × QT [$F(4, 366) = 1.13$].

The failure to obtain significant results for the Game Status (GS) main effect means that playing versus not playing the simulation game did not help or hinder a student's overall performance on the final exam. The failure to find a significant Game Status (GS) × Mid-Term Exam Performance Level (MTL) interaction means that a student's overall final exam performance was independent of the combined effect of the student's mid-term performance and game participation. The nonsignificant results for the Question Type (QT) main effect was expected because of the nature of the transformations carried out on the data. Finally, the failure to find a Game Status (GS) × Mid-Term Exam Performance Level (MTL) × Question Type (QT) interaction means that a student's performance on the final exam is independent of the combined effect of the level on each of these variables.

The significant Mid-Term Exam Performance Level (MTL) main effect indicates that there is a relationship between a student's performance on the mid-term exam and his/her performance on the final exam. Specifically, the investigation of the nature of this main effect by means of a follow-up analysis focusing on each pairwise contrast finds that those who performed better on the mid-term exam also performed better on the final exam (see Table 1). It was because of the expectation of this result that using MTL as a

TABLE 1 *Multiple Comparisons (via Tukey WSD Technique) of Final Exam Scores for the Mid-Term Exam Performance Level (MTL) Main Effect*

MTL pairwise contrast	Means compared	Obtained t-value
High – Low	0.460 – (– 0.477)	7.27*
High – Moderate	0.460 – (– 0.020)	3.41*
Moderate – Low	– 0.020 – (– 0.477)	3.85*

Notes. df = 183, critical t-value = 2.36, and FWI = .05. The $MS_{eb} = 1.57$ was used as the error term. All means are Z-scores.
*indicates pairwise contrast in significant.

blocking variable was deemed an appropriate means by which to reduce the experimental error.

While the results of the follow-up analysis of the MTL main effect are interesting, more insight can be gained by investigating the significant Mid-Term Exam Performance Level (MTL) × Question Type (QT) interaction. The results of the follow-up analysis for the applied and theory questions are consistent with the results for the MTL main effect. That is, those who performed better on the mid-term exam also performed better on the final exam in these areas. However, with respect to the quantitative questions, those who performed at an overall moderate or high level on the mid-term exam performed equally well on the quantitative questions and both of these groups performed significantly better than those who were classified as low performers on the mid-term exam (see Table 2).

TABLE 2 *Multiple Comparisons (via Tukey WSD Technique) of Final Exam Scores for the Mid-Term Exam Performance Level (MTL) × Question Type (QT) Interaction Effect*

Question type	MTL pairwise contrast	Means compared	Obtained t-value
Applied	High – Moderate	0.652 – (– 0.579)	7.59*
	High – Low	0.652 – (– 0.073)	4.47*
	Moderate – Low	– 0.073 – (– 0.579)	3.12*
Theory	High – Moderate	0.577 – (– 0.527)	6.80*
	High – Low	0.577 – (– 0.049)	3.86*
	Moderate – Low	– 0.049 – (– 0.527)	2.95*
Quantitative	Moderate – Low	0.182 – (– 0.323)	3.11*
	Moderate – High	0.182 – (0.152)	0.19
	High – Low	0.152 – (– 0.323)	2.93*

Notes. df = 549, critical t-value = 2.35, FWI = .05, and $MS_{ew.cells}$ = 0.83 was the error term for the set of pairwise contrasts for each Question-Type analysis. All means are z-scores.
*indicates pairwise contrast is significant.

The final, and perhaps most interesting, significant result to investigate is the Game Status (GS) × Question Type (QT) interaction. The follow-up analysis of this effect indicates that performance levels on the applied and theory questions on the final exam were the same for those who played the game versus those who did not play the game. However, with respect to the quantitative questions, those who played the simulation game performed significantly better than those who did not play the game (see Table 3).

Discussion

The purpose of the present study was to determine if playing a marketing simulation game in a principles of marketing course improves a student's acquisition of marketing knowledge as measured by the student's final exam performance. The results of this study show that performance on the final

A study of the relationship 143

TABLE 3 *Multiple Comparisons (via Tukey WSD Technique) of Final Exam Scores of The Game Status (GS) × Question Type (QT) Interaction Effect*

Question type	GS pairwise contrast	Means compared	Obtained t-value
Applied	Played – Did Not Play	0.112 – (– 0.064)	1.28
Theory	Did Not Play – Played	0.032 – (– 0.055)	0.63
Quantitative	Played – Did Not Play	0.206 – (– 0.113)	2.32*

Notes. df = 549, critical t-value = 1.96, and FWI = .05. $MS_{ew.cells}$ = 0.83 was used as the error term. All means are z-scores.
*indicates pairwise contrast is significant.

exam overall, and on the applied and theory questions in particular, did not vary as a function of game status. However, with respect to the quantitative questions on the final exam, those who played the game performed better than those who did not play the game.

There are two possible explanations for the lack of difference between the game status groups on the applied and theory questions. One is that the nature of the simulation game played was such that there was no need for the student to significantly draw on the applied and theoretical material covered in the course. All that was required of the participants was for them to make decisions that were primarily of a quantitative nature (e.g. price, sales force size, shipment quantities, level of advertising). Further, no strategy or post-game performance reports were required. The preparation of such reports would have made it necessary for the students to refer to the appropriate applied and theoretical course material. Reports of this nature might have enhanced the final exam performance on these question types.

Another possible explanation for the lack of difference between the game status groups on the applied and theory questions is that only a limited number of decisions were required. The participants in the game only had to make 4 weekly decisions. Had more decisions been required, the results might have been different. A longer game could have given the students a greater opportunity to draw on the applied and theoretical course material. The limit on the number of decisions was necessary, however, as game participation could not begin, due to the study design, until after the mid-term exam.

The finding that those who played the game performed better on the quantitative questions than those who did not play the game also can be explained in either of two ways. One explanation is that participation in the game allowed the students to become more skilled in the quantitative techniques covered in the course. By playing the game, the students had the opportunity to practice and apply techniques such as sales forecasting, return on investment, markups, average cost pricing, and breakeven analysis. Those who did not play the game did not have this opportunity. As a result, those who played the game performed better on the quantitative questions than those who did not play the game. Based on the raw exam scores for the quantitative questions, participation in the game resulted in a grade of 54.6% versus 48.2% for the non-players, a 13.3% improvement.

While the above explanation for the difference in performance is reasonable, it is also possible that those who chose to play the game had better quantitative skills than those who chose not to play. The plausibility of this explanation is limited, however, because it leads to the conclusion that those who chose to play had better quantitative skills but only comparable skills in the applied and theoretical areas. (An analysis of variance based on the midterm grades found no difference between the game status groups overall or on the applied and theory questions.) Since there is no reason to believe that only those who had better quantitative skills were likely to participate in the game, the validity of this explanation is open to question.

Conclusion

The evidence from past research as to whether business simulation games are a more effective teaching tool than other instructional approaches is inconclusive. This state of affairs may be due to the limitations associated with the experimental designs, the sample sizes, and the measurement approaches used in the various studies. The results of the present study show that simulation games are an effective means by which to improve quantitative skills but not an effective means by which to improve the acquisition of applied or theoretical knowledge.

Session Review

The 15 conference delegates at the LAPTOP session were given some background information about the game before they had the opportunity to actually run the game on the personal computers provided. In their initial role as game co-ordinators, the participants, following the step-by-step, on-screen instructions, had to determine the importance of the various game parameters (e.g. price, size of sales force). The number of industries and companies per industry also had to be determined. In most cases, one industry with two companies was set-up. Once these preliminary steps were accomplished, the participants took on the role of game players. As company managers, they made a number of marketing decisions for their companies (e.g. price, advertising, sales force compensation, shipments). The decisions were then analysed and the results printed. In most cases, the decision process was repeated for two or three periods. Overall, the session was successful in demonstrating the ease with which one can learn how to set-up and run the LAPTOP simulation game.

References

Biggs, W. D. 1975. Some impacts of varying amounts of information on frustration and attitudes in a finance game. *In* Buskirk (1975).

Biggs, W. D. and Greenlaw, P. S. 1976. The role of information in a functional business game. *Simulation & Games* 7.
Brenenstuhl, D. C. and Badgett, T. F. 1977. Predictions of academic achievement in a simulation game via personality constructs. *In* Neilsen (1977).
Brenenstuhl, D. C. and Biggs, W. D. (eds.) 1980. *Proceedings of the Seventh Annual Conference of the Association for Business Simulation and Experiential Learning.*
Burns, A. C. and Kelley L. (eds.) 1986. *Proceedings of the Thirteenth Annual Conference of the Association for Business Simulation and Experiential Learning.*
Buskirk, H. (ed.) 1975. *Proceedings of the Second Annual Conference of the Association for Business Simulation and Experiential Learning.*
Certo, S. C. and Brenenstuhl, D. C. (eds.) 1979. *Proceedings of the Sixth Annual Conference of the Association for Business Simulation and Experiential Learning.*
Chisholm, T. A. 1979. An examination of the perceived effectiveness of computer simulations in a classroom setting as affected by game, environmental and respondent characteristics. *In* Certo and Brenenstuhl (1979).
Currie, D. M. and Gentry, J. W. (eds.) 1984. *Proceedings of the Eleventh Conference of the Association for Business Simulation and Experiential Learning.*
Edwards, W. F. 1987. Learning macroeconomic theory and policy analysis via microcomputer simulation. *In* Kelley and Sanders (1987).
Faria, A. J. 1986. A test of student performance and attitudes under varying game conditions. *In* Burns and Kelley (1986).
Faria, A. J. 1987. A survey of the use of business games in academia and business. *Simulation & Games* 18.
Games, P. A., Gray, G. S., Herron, W. L. and Pentz, A. 1979. *ANOVR*. University Park, Pa.: The Pennsylvania State University Computation Center.
Gentry, J. W. 1980. The effects of group size on attitudes toward the simulation. *In* Brenenstuhl and Biggs (1980).
Greenlaw, P. S. and Wyman, F. P. 1973. The teaching effectiveness of games in collegiate business courses. *Simulation & Games* 4.
Hall, D. R. 1987. Developing various student learning abilities via writing, the stock market game, and modified marketplace game. *In* Kelley and Sanders (1987).
Hsu, T. 1984. A further test of group formation and its impact in a simulated business environment. *In* Currie and Gentry (1984).
Kelley, L. and Sanders, P. (eds) 1987. *Proceedings of the Fourteenth Annual Conference of the Association for Business Simulation and Experiential Learning.*
Keys, B. 1976. A review of learning research in business gaming. *In* Sord (1976).
Miles, W. G., Biggs, W. D. and Schubert, J. N. 1986. Student perceptions of skill acquisition through cases and a general management simulation. *Simulation & Games* 17.
Neilsen, C. E. (ed.) 1977. *Proceedings of the Fourth Annual Conference of the Association for Business Simulation and Experiential Learning.*
Neter, J. and Wasserman, W. 1974. *Applied Linear Statistical Models*. Homewood, Ill.: Irwin.
Sord, B. H. (ed.) 1976. *Proceedings of the Third Annual Conference of the Association for Business Simulation and Experiential Learning.*
Walker, C. H. 1979. Comparing performance during three managerial accounting simulation schedules. *In* Certo and Brenenstuhl (1979).
Wolfe, J. 1985. The teaching effectiveness of games in collegiate business courses: a 1973-1983 update. *Simulation & Games* 16.
Yancey, J. M., Howell, J. F., James, P. A. and Serapiglia, T. 1977. *FOLUP*. University Park, Pa.: Pennsylvania State University Computation Center.

Simulation References

LAPTOP. Faria, A. J. and Dickinson, J. R. 1987.

Dysfunctionalism in American management systems: management mania in corporate America

John F. Lobuts, Jr.

George Washington University Washington, D.C., USA

ABSTRACT: The Escalation Simulation exemplifies destructive competitive behaviour which often occurs in society, and in fact, is reinforced by society. We have learned to accept, if not expect, competitive behaviour, in which individuals pit themselves against each other for the grand "prize", even though the "prize" may actually be gained easily through collaboration. The attitude of "winning at any cost" which society fosters, is one variable among many serving to reinforce a behavioural attitude in which ethics and morality are often forgotten in the fight to win and which ultimately leads to the lack of ownership of behaviour. As this simulation demonstrates, however, we all stand to lose from this type of attitude, whether we are talking about bidding for nickels or organizational management.

The writer expresses appreciation to the following authors, without whose work this exercise would not have been possible: Aronson, E.; Fisher, A.; Harvey, J.; Kohn, A.; Kolb, D. *et. al.*; Rappaport, A. and Raiffa, H.

KEYWORDS: Competitive conflict, disruptive conflict, destructive competition.

ADDRESS: School of Government and Business Administration Monroe Hall, Room 203 2115 G Street, N. W. Washington, D.C. 20052, USA.

Introduction

This paper is about one rather special style of managerial behaviour that is widespread and termed "dysfunctional". For lack of a better term, it shall be referred to here as "Perturbatory Management." This paper discusses some managerial behaviours that perturbs people causing anxiety, frustration, and dysfunctionalism which affects organizational operations. The resultant manifestation often leads to destructive competitive behaviours which work against the achievement of goals. These managerial actions could be called "Games Managers Play" a variation on a concept developed in the late 1960s by Eric Berne in his classic work "The Games People Play." The "Games Managers Play", relates generally to those managerial actions, which by design place subordinates and the organization in general, into a state of "perturbation", and interfere with the efficient accomplishment of tasks and responsibilities. Any action of a manager which perturbs those around him and does not aid or assist subordinates to work more effectively is defined here as perturbatory management action.

While Victor Thompson is credited with the descriptive characterization of

Dysfunctionalism in American management systems   147

dysfunctional behaviours there are many theories about managerial dysfunctionalism. Sayles and Strauss talk about the costs of "coordinating complexity", Parkinson postulates his first law, Laurence J. Peter hypothesis the "Peter Principle", Bennis cites the dysfunctional result of the competitive situation and Sterling Livingston describes how management education fails to relate to the real world. All of these constructs contribute to the phenomenon I am calling Perturbation Management.

Narration

A new era is dawning in corporate America. It is a world of buy-outs, take overs, downsizing and restructuring. Mergermania and corporate cannibalization is the new way of business life. These corporates actions/behaviours often destroy rather than build organizational effectiveness. They cause intergroup competition and design into the organization destructive conflict versus intergroup collaboration which relates to total organization effectiveness, individual productivity and high morale.

All competition is not bad! The merits of competition are readily seen in the free industrial world of work and life in general. However, an effective manager must recognize when competition needs tempering, and be able to coordinate and integrate the organization into a total functional system. This collaborative behaviour is manifesting itself within the European community, as the 12 countries count down to the day in 1992 when trade barriers fall.

Presented here is a simulation, presented to ISAGA 88, which I refer to as the Escalation Simulation. The simulation cites an overwhelming example of destructive competition and addresses an area of competitiveness that all managers should readily be aware of. To ignore the destructive nature of this type of competitiveness, which manifests itself in the following simulation, is to collude with environments that allow for management by perturbation.

Refutation

In the real world of business "Perturbatory Management" is practiced by individuals who have risen to a level where they are no longer competent to perform effectively. These individuals quite often have no apparent understanding of how their organization functions or how to integrate all its parts. These people behave in a way that depicts no understanding of how the organization functions as an interaction-influence system. Least of all, their actions indicate that they do not understand basic organizational behaviour, the dynamic that turns an abstract formal organization into a functioning and viable organism. Perturbatory Management practices are prevalent in many organizations with a hierarchical management structure. Large numbers of managers, regardless of the degrees held, grades received in school, or the formal management education programmes attended, have aspired to and risen to a level of inability to manage. This inability is not always the fault of incompetence or malice on the part of a manager, but is tied to inflexible structures and rigid organizational designs. The structures born of the industrial revolution were not designed to operate as a colaborative and interactive system.

Some Perturbatory Management techniques are described in the following examples. Paul Pickle Pepper went to work for a university. Paul attended all the right schools, held the right degrees and was young, bright, and aggressive. However, what Mr. Pepper failed to gain in his formal education was soon to become very evident. In short, Mr. Pepper was considered a threat by certain superiors due to his formal training and background and with that, the perturbatory management practice was put into action. Inside of 2 years, he occupied eight different offices. His superiors ordered moves on the average of every 3 months. With job responsibilities as such, no sooner would he get settle and have some time to think and analyse his new situation when it was moving time again. Since office space was always at a premium, any move involved a second, and sometimes a third or fourth party. These disruptions not only had an affect on his life but the lives of all those involved. Among the employees it became a sardonic comment: management sings to the tune of "musical chairs" and thereby forces the staff to dance to what was commonly referred to as the "office shuffle".

Another management by Perturbation technique Mr. Pepper recalls was the unstructured staff meeting. Staff meeting were held every Monday morning, religiously, from 9:30 until . . . ? During these meetings all employees were entitled to speak and were called upon for contributions. Two or three hours later everyone had spoken but no one knew exactly about what. Many times the staff would ask for a summation of the discussion and quite often management was unwilling or unable to give a recapitulation.

An additional management by perturbation technique is described in the actions of managers who repeatedly ignored recommendations sent forth from the staff to superiors. Unless the recommendation crossed the manager's desk at least three times, the request would go completely unacknowledged. This oversight was viewed by the subordinates within the organization as a deliberate action by management. In defense of their actions, often times management would say: "The problem will solve itself." This conflict avoidance technique allowed the manager to avoid facing the issue. However, such behaviour avoids dealing with substantive issues, further creating more perturbation.

The phenomenon of perturbation management often gives rise to conflict. Which in turn can create a win-lose environment or a "we" versus a "they" thinking. These actions lead to a "we are right, they are wrong" mentality. When the intensity of these environments grow, collaboration becomes an impossible dynamic. Collaboration is the very foundation of the interacting and high performing systems. Collaboration is the foundation for the European Communities "single market" concept.

Take five people who have never before been together and give them a task to be judged by an impartial observer. In less than half an hour watch how the intensity of competitiveness grows and destructive actions occur. When these destructive dynamics build collaboration becomes impossible.

To see how these dynamics manifest within groups the writer presents the following simulation for the readers consideration, as presented to ISAGA 88 in Utrecht, The Netherlands.

The Escalation Simulation[1]

The Escalation Simulation is an auction of a $2.00 roll of nickels and is an experiential learning method which illustrates the detrimental effects of destructive competition. It is easily executed in a group or classroom environment. The professor plays the role of auctioneer and selects five persons or volunteers from the class to serve as bidders. He then auctions off the roll or $2.00 worth of nickels, one at a time, so each bidder must have enough pennies to participate in the auction. Two regulations must be followed during the auction: (1) neither the participants not the audience are permitted to talk, (2) the auctioneer must receive a minimum opening bid of one penny for each nickel. I have observed that the intensity of the competition usually builds by the time the bidders reach the price of five cents, and surprisingly, it is not unusual for a participant to bid six or seven cents for a nickel. (Always in the spirit of competition.)

The participants would stand to gain if they agreed (nonverbally) to collaborate with each other by allowing each to bid a penny for each nickel. By doing so, they could make a profit of 32 cents. However, collaboration, as this simulation illustrates, is rarely, if ever, achieved. In fact, I have yet to observe a Nickel Auction in which it has occurred.[2]

The Simulation exemplifies destructive competitive behaviour which often occurs in society, and in fact, is reinforced by society. We have learned to accept, if not expect, competitive behaviour, in which individuals pit themselves against each other for the grand "prize," even though the "prize" may actually be gained more easily through collaboration.

This attitude of "winning at any cost" which society fosters, is one variable among many serving to reinforce a behavioural manifestation I have termed "risk-free decision" phenomenon.[3] It is an attitude in which ethics and morality are often forgotten in the fight to *win* and which ultimately leads to the lack of ownership of behaviour. As simulation demonstrates, however, we all stand to lose from this type of attitude, whether we are talking about bidding for nickels or organizational management.

Alfie Kohn writing in *Psychology Today*, states the following:

Social psychologist Elliot Aronson has observed that "The American mind in particular has been trained to equate success with victory, to equate doing well with beating someone." Our society so values competition that it seems almost blasphemous to doubt its supposed connection to achievement. The relationship is, to many people, self-evident. But what does the evidence show? Do we really perform better when we are trying to beat others than when we are working with them or alone? Many psychologists, instead of taking competition's reputed benefits for granted, have put them to the test. And with astonishing regularity, they have found that making one person's success depend on another's failure – which is what competition involves by definition – simply does not make the grade.

Confirmation

Sayles and Strauss, in their book state:

> Managers sometimes believe they can get something for nothing. They can get the advantages of increased specialization by dividing up more complex jobs into simpler jobs or by hiring additional trained specialists. But they ignore the possibility that the problems or coordination so created may more than outweigh the advantages gained. . . . No matter how hard an employee works, his efforts are wasted unless they integrate with those of his fellow workers (Sayles and Strauss 395).

The classic Peter Principle makes the following assumption: 'In a hierarchy, each employee tends to rise to his level of incompetence: Every post tends to be occupied by an employee incompetent to execute its duties" (Peter 339).

Another example to be examined is C. Northcote Parkinson's Law. Parkinson's axiom is that the number of people employed in any organization, be it government, business, industry or education, has no relationship to the amount of work to be accomplished. In an attempt to improve their own position in the organization, incumbent superiors are motivated to expand their subordinates. Parkinson attests to how seven individuals are kept busy doing work formerly accomplished by one Person, Mr. A.

> For these seven make so much work for each other that all are fully occupied and A is actually working harder than ever. An incoming document may well come before each of them in turn. Official E decides that it falls within the province of F, who places a draft reply before C, who amends it drastically before consulting D, who asks G to deal with it. But G goes on leave at this point, handing the file over to H, who drafts a minute that is signed by D and returned to C, who revises his draft accordingly and lays the new version before A. . . . (Parkinson 5).

If things were not bad enough, conceivably we're all doomed. At least according to J. Sterling Livingston:

> There is no direct relationship between performance in school or training programs and records of success in management. . . . How effectively a manager will perform on the job cannot be predicted by the number of degrees he holds, the grades he received in school, or the formal management education programs he attends (Livingston 79).

What then remains for a management style? The Perturbation Management Phenomenon.

Conclusion

To make any concrete suggestions for changing the organizations, we must study organizational management style: how it is used, by whom, for whom, against whom, for what purpose, and in what context.

Our institutions and organizations are in need of collaboration and co-operation from all persons involved; an integration of ideas and activities at all levels is eminently necessary, to be properly managed. The lack of collaboration and co-operation creates a void of mistrust, ambiguity, anxiety, defensiveness, internal frustration, competitiveness, intolerance, poor communication and a decrease in individual levelling, coping and interfacing. Needless to say, when these dynamics are present, an organization can be neither viable nor functional, these dynamics lead to a competitively destructive environment. People who work in organizations, at all levels, should know how easy it is to induce conflict and how difficult it is to arrest it.

The concern for efficiency in this context is not aesthetic, economic, or even humanistic. It stems from the maxim that no system can long survive at either input or output levels that consistently or substantially deviate from an optimum range. As their data grows increasingly sophisticated, the management system and its functions, be they educational, governmental, business or industry, are increasingly endangered by such deviations. The deterioration and ultimate demise or death of systems, as presented here, is well documented in Allen Bloom's book: *The Closing of the American Mind*.

Summary

A management example which exemplifies the co-operative/collaborative model which is the thesis of this paper is the former economic success of People Express Airline, under the former Chief Executive Officer, Donald Burr.

When Don Burr left Texas International in 1979, he had no specific direction in mind. The US Congress had passed the Airline Deregulation Act of 1978, which freed existing airline to compete aggressively in both route selection and pricing and encouraged new entrants. Burr sensed that the end of 40 years of control by the United States of America Civil Aeronautics Board (CAB) created a historic moment exactly when he was looking for a new challenge.

In one furious acceleration akin to the lift-off thrust of Boeing 737, People Express rose in just two and a half years, from relative obscurity to international prominence. From three planes serving three cities with 24 flights a day to a fleet of 22 Boeing 737s, 10 Boeing 727s and one 747 making 264 nonstop flights daily to 20 destinations, including London. From a loss of $9.2 million, including start-up costs, on $38.4 million in revenues to earnings of $1 million on revenues of $138.7 million and net income of $6.3 for the first 6 months of 1983 (Rhodes 42).

"Most organizations frustrate people who really want to work," Burr says. "They control them, they watch them, and check up on them. They subsume the individual. They consider them guilty until proven innocent" (Aplin 341).

In Burr's view, the only way to develop a company is to develop the people within it. He created a working environment in which there was no distinction between company and employee because the two have to become one. Burr's attempt to reach this unity, yielded a system in which each employee was given ownership in the firm, participation in decision making, freedom to create an individual contribution, and opportunity for personal growth.

If Mr. Burr is correct in his assessment about how organizations build in frustration for their people (and I believe he is) then it's no wonder why management by perturbation lives and operates in our formal organizational systems.

People Express success was due to an environment built in which there was no distinction between organization and employee because they are one. This and numerous other similar examples tells us that the day of the "Imperial Organization" is over. The imperial organization as used here is a hierarchical structure with the great Messiah residing at the pinnacle of power. One who knows all, sees all, and tells all! This is an obsolete principle. It has never been congruent with the democratic processes, so cherished by so many. The imperial organization was *not* built on a premise of co-operation and collaboration. The Imperial model served us well during the industrial revolution and brought us to a new dawn requiring the development of a new management model and system of operation. The writer is certain the new management model has two dynamics: 1. Cooperation and 2. Collaboration built from dialogue.

Recommendation

If indeed the day of the imperial management model is over, what is the new direction? This writer believes the new model will be titled the Multi-C Managerial Model, the co-operative collaborative model. Rensis Likert referred to this as the "interaction influence model." In addition the new management model will be shaped from some of the following C's: caring, congruence, calmness, calumniation, chanceful, changeable, cohesion, collectivism, competent, complementary, comprehensible, concur and constancy.

The industrial revolution was born from the competitive imperial model and it served us well. Learning from the old and giving birth to a new model will not be without pain and pitfalls. One of those pitfalls will be the giving up of the Imperial "I", deeply rooted in the American psyche. People today are encouraged to think in terms of "selfishness" rather than "self-fullness" – in terms of competition rather than co-operation. This attitude is reflected everywhere, from Little League baseball to popular books, such as *Looking Out for Number One*, to the nuclear arms race between the United States and the Soviet Union. Max Lerner points this out in his article, "Business Ethics at Home and Abroad":

I think the great American malady today, by the way, is not necessarily corruption, not even power corruption, although these are real problems. I think the real malady is what I call the Imperial I.... The Imperial I – gimme, gimme, grab, grab. We have become a polity of pressure groups, not just individuals, that talk in terms of the Imperial I, with everyone thinking in terms of what they get from the public treasury, not thinking in terms of what the total bill is, or in terms of priorities or in terms of what it means for the tranquility and the cohesiveness of the nation. What that means, of course, is that the ethic that has been emerging is that of following the line of least resistance – what everyone can get away with.

The Escalation Simulation exemplifies destructive competitive behaviours (or what Harvey refers to as a collusion with mediocrity) which often occurs in organizations and society, and in fact is reinforced by these institutions.

This attitude of "winning at any cost" which has been fostered in the old is one variable among many serving to reinforce a behavioural manifestation I term the "escalation game." It is an attitude in which ethics and morality are often forgotten in the fight to win and which ultimately leads to the lack of ownership of behaviour. As the simulation demonstrates, however, we all stand to lose from this type of attitude, whether we are talking about bidding for nickels or organizational management.

To compete economically and remain viable in this global economy, organizations through their management practices must remember what Harmon and Mayer purport in their research, they state:

Without being consciously aware of it, we think and act as if organizations were real entities. Organizations and institutions are constructs, products of the mind. . . .

Thus, the only way to develop the new management model, for organizations today, is to develop the people within them. Organizations must create working environments in which there are no distinctions between company and employee, for they must collaborate and co-operate, because the two are one, and the dependency is mutual.

Notes

1. The writer expresses appreciation to the following authors without whose work this exercise would not have been possible: Aronson, E.; Fisher, A.; Harvey, J.; Kohn, A.; Kolb, D. *et al.*; Rappaport, A. and Raiffa, H.
2. This writer has illustrated this simulation for the past 15 years at least three times a year and the collaboration has never occurred, in the 45 tries.
3. Lobuts and Pennewill, "The Risk-Free Decision Making", *Journal of Business Ethics*, February 1986.

References

Aplin, J. C., Cosier, R. A. and Schoderbeck, P. P. 1988. *Management*. New York: Harcourt Brace Jovanovich, Inc.
Aronson, E. 1980. *The Social Animal*. San Francisco: W. H. Freeman & Co.
Berne, E. 1964. *Games People Play, The Psychology of Human Relations*. New York: Grove Publishing.
Bloom, A. 1987. *The Closing of the American Mind*. New York: Simon and Schuster.

Campbell, P. and Dunette, M. D. 1972. Laboratory Education Impact on People and Organizations. *In* Nord (1972).

Fisher, A. B. 1980. *Small Group Decision Making*. New York: MacGraw-Hill Book Co.

Harmon, M. M. and Mayer, R. T. 1986. *Organizational Theory for Public Administration*. Boston: Little Brown & Co.

Harvey, J. B. 1974. The Abilene Paradox. *Organizational Dynamics*. American Management Association.

Kohn, A. 1986. How to Succeed Without Even Trying. *Psychology Today*. **20**:9.

Kolb, D. A., Rubin, I. M. and McIntyre, J. M. 1971. *Organizational Psychology: and Experiential Approach*. Englewood Cliffs, M.J.: Prentice-Hall, Inc.

Learner, M. 1977. Business Ethics at Home and Abroad. *The Personnel Administrator*.

Liker, R. 1961. *New Patterns of Management*. New York: MacGraw-Hill Book Co.

Livingston, J. S. 1971. Myth of the Well-Educated Manager. *Harvard Business Review*.**49**:1.

Nord, W. (ed.) 1972. *Concepts and Controversy in Organizational Behavior*. Santa Monica, CA.: Goodyear Publishing Co., Inc.

Parkinson, C. Northcote. 1957. *Parkinson's Law*. Boston: Houghton Mifflin Co.

Peter, L. J. 1967. The Peter Principle: We're all incompetent. *Phi Delta Kappa*. Indiana: Phi Delta Kappa, Inc.

Raiffa, H. 1982: *The Art and Science of Negotiation*. Cambridge, Mass,: Harvard University Press.

Rappaport, A. and Chammal, A. 1970. *The Prisoner's Dilemma*. Chicago: The University of Michigan Press.

Rhodes, L. 1984. That Daring Young Man and His Flying Machines. *INC*. January 1984.

Sayles, L. R. and Strauss, G. *Human Behaviour in Organizations*. Englewood Cliffs, N.J.: Prentice-Hall, Inc.

Thompson, V. A. 1961. *Modern Organization*. New York: Alfred A. Knopf, Inc.

Australian rehearsal technique

Erica Bates, Elizabeth Christopher and Barry Moore

Gretam Pty Ltd

ABSTRACT: Simulation-gamers in Australia use the Australian Rehearsal Technique (ART) to improve the competence of decision makers in business and government. In ART the game producer develops a scene which models a decision-making process, selects volunteers to play the characters in that scene, and sometimes appoints a director to organize the cast and present the drama. The director and players stage a rehearsal of the scene. The other participants form an audience and are given cards setting out some aspect of the scene that they specifically observe. The producer returns to the stage at critical moments to stop the rehearsal and involve the audience as commentators. The debriefing process thus takes place throughout the simulation, which gives it a greater impact. Because the theatrical nature of the simulation is emphasized, players can distance themselves more readily from the characters they play and are less defensive about any analysis of their decision-making behaviour. People who prefer to be speculative observers rather than active participants have important roles to play as members of a critical audience.
KEYWORDS: Drama; decision making; ethics; games; gaming; management; role-plays; simulations; training.
ADDRESS: Gretam Pty Ltd., 75 Albert Drive, Killara NSW 2071, Australia.

Introduction

For all the arts, the existence of an audience is essential. In drama its contribution to the quality of performance is very obvious. An unresponsive audience discourages the actors, the performance suffers, and everyone goes away feeling depressed. A small audience is also discouraging, both to actors and to members of the audience itself. A critical mass is necessary to produce a chain reaction of morale and competence.

Simulation games are a form of drama, which suggests that an audience is an essential part of the action. There has been a rediscovery of this connection during the 1980s, particularly in Australia (Christopher, 1983; Moore, 1988). This rediscovery has led to the development of an educational simulation technique; a technique which did not have a label until it began to be discussed outside Australia. So it has become necessary to name it the Australian Rehearsal Technique (or ART). The "rehearsal" part of the name will soon become apparent.

ART is based on the assumption that observers are a vital element in simulation. ART requires that the trainer (or "producer" in ART terminology) set a scene and describe the characters in a role-play of a "rehearsal" of a real-life situation, event or dilemma. The producer may then step into the background, leaving one or more of the participants in control of the situation, with instructions to act as drama director (or joint directors), but when dealing with inexperienced groups, the producer will play the drama director's role in addition to the standard producer's role of Game Overall Director.

In the drama director's role, the task is to organize the actors, brief them on how to play their roles, and organize the rehearsal of the simulation-drama. This dramatic device would be familiar to anyone who has seen Pirandello's play *Six Characters in Search of an Author*.

The action on stage develops, with consultations between directors and actors and false starts by the actors. Eventually the scene develops, and actors begin to immerse themselves in the roles of their characters. The action is watched closely by the other participants, that is, those who are playing the role of members of the audience. Ideally the "watchers" would outnumber the actors, in order to create the sense of an audience watching a stage performance.

Participants in the audience are informed by the producer that they are members of a studio audience privileged to watch a master class of actors rehearsing a scene. The producer instructs them to observe carefully and be ready to comment on the performance that is unfolding before them.

At critical moments in the rehearsal the producer returns to the stage, stops the action and involves the audience as commentators. The actors respond to these comments, and may change their interpretation of their roles accordingly.

Ethical Competence

Consider the problem of teaching ethics to managers in business and government. Ethics is shaping up as one of the growth areas in business education for the 1990s. The tough-minded 1980s have been all about financial competence and the virtues of the "bottom line" of a profit-and-loss account. However, nothing lasts forever and now we are moving into an era where the undesirable consequences of yesterday's priorities are being sheeted home.

A segment on ethics is a compulsory part of the Master of Business Administration course at Harvard. Undergraduate and graduate students of health administration, in Australia, are also being exposed to compulsory courses dealing with the ethical implications of management decisions. In our country the challenge for teachers of ethics has been to develop strategies which will ensure the ethical competence of health administrators in an environment in which the financial outcome is still considered the most important aspect of any decision.

However, in the health arena there are a number of compensating advantages. For example, it is relatively easy for a teacher of ethics to dramatize the painful decisions that have to be made by a health service manager as the result of technological changes in the delivery of health care.

Lady Jane Hospital

We wrote the scenario of LADY JANE HOSPITAL to highlight some of the agony of decision making in a health care facility. It is a fact of everyday life in health administration that there is not enough money, nor trained staff, to do as much as can be done for all who have health problems. Choices have to be made, between preventive and curative services, and within each of these categories.

Each time we play LADY JANE HOSPITAL, some actors try to avoid their ethical problems by expanding the hospital's resources, by such means as sending some patients away to another hospital, by threatening politicians with loss of votes, or by "inventing" a Rotary Club or some other dispenser of benevolence in the community. It requires considerable skill to design a game that closes off all of these options, by writing bad news into actors' scripts ("all other hospitals are having the same problems, they want us to take some of their patients!") or by having the producer suddenly leap up on stage to stop the action and obliterate some lateral thought on political action ("I'm sorry, it is no use harassing the Minister; the Government's resources are fully committed to other essential purposes like law and order, and keeping your superannuation fund solvent.")

ART Effects

What effect does ART have on the usual outcomes of a simulation game? Our answer is that there seem to be at least three ways in which favourable outcomes are enhanced, and unintended consequences suppressed. These are related to the use of a *rehearsal*, the strategy of *producing a drama*, and the role of the *audience*. Let us take them one at a time.

The rehearsal

By defining the first run-through as a rehearsal, the actors are able to try out different types of performance. In writing the scenarios and role-plays, we avoid very definite character descriptions so that the actors can change their interpretation of the role, depending on the responses they get from other characters, the director, or the audience. Consequently there is less pressure on the actors. In particular there is not the fear that they may expose themselves as autocratic, heartless or weak; it is clear that the characteristics they portray are not their own but those of a character in the drama. Nevertheless, the players are well aware that much of their acting fits aspects of their own personalities, and although they may never acknowledge this publicly, they do learn quite a lot about themselves, and often refer to their own covert responses later.

Producing a drama

In ART, an important role is that of directing the actors on stage. This is a very skilled role which may be allocated to a particularly competent student or be played by the Game Overall Director personally.

A great deal of responsibility falls upon the person who plays the drama director. The drama director is responsible for the guidance of the cast, the replay strategies, and consequently for the degree of interest generated amongst the actors and the audience. The drama director sets up the first rehearsal by telling various members of the cast how to play their roles, and may even develop character sketches for those who cannot work these out for themselves. The drama director decides the stage at which the first try-out has been completed, and whether a different set of strategies should then be rehearsed. The drama director should act as if he or she is the compere of an event and call on members of the audience to act as chorus to the play, or as a double to a character, or to reflect on some part of the action.

In the role of producer, on the other hand, the Game Overall Director will decide when to bring the audience into the discussion and ensure that their interest is being maintained. Indeed, the producer, using discretion, may make awards to members of the audience, such as gold stars, or chocolates, for guessing what a character has in mind, advising what a character should do, or analysing an important theoretical point.

The audience

Members of the audience play out roles that are far more active than those of a normal theatre audience. Nevertheless, their roles are more passive and less visible than those of the actors or directors. Participants in a training session who prefer the role of a reflective outsider rather than that of an active experiential learner, can participate in a legitimate studio-audience role and achieve their own learning goals as well as assisting the actors to achieve theirs.

Like audiences in general, the audience here watches the unfolding drama and takes its cue from the drama director and the producer, becoming involved mainly, but not exclusively, at the producer's request. When the producer asks for audience feed-back, the well-managed audience should give it, in the form defined by the producer, so that members of the audience will comment on the acting skills of the cast or the appropriateness of their portrayal of the characters, and what seems to be in the mind of a particular character being represented on stage. Sometimes a volunteer member of the audience may be called on stage to act as an *alter ego*, or an interpreter of the drama.

But the main role of the audience occurs at the end of the dramatic action. The producer has a set of cards, on each of which there is one criterion for the audience to observe. For example, in LADY JANE HOSPITAL the actors are contending with problems of life and death, each following an agenda prescribed by the script, and each relying on a set of internal values not prescribed by the script. One member of the audience receives a card reading "need", meaning "watch for examples of actors arguing on the principle that patients who need the most should get the most". Other members of the audience receive cards reading "existing entitlement", "efficiency", "social merit", or any other ethical basis for allocating scarce resources within a hospital.

At the end of the rehearsal, members of the audience each describe what they observed, in terms of a principle, a character trait or a theoretical quality. Hence the audience and not the producer dominate the debriefing session, and produce considerably more variety in their observations. Their comments are also more likely to be accepted because the members of the audience are part of the learner group.

Implications

Not all simulations are suitable for ART. More work still needs to be done to establish the kind of simulation game that works best, and the proportions of time that should be devoted to rehearsal and performance. There is a risk that the performance may be too similar to the rehearsal and that everybody's interest may therefore decline. If that proves to be the case then there seems to be little alternative to abandoning the idea of a replay and using the rehearsal as an end in itself.

One significant advantage of simulation gaming is the happy combination of education and entertainment, and this advantage should not be discarded light-heartedly or wantonly. To date we feel that ART has certainly proved its value both as a teaching technique and as a crowd-pleaser.

Session Review

At the ISAGA 88 conference we had a group of 15 participants, of whom seven took acting roles and eight formed the audience. The three authors of this paper took complementary roles. Elizabeth Christopher was the drama director, Barry Moore the producer and Erica Bates the theoretical analyst at the beginning and the end of the play. Our subjective impression of a successful session was supported by the general comments of the participants, and Ken Jones, who had been a member of the audience, cheered us up by saying that we had given him a unique experience that he would not forget. Important comments were that the roles of drama director and Game Overall Director might be combined rather than separated, and that the idea of debriefing a rehearsal rather than a performance should lead to interesting developments in the educational technology of simulation-gaming.

References

Christopher, E. M. 1983. Representation and impersonation, action and observation: a study of the dramatic environment created by role-playing simulation as a directed group learning method. *Ph. D. Thesis*. Kensington, N.S.W., Australia: University of New South Wales.

Crookall, D. (ed.) 1988. *Simulation-Gaming in Education and Training: Proceedings of ISAGA 87*. Oxford: Pergamon.

Moore, B. 1988. Management training by simulation-gaming. *In* Crookall (1988).

Simulation References

LADY JANE HOSPITAL. Bates, E. M. 1987. Gretam Pty Ltd.

Talking heads

Elizabeth Christopher
Mitchell College of Advanced Education, Bathurst, NSW 2795, Australia

ABSTRACT: The purpose of the game is to make players more aware of the range of factors that affect the conduct of a business meeting of people from different cultures. It is intended to take several hours and debriefing runs throughout the game. The number of players may vary from about 6 to 13 and "extras" may function as critical observers. Players assume roles as national managers of an international company. They have two tasks: (1) to make recommendations concerning proposed changes for Saito corporation; (2) to recommend a single corporate language for Saito's international dealings. All players are given the same general scenario and individual role instructions which require them to adopt certain positions. Thus ostensibly at the conference table they discuss ramifications of the proposed organizational changes – but they all have hidden agendas.

KEYWORDS: International negotiation; game; intercultural conflict; organizational behaviour; management training.

ADDRESS: 19 Ryries Parade, Cremorne, NSW 2090 Australia.

Designing an intercultural business simulation

Richard D. Teach
Georgia Institute of Technology

ABSTRACT: The theme of this paper is to design the framework for a computer run business simulation that incorporates both corporate and national cultural variables in a way that dominates the more traditional bottom line outcomes. The major issues are to define the cultural variables, determine how to measure the different responses to the cultural variables and how to structure this simulation in order that the cultural sensitivities can play the dominant role. The game is to be computerized and can be played in a series of distinct cycles over some length of time in several short periods. Thus, the game could be used in a university course.
KEYWORDS: Business simulations; corporate cultures; national cultures; simulation design;
ADDRESS: College of Management, Georgia Institute of Technology, Atlanta, Georgia 30332-0520 USA.

The educational challenge of business simulations

Ronald Brech
RBRC Software, London

ABSTRACT: Computer-assisted learning has added a new dimension to education. Students are brought into the real world, where uncertainty is the norm rather than the exception. Simulations provide students with "live" experience and enable them to understand and handle risk, so that they can come to terms with uncertainty.

Students are made to realize that there are no pre-determined answers to problems. To take a right decision implies that you have *either* an accurate forecast *or* luck. Computer simulations are in fact the only way where students can be taught to make decisions *come* right . . . as in life itself. They help students to learn from mistakes, first by teaching them to recognize them and correct them, and later to foresee them and prevent them. The students in fact learn by doing . . . from experience.

Business simulations have a particularly important role to play here, because business concepts are similar to those of everyday life. Good business, after all, is nothing more than sound common sense.

But business simulations are also a means for tutors to adapt themselves to the modern idiom of education, by helping them to switch from the actor dynamic to the spectator dynamic. Instead of providing the answers, the tutors have to ask the pertinent questions.

KEYWORDS: Computer-assisted learning; "live" experience; learning from mistakes; spectator-dynamic; handling uncertainty.

ADDRESS: RBRC Ltd, The Guild House, 32, Worple Road, Wimbledon, London, SW19 4EF, The United Kingdom.

Student- and instructor-oriented features in a business simulation

Precha Thavikulwat and Jimmy M. T. Chang

Towson State University and Hong Kong Polytechnic

ABSTRACT: Simulations for the classroom must account for two users: the student and the instructor. For the student, a simulation should be challenging, procedurally intuitive, and conserving of student's time. For the instructor, it should be flexible, administratively facile, and resistant to students' mischievous propensities. MANAGEMENT 500 is a prototype microcomputer-based, total-enterprise business simulation designed to meet these requirements. To be challenging to students, the simulation produces complex results with simple relationships. To be procedurally intuitive, the simulation programme incorporates Lotus-like menus and error-tolerant routines. And to be conserving of student's time, the simulation includes a task-managing master programme that enables several programmes to be simultaneously retained in memory. To be flexible for the instructor, the simulation programme is supported by a configuration programme that allows the instructor to alter the parameters of the simulation. To be administratively facile, the simulation is of an independent-across-players design, which does not require the instructor to co-ordinate programme inputs. To be resistant to students' mischievous propensities, the simulation programme encrypts its data files and stamps printed outputs with tamper-evident identifying information. Incorporating student- and instructor-oriented features in a simulation improves competence in user-oriented simulation design.
KEYWORDS: Business; management; total-enterprise; simulation; microcomputer; 500; instructor-oriented; student-oriented.
ADDRESSES: Precha Thavikulwat, Department of Management, Towson State University, Towson, Maryland 21204, USA.
Jimmy M. T. Chang, Department of Management Studies, Hong Kong Polytechnic, Hung Hom, Kowloon, Hong Kong.

Adapting a wargame for multiskill tasks

David Crookall and Danny Saunders
*The University of Alabama, USA and
The Polytechnic of Wales*

ABSTRACT: The workshop described here aimed to look at ways in which a wargame could be adapted for other uses. An outline is provided of the original objectives of the workshop and a brief account is given of what happened during play and the debriefing. The game materials as used during the workshop are included here to enable others to pursue ideas.
KEYWORDS: Wargame; simulation/game; board game; conflict.
ADDRESSES: David Crookall, Department of Speech Communication, The Pennsylvania State University, 305 Sparks Building, University Park, PA 16802. USA.
Danny Saunders, Department of Behavioural and Communication Studies. The Polytechnic of Wales, Pontypridd, Mid Glam CF37 1DL, Wales (UK).

Section Four

Policy Exercise

Editor: **Cees A. Th. Takkenberg**

Policy exercise, (group) decision support

Cees A. Th. Takkenberg
Utrecht University, The Netherlands

Policy will be considered as a sort of planning. Planning is the most basic of management functions and requires the designation of goals and the selection of alternatives throughout all levels and divisions of the organization. Structuring the organization as such, activating human resources within the framework of this structure, and the subsequent control processes require all planning. When we plan we decide in advance what must be done. The alternative to planning could only be something akin to random behaviour with frequent shifts in direction and inconsistent activity. A lack of planning on the part of management would be seen as surrendering to excessive attention to immediate problems – putting out fires – and erratic, *ad hoc*, and perhaps inconclusive decisions. Plans may be firm or flexible; they deal with goals and the means of achieving them. Objectives will be seen as special goals and to accomplish them, policies, rules and procedures may be used. These are all constraints or guidances that border the path leading to the objective. Policies are likely to take the form of general statements or understandings that serve to channel activity toward the objective. Rules and procedures are more rigid and more suitable for guiding short term decision making.

Policy exercise is a field where simulation and gaming come into action; often some form of decision support system will be used for interactive simulation. The gaming context for DSS is an effective one.

Steven Underwood presented a paper describing a policy exercise in technology assessment. The policy exercise is a scenario development procedure that brings together stakeholders, policy experts and institutional experts to synthesize and explore collective knowledge for policy making. The core of the method is a workshop where participants develop scenarios to forecast technological, environmental and institutional events and to assess their potential impacts. The paper presents the concept for the policy exercise, describes the current research on methodological developments, and considers the implementation in the context of technology assessment. The improvement of communication among scientists and policy makers is seen as the most important function.

Mario Polic and Ivo Wenzler use gaming-simulation for a definition of a project on the formulation of strategies of the Yugoslav Bank for International Economic Cooperation (YUBIEC). The formulation of the strategies is a complex problem that will be seen from multiple perspectives and gaming-simulation was used to define the project outline and organizational structure of the project. Based on the workshop and the different approaches taken, four distinct perspectives on the competitiveness of Yugoslav exports were formulated at the beginning of the game session. Effective communication leads to consensus.

Marian Sackson introduced a new topic: the simulation of group decision making in a stochastic environment. This will be seen as a new application of expert systems that also exhibits learning. The paper describes an experiment conducted to test a model developed to simulate group decision makers in a strategy and policy-making environment. First a comprehensive expert system model is built. The intent of this model is to provide an experimental environment to test the use of an expert system as a tool to help group decision makers to learn more about the decision process that is used to evaluate strategic scenarios. This methodology is applied in a dynamic business environment.

Henk Sol and Michael van der Ven presented a group decision support system for international transfer pricing decisions. This experimental system is developed in close co-operation with a European based pharmaceutical company and is in real use since spring 1988! The authors state some interesting conclusions; one of these is the added value of gaming with respect to the use of computerized decision support tools. These tools in themselves are not sufficient for improving the decision process. With gaming, problem decomposition, use of the support tools and the role of organizational members are integrated in the decision process structure.

Pieter Bots, Frans van Schaik en Henk Sol used gaming as an environment for testing Decision Support Systems. A crucial question is the added value of DSS. Do decision support systems lead to better decisions? The paper presents a gaming approach towards testing the effectiveness of a decision support system and a so-called task structure. This approach enables the assessment of their effectiveness before they are actually introduced in a practical situation. A management game is developed to simulate an actually existing company in which the introduction is considered. Experiments show that a DSS does not cause decision makers to make any better decisions. Decision quality does improve dramatically, though, when decision makers are guided in the way they structure their problem situation by providing them with a task structure.

Eduard Rădăceanu presented a paper concerning "the Value Analysis of Management Information Systems, Problem Analysis and Gaming".

The development of a Management Information System is guided by a game. The game uses value analyses and a way of problem analysis that brings managers and system development people together. The analysis leans upon a complex model of the enterprise concerned. The use of an optimization model with an object function enables the evaluation of alternative scenarios.

Structured participation in technology assessment: the policy exercise

Steven E. Underwood

The University of Michigan, USA

ABSTRACT: This paper presents a new hybrid method with potential benefits for technology assessment. The Policy Exercise is a scenario development and assessment procedure that brings together stakeholders, policy experts, and institutional experts to synthesize and explore collective knowledge for policy making. The core of the method is a workshop where participants develop scenarios to forecast technological, environmental, and institutional events and to assess their potential impacts. This paper presents the concept for the Policy Exercise, describes the current research on methodological developments, and considers the implementation and possible benefits in the context of technology assessment.
KEYWORDS: Technology assessment, decision conferencing, policy exercise, scenarios, gaming-simulation, forecasting.
ADDRESS: University of Michigan Transportation Research Institute, University of Michigan, Ann Arbor, MI, 48109, USA.

Introduction

Addressing difficulties in adjusting to technological change, William Ogburn wrote in 1937 that "we seem to be looking backward as we drive rapidly through the fog across open country on our course toward the future" (see Ogburn, 1937). I doubt that today his impression of technological forecasting would be any less dramatic.

In recent years we have witnessed a significant increase in the rate and impact of technological change. Current advances in energy, chemical, industrial, and biological technologies, to name a few, are a source of threat as well as promise for future generations. Responding to these developments, institutional and methodological improvements in technology assessment (TA) have enhanced our potential for assessing environmental impacts and informing the public decision process. Notwithstanding this evident progress, "the need for environmental action is at least as great as it has ever been. Problems long recognized remain unresolved, and new ones continually appear. Public understanding of environmental threats lags well behind reality, and political consensus on how to meet new needs is not apparent (see Conservation Foundation, 1987)." Although this statement describes environmental action in the USA, it might be even more descriptive of the European context where TA is conducted in a multitude of organizational settings and must respond to a growing diversity of interests (see Leyten, Smits and Geurts, 1986). These problems reflect the increasing institutional complexity of most technology-environment issues and the growing need to address pluralistic policy considerations in TA.

Recent methodological improvements are directed at integrating scientific and technical knowledge with institutional and policy considerations in TA. The International Institute for Applied Systems Analysis (IIASA) development of the Policy Exercise (PE) is one such effort. IIASA's research has advanced a new participative approach that is aimed at synthesizing technical and institutional knowledge while building alternative scenarios to guide an assessment. The PE is used to explore the long-run consequences of technology-related policies from a number of perspectives while providing a structure for project co-ordination. Through the interactive participation of technical experts and social leaders, the PE is designed to raise public consciousness about technological choices and their possible impacts. It is also designed to guide TA research in a path that is sensitive to public interests.

Technology assessment is "a class of policy studies which systematically examine the effects on society that may occur when a technology is introduced, extended, or modified. It emphasizes those consequences that are unintended, indirect, or delayed" (see Coates, 1976). As such, a major function of TA is to increase the public consciousness about societies' choices regarding science and technology. This function is best served when technical considerations are framed in the context of realistic institutional constraints. However, large-scale TAs have traditionally focused on technical analysis and expertise, addressing institutional issues in a secondary and/or separate manner. It is difficult enough to integrate the technical analyses around a central theme, let alone respond to a diversity of public interests. Integration of institutional and policy concerns has always been a problem in TA. As a result, the public is often left wondering "what are the choices we face?" Or worse, "what does this study have to do with the real issues that we are addressing."

Concept for the Policy Exercise

Facing similar prospects, William Clark decided to address this problem directly in his design of Biosphere Project at IIASA, noting that the conventional methods for making the policy-science link (i.e. formal models and expert panels) were not up to the task. He proposed a new approach – the Policy Exercise (PE) – for "synthesizing large amounts of uncertain data into useful scenarios for environment-develop interactions and social response" (see Clark, 1986, p. 39).

Clark identified two communities who could affect change: the technical/scientific community, and the policy/institutional community. Both of these

174 *Steven E. Underwood*

	Institutional interests	
	Unified	Fragmented
Scientific knowledge — Unified	Extrapolation Formal models	Gaming-simulation Scenarios Panels
Scientific knowledge — Fragmented	Integrative models Cross-impact Delphi	Policy exercise

FIGURE 1   Futures Methods in TA Contexts

communities are nonmonolithic. That is, the technical community is actually an aggregation of a number of different and sometimes conflicting perspectives in the issues, and the institutional community is also an aggregation of differing interest groups. Synthesis in this context refers to the integration of multiple technical perspectives, multiple institutional perspectives, as well as combining the perspectives of the technical and institutional communities. Figure 1 shows four possible contexts for TA and the relationship of the PE to other TA methods in this scheme.

Gary Brewer (1986) extended this notion, providing the conceptual foundation for what he viewed as a "deliberate procedure in which goals and objectives are systematically clarified and strategic alternatives are invented and evaluated in terms of the values at stake" (p. 468). Brewer's original concept was that of a hybrid procedure which merged free-form gaming and scenario writing with other methods where appropriate. Brewer felt that the PE was needed as a "preparatory activity for effective participation in official decision processes" (p. 468). The essential features of the PE in these early formulations were (1) structured participation of experts, (2) policy-science linkage, and (3) scenario generation for exploring unconventional futures (see Fig. 2).

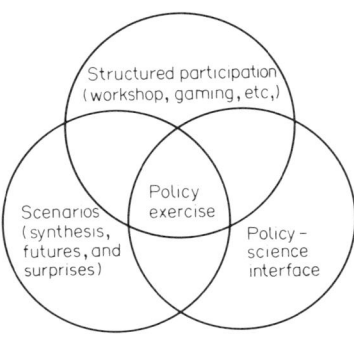

FIGURE 2   Methodological Elements of the Policy Exercise

Action Research: Developing the Policy Exercise

Current research at IIASA, the University of Michigan (UM) in the United States, and Environmental and Social Systems Analysis Ltd. (ESSA) in Canada centres on developing an operational procedure as specified by Clark (1986) and Brewer (1986). Current applications include European forest sector (IIASA), the European environment (IIASA), Great Lakes fisheries (UM), and global warming effects on the Great Lakes (ESSA).

Recent test applications have been encouraging. Based on an initial operational procedure developed by Toth (1986) and the results of experimental runs (see Duinker, 1987; Toth, 1987; Underwood and Toth, 1988) the actual design of the PE has evolved into a staged procedure that produces integrative scenarios of long-run scientific-institutional developments. It combines the methodological features expert opinion, formal modelling, and scenario writing, into a single hybrid approach. A key component is the *scenario development workshop* where scientists and policy representatives have an opportunity to interact directly while constructing and evaluating scenarios of long-run technological-environmental interactions.

Favourable results in the early experiments include: (1) high participant satisfaction with the workshop, (2) desirable levels of future-oriented policy-science discussion and analysis, and (3) successful development of scenarios that reflect both technical and institutional considerations. Nevertheless, a number of operational questions remain. Further trials will examine the appropriate number and mix of participants, the appropriate timing and duration of the workshops, effective integrative modelling techniques, and alternative approaches to debriefing the exercise and summarizing the results for dissemination. Although the research teams have made significant advances in the development of the approach, the procedure requires further research and testing before it can be applied in a systematic fashion.

The Policy Exercise and Technology Assessment

Although the PE grew out of needs specific to large-scale long-term environmental impact studies (i.e. specifically the Biosphere Project at IIASA), the assessment requirements of a long-term TA are not much different from those of this original application context. Especially in the European context where decisions related to the development of environmentally sound technology have interjurisdictional elements, the PE could offer an approach for bringing together representatives of various interest groups in a constructive atmosphere. The assessment functions of the PE, that is the synthesis of knowledge and the exploration of alternative development scenarios, is essentially the same in TA as in an environmental impact study. These functions are served through the structured interaction of the participants while building the scenarios and evaluating alternative development strategies (see Underwood, 1988).

Beyond the assessment function of the PE, the procedure can also be used as a project management tool to help scope and bound the assessment problem and to co-ordinate the research effort. Most large assessments are composed of a number of smaller studies conducted by individuals or teams of experts. Combining and co-ordinating the efforts of these studies is not an easy task. The scenarios from the PE can serve as a basis for this co-ordination. Moreover, large-scale TAs frequently involve some form of steering committee or governing board that is composed of sponsors or selected interest representatives. The PE can provide structure for this kind of study while offering a means for soliciting direction from the governing board and for presenting information from the technical components of the study. This type of activity can be especially effective for initiating and completing large projects that require extensive co-ordination of several smaller studies. In fact, the PE can serve as an overall planning and co-ordination mechanism for the large TA.

Perhaps the most important function of the PE is improving communication among the experts (i.e. scientists and policy makers) who participate in the exercise and dissemination of information to the wider community. The communication function of the PE reaches beyond creating a synthetic information base for the assessment. Through involvement of noted experts and community leaders, the PE also draws attention to the interrelationship between important scientific and social issues in a manner that is digestible to the scientific and nonscientific community alike. Critical information and its implications for both communities surfaces in the co-operative dialogue required for construction and assessment of the scenarios. This dialogue is captured through the recording process in the PE and summarized in a briefing document. Furthermore, an intuitive understanding of the situation is conveyed to the participants through their experiences in the exercise. The PE becomes an opportunity for the technical experts and the community leaders to address the issues in a common language with a common referent system.

Implementing the Policy Exercise for Technology Assessment

This section describes one possible implementation of the PE in the context of a large TA. Other implementation strategies may be used to greater effect depending on the circumstances. In a large TA, the PE should be an integral part of the study. From the initial conceptualization and bounding stages, to the final dissemination of the results of the study, a PE can contribute to the synthesis, assessment, and communication functions.

In the interim stages the PE makes significant contributions to the underlying framework and projections and to the integration and co-ordination of the components of the study.

The activities of a PE coincide with the sequence of activities in a TA, with many of the PE activities making a direct step-by-step contribution to the phases of the TA.[1] I recommend that at least two scenario workshops be held in connection with a large-scale TA, one toward the front end for scoping and bounding purposes, and one toward the back end for assessment and communication purposes.

Shortly after the project start-up, an initial PE workshop will bring the selected participants (e.g. sponsors, steering committee, etc.) together for the first time, orienting them to the approach and bounding the problem for assessment. The participants are carefully selected for their knowledge of the technical and/or social system and for their ability to disseminate the information to a broader audience. This workshop is conducted to formulate the situation, asking the participants to identify key issues, stakeholders, and factors that must be addressed in the assessment. It is helpful at this stage to develop an integrative cross-systems model of the situation that could provide a basis for developing a set of appropriate forecasting models.[2] This workshop, and the tasks in preparing for it, correspond to the problem definition, technology description, and social description components of the TA. The products of this workshop – the integrative model and the listing of issues and interest groups – provide a co-ordinating structure for later stages in the study.

Following the initial workshop, the TA will set out to develop and calibrate a set of appropriate forecasting and impact models (or other methods, e.g. cross-effect matrices) which will capture the system as defined in the conceptual cross-systems model. The more integrated these models are, the better they will fit the requirements of the scenario construction workshop. Despite their well-publicized drawbacks[3], systems dynamics models are one possible approach for this type of modelling task. However, the workshop constraints should not be the determining factor in selecting the modelling approach. In fact, the models should be selected on their appropriateness for the assessment task modified to support of the workshop. As in most assessment practices, these methods and models will be used to forecast the state of the social and technological systems and to determine the related impacts as specified in the first workshop. This initial forecast and impact assessment is based on conventional assumptions on the state of the future. Scenarios generated in the second workshop will be compared with this baseline assessment as a form of impact evaluation and institutional/policy analysis.

The purpose of the second workshop is to develop a set of scenarios that will synthesize contributing perspectives (i.e. forecasts, impacts, and evaluations of the impacts) with some not-so-conventional institutional factors and policy moves. The workshop will apply the forecasting and impact models, which hopefully have been prepared in a workable format, in concert with simulated institutional responses from the workshop participants, to develop a set of technology development/social impact scenarios that are based on knowledge of the technological and institutional systems. The events and themes of the scenarios will be guided by the desired form of analysis. More than one workshop scenario is probably desirable. Some of the scenarios may include plausible surprises, again the decision on including surprises being based on the desired form of analysis. The workshop will conclude with a group evaluation and analysis of the scenarios. Alternative mechanisms (e.g. public policies, research programmes, regulations, monitoring efforts, etc.) for controlling the impacts of the technologies is a point of departure for the debriefing. The results (i.e. scenarios, analysis, workshop protocols, etc.) of the workshop will be summarized in a briefing document that will be distributed to all of the participants and may be included as an appendix to the final assessment report.

Following the scenario generation workshop, the analysts may wish to go back and revise their baseline scenario or examine new contingencies that surfaced during the workshop. Further analysis of the institutional and policy mechanisms considered in the workshop is also desired at this point. This stage roughly parallels the policy analysis in conventional TA. This analysis of the workshop results and revision of the baseline assessment is also communicated to the workshop participants.

A key element of the PE approach is the extensive communication between the study groups and the workshop participants. The usual means of disseminating the study results is bolstered by the constituency of PE participants and their ongoing involvement in the assessment. Hopefully, through the continuing dialogue between the technical experts and community leaders, the assessment will reflect the concerns of interest groups and the policy implementers. The assessment should reach beyond a single conventional and technocratic image of the future and address alternative policies, alternative images, and the means of getting from the present to alternative desired futures. The PE approach is designed to induce people to talk and to *look forward* "as we drive rapidly through the fog across open country on our course toward the future" (see Ogburn, 1937).

References

Becker, H. and Porter, A. (eds.) 1986. *Impact assessment today, Vol. 2*. Utrecht: Uitgeverij Jan van Arkel.
Brewer, G. D. 1986. Methods of synthesis: The policy exercise. *In* Clark and Munn (1986).
Cecchini, *et al.* (eds.) 1988. *Simulation/gaming in education, training, and development.* Oxford, UK: Pergamon Press.
Clark, W. C. 1986. Sustainable development of the biosphere: Themes for a research program. *In* Clark and Munn (1986).
Clark, W. C. and Munn, R. E. (eds.) 1986. *Sustainable development of the biosphere.* Cambridge, UK: Cambridge University Press.
Coates, J. F. 1976. Technology assessment: A toolkit. *Chemtech* 6.
Conservation Foundation. October, 1987. *State of the environment: A view toward the nineties.* A report from the Conservation Foundation. Washington, D. C.: The Conservation Foundation.
Crookall, *et al.* (eds.) 1987. *Simulation-gaming in the late 1980s.* Oxford, UK: Pergamon Press

Duinker, P. December 1987. *Policy exercises and IIASA*. Lecture delivered at the International Institute for Applied Systems Analysis, Laxenburg, Austria.

Holling, C. S. (ed.). 1978. *Adaptive environmental assessment and management*. Chichester, UK: John Wiley and Sons.

Leyton, J., Smits, R. and Geurts, J. 1986. The organization of technology assessment: A comparative analysis of five European countries. *In* Becker and Porter (1986).

Ogburn, W. F. 1937. Technology and national policy. *Plan Age,* **9**.

Porter, A. L., Rossini, F. A., Carpenter, S. R. and Roper, A. T. 1982. *A guidebook for technology assessment and impact analysis*. New York, NY: North Holland.

Sonntag, N. C. 1986. Commentary on "Methods for synthesis: Policy exercise." *In* Clark and Munn (1986).

Toth, F. 1986. Practicing the future: Implementing the policy exercise concept. *Research report WP-86-23*. Laxenburg: IIASA.

Toth, F. 1988. Practicing the future part II: Lessons from the first experiments with the policy exercise. *Research report WP-88-12*. Laxenburg: IIASA.

Underwood, S. E. 1988. The policy exercise: Cooperative learning for long-run policy assessment. *In* Cecchini, *et al.* (1988).

Underwood, S. E. and Duke, R. D. 1987. Decisions at the top: Gaming as an aid to formulating policy options. *In* Crookall *et al.* (1987).

Underwood, S. E. and Toth, F. 1988, in preparation. Improving the science-policy interface: A teleconferencing exercise for managing long-term large-scale issues. Draft of an IIASA working paper.

Notes

1. This section draws on Porter *et al.* (1982) for the steps of a TA and Toth (1986) and Underwood (1987) for the steps of the PE.
2. One possible approach for developing a cross-systems model would be to employ a simplified version of the AEAM procedure (Sonntag, 1986; Holling, 1978).
3. Most modelling approaches have benefits and drawbacks. In the case of systems dynamics, criticisms are that the models (or modellers) have made insufficient use of known empirical data, excluded unprecedented events, been inadequately verified, overemphasized the mechanistic view of systems, and have not adequately handled random events. Nevertheless, this is in integrated form of modelling and may be useful in a PE when supplemented with other forms of forecasting and assessment, especially expert-based approaches.

Project definition gaming/ simulation exercise

Mario Polic and Ivo Wenzler

Institute for Developing Countries, Zagreb

ABSTRACT: With the goal of contributing to the export competitiveness of Yugoslav economy, the Yugoslav Bank for International Economic Cooperation (YUBIEC) has initiated a project intended to serve as a permanent basis for the formulation of YUBIEC strategies. In view of the great complexity of the problem and the need for a multiple-perspective approach, a gaming/ simulation exercise was designed with a goal to define the project outline and organizational structure of the project. The exercise included a Problem Definition Workshop and the running of the YUBIEC game. Both were attended by a team of YUBIEC executives, and by experts from major research, business and government institutions. Based on the Workshop results and the different approaches taken, four distinct perspectives on the competitiveness of Yugoslav exports were formulated at the beginning of the game run. Because of effective communication among participants an appreciation for, and a consensus of the problems were generated. In addition joint proposals for possible solutions to certain key elements of Yugoslav export competitiveness were developed.
KEYWORDS: Managing complexity; game design; multiple perspectives; group problem solving; impact assessment; enhancing communication; project development.
ADDRESS: Institute for Developing Countries Ul. 8. maja 82/II 41000 Zagreb, Yugoslavia.

The Problem

Seeking to improve the competitiveness of the Yugoslav export industries, the Yugoslav Bank for International Economic Cooperation (YUBIEC), in co-operation with the World Bank, has initiated a project entitled "The Competitive Strategy for Yugoslav Exports". The project is intended to serve as a permanent basis for the formulation of YUBIEC strategies in providing financial support for those industrial sectors that could contribute to the faster development and greater integral competitiveness of the national economy. The Institute for Developing Countries (IDC) was given the task of developing the project outline and its organizational structure.

The problem was how to develop the outline and the appropriate organizational structure of a project whose outputs could effectively guide YUBIEC in formulating its short- and long-term strategies. For developing such a project a better understanding of the issues of Yugoslav export competitiveness was needed.

A joke told by one of the participants in the project, very closely represents some of them:

Two guys, a South Korean and a Yugoslav, were sitting in the desert when encountered by a hungry lion. The South Korean immediately jumped to his feet and started to warm-up. The Yugoslav continued to sit and asked: "why do you bother to warm-up when you can't possibly outrun the lion?" The South Korean answered: "I am not warming-up to outrun the lion, I am warming-up to outrun you."

The issues of Yugoslav export competitiveness are characterized by too many interconnected and unidentifiable variables of a highly qualitative nature and there is no proven conceptual model for decision making on the key elements upfront.

Besides their overall complexity (economics, finance, technology, science, bureaucracy, politics, etc.), one of the most important characteristics that has determined our approach is the expected decision-making context during the project development and implementation. In a situation where huge resources are in question and whose allocation can significantly influence whole sectors of the Yugoslav economy, socio-political determinants dominate the decision-making process. Participants' actions are expected to be largely intuitive, irrational, idiosyncratic, and politically motivated.

Because of the complexity of the issues and since there are no single correct approaches or answers to them, the project outline and its organizational structure have to be precise, yet flexible and acceptable from the very beginning to all actors included in the process. Since these actors hold different perspectives and are motivated by different goals relative to the project and its outcomes, it was important to employ an approach which could effectively deal with multiple perspectives and provide an environment for participatory, small-group decision making. The approach selected for developing the outline and organizational structure of the project was gaming/simulation.

The following figure indicates the phases of the IDC – YUBIEC gaming/simulation exercise.

Pre-game Activities

Pre-game activities lasted for 3 weeks, including preparation and running of the Issues Definition Workshop. After that, on the basis of the Workshop results, the YUBIEC game was designed.

Issues definition workshop

A one-day Issues Definition Workshop was held in the YUBIEC headquarters. The workshop was attended by a team of YUBIEC experts and representatives of various research, business and government institutions.

During the course of the day there were more than thirty people involved in discussing and defining the issues of Yugoslav export competitiveness. Their interest in participating was piqued by two things. Since gaming/simulation is just beginning to be used in the Yugoslav business environment, none of the

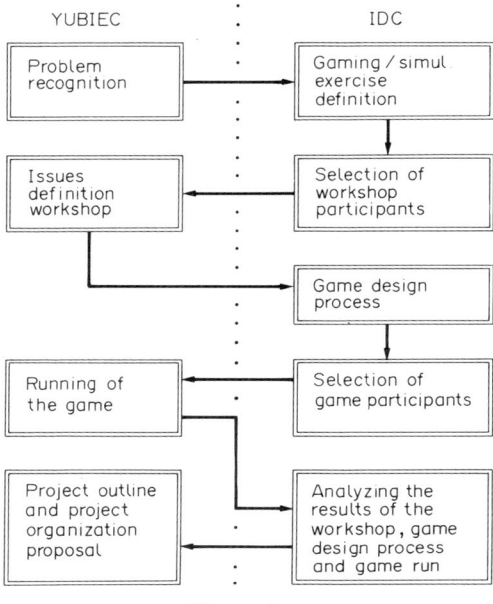

FIGURE 1

potential participants had ever experienced such an exercise. Secondly, due to the increasing need for significant socio-political and economical changes, there is a growing interest for new and different problem-solving and decision-making methodologies.

Workshop Objectives – (a) To provide a setting for an argumental and fruitful discussion on all aspects of the Yugoslav export competitiveness, as perceived by the perspectives represented; (b) To identify and record as many ideas and issues as possible; and (c) To provide a setting for gaining a comprehensive understanding of the Yugoslav export competitiveness problem environment.

Running of the Workshop – The workshop started early in the morning with an introduction of the issues under consideration and how the group was expected to discuss them. Knowing the flaws of standard discussions, a different strategy for organizing it had been prepared.

Participants were not provided with fully elaborated objectives and procedures of the Workshop. It was emphasized that no written materials or special preparations were needed. They were expected to freely discuss any issue they considered to be important. The only exception was the occasional introduction of a provocative statement by a moderator, to which the participants were asked to react. These statements were biased enough to trigger vastly different responses and thus enrich the discussion. During the discussion the participants were also asked to write as many ideas or issues as possible on small slips of paper ("snow-cards") which were to be used later in the process.

During the morning session of the Workshop, the habit of monology as a way of communicating ideas was so strong that, despite our efforts to break it, communication among participants was not enhanced to the extent we hoped for. Another reason for the lack of communication was their anticipation of something completely different from the Workshop. They all knew they would be participating in a gaming/simulation exercise, but since they had never experienced one, they expected to start "playing" from the very beginning. When it didn't happen they felt somewhat disappointed and slipped further into the usual way of communicating and discussing ideas. The domination of monologue hindered their communication, thus resulting in a lack of associative thought and use of fewer "snow-cards" in expressing themselves.

After lunch participants were divided into two groups. They were asked to organize their "snow-cards" and build a schematic presentation of the Yugoslav export competitiveness problem environment. In the process both groups were expected to link ideas on their schematics with coloured threads representing different influences and impacts. At the end, each group had to explain the logic behind their approach to the problem.

Although some of the atmosphere from the morning session remained, the involvement of the participants during this part of the Workshop increased significantly. Through the process of selecting, organizing and clustering "snow-cards", and identifying their relationships, participants began to develop a new understanding of the issues discussed. The main reasons for increased involvement were the opportunities for teamwork, informal communication, and creative involvement in transforming ideas into a schematic presentation. During this process leaders started to emerge, organizing activities within each of the groups. Presentations were listened to more closely and communication was enhanced substantially through many questions, comments and disagreements.

Despite the problems of communciation during the morning session, the objectives of the Issues Identification Workshop were primarily met.

Outputs of the Workshop – In addition to the tape-recorded discussions, the workshop resulted in two different schematic presentations, with about 500 ideas and issues written on the "snow-cards". These schematics and their elements, as well as transcripts from the discussions, served as input to the game design process.

Designing the YUBIEC game

The YUBIEC game had to be designed and run two weeks after the Workshop. In designing the game we relied primarily on the results from the

Workshop. However, it was not always possible to rely exclusively on the ideas and issues raised during the Workshop. We had to resolve some of the conflicting issues and prepare more information on several key elements before using them for the game design.

Game Design Objective – We sought to design a game that would assist in: (a) Identifying and analysing the key elements of the Yugoslav export competitiveness and determining their relationships; (b) Improving and facilitating communication, as well as supporting the exchange of different perspectives on the issues discussed; and (c) Conveying the complexity of the issues to all participants in the process, as well as facilitating a multiple-perspective overview of the issues and their impacts on the project development.

Game Design Process – In designing the game, a specific set of procedures were used to ensure achievement of the game's objectives (see Duke, 1974, 1980). The process began with the selection and occasional re-definition of the ideas/issues raised during the Workshop, and/or used in building the schematics. Through systems analysis and from the selected elements, a new and detailed schematic presentation was made. This was an attempt to combine the participants' approaches from the Workshop with ideas we brought into the process. The final schematic visually presented the elements of Yugoslav export-industry competitiveness that were considered to be important, together with the key interrelationships among them. The final number of elements used in the schematic was 370. The following figure is a simplified version of this schematic.

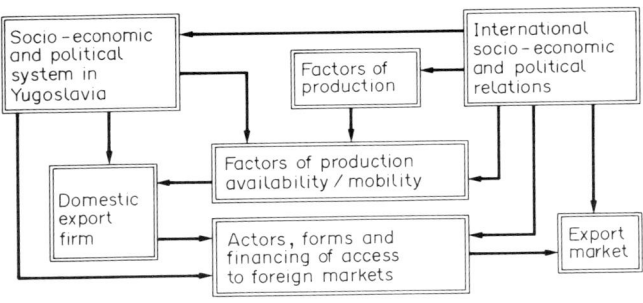

FIGURE 2

The next step in the process was the building of a system components/gaming elements matrix which guided us in formulating the gaming elements such as the format of the game, scenario, pulse, game sequence, steps of play, roles, decisions, etc. After all of the gaming elements had been defined, a game objectives/gaming elements matrix was developed in order to ensure that the game, when run, would meet the stated objectives.

Game participants selection

In selecting the participants for the YUBIEC game run, several criteria were used to ensure the best possible results during and after the game run. The participants had to: (a) be proven experts in the fields relative to the issues discussed; (b) come from differing educational and professional backgrounds; (c) hold different perspectives on the issues discussed; and (d) be willing to spend a whole day participating in the gaming exercise.

From the group of about 30 experts that participated in the Issues Definition Workshop, 16 were selected and invited to participate in the game run. Six participants were YUBIEC vice-presidents and members of the YUBIEC Management Board, four participants were directors of large business corporations, three participants came from government institutions, and the remaining three came from major research institutes.

Running of the Game

The YUBIEC game had four cycles. It started at 8 am and debriefing ended after 5 pm. Each cycle consisted of an objective to be met and several questions to be answered. A set of procedures were followed to ensure that all objectives were met.

Game Objectives

Cycle I – The objective was to reach a decision about which internal and external elements of an export business enterprise are critical for Yugoslav export competitiveness.

Cycle II – The objective was to rank the elements that were selected by the groups in the previous cycle according to their criticality in relation to the Yugoslav export competitiveness.

Cycle III – The objective was to select the key elements of the Yugoslav export competitiveness, and determine their interrelationships.

Cycle IV – The objective was to select and analyze those key elements for which there is an immediate need for further analysis, and those elements that require a significant structural change to meet the goal of improving the Yugoslav export competitiveness.

Steps of Play

Cycle I – Participants were divided into four groups each representing a role (Government, Economy/Business, Science, Bank). They were given a small schematic of the problem environment, a role description, a list of their role goals, and all other gaming materials including 370 cards with elements of the problem environment written on them. Moderators then explained the schematic presentation that resulted from the Issues Definition Workshop experiences, and the ideas generated during the game design process.

From the 370 elements each group had to select those which they felt to be most critical for Yugoslav export industry competitiveness. In the process of selection cards (elements) were to be sorted into two categories – those considered more and less critical. For elements selected as more critical the process was repeated until the number of elements that remained were one hundred or less. Each member of the group had a right to add two elements that he/she considered critical and that had been rejected by the others, and two completely new elements which they thought had been omitted from the given schematic. During the selection process participants were encouraged to analyse the schematic so they could establish a better understanding of the relationships among the elements chosen.

Using their selected elements, each group had to express their perspective on the issues discussed by building a new schematic, thereby establishing the key relationships among the elements. They then had to explain the logic behind their schematics, and emphasize the elements they considered to be most critical to the Yugoslav export competitiveness and explain the reasoning behind their choices.

Cycle II – At the beginning of this cycle every participant (independently from the group) had to vote for, and rank, those elements that he/she thought were important enough to be considered in further analysis. This was accomplished by placing voting stickers on the edges of the selected elements. Stickers were coloured and numbered, with each colour representing the group the participant was a member of, and numbers representing priority.

Moderators then counted the votes and chose the highest ranking 32 elements. If several elements had the same number of votes, those which more participants or groups had voted for preceded the others in being selected. New cards for the elements selected were placed on the prepared wall matrix and, parallel to that, moderators prepared small copies of the matrix to be given to each group as a reference.

Cycle III – Before analysing the 32 most important elements, each group had an opportunity to add a new element, which they felt to be crucial. From the 36 elements on the matrix (both in rows and columns) each group could select 12 (in rows), and three elements (in columns) influenced by every one of the 12 selected.

Each group used the small matrices to record the influences. When all the interrelationships were agreed upon within the group, a representative recorded them on the big matrix. Decisions were recorded by using coloured pins with small coloured flags which indicated the nature of the influence. After this was completed, each group had to explain the logic behind their selections and the relationships they identified, emphasizing those of a multiple nature.

Cycle IV – Each group was then asked to select up to four elements from the master wall matrix which they believed merited further analysis or required significant structural change.

For each element selected, the group had to prepare written answers to the following questions: (a) Why is there a need for further analysis of this element? (b) Why is significant structural change required? (c) What are the ways in which the analysis could be conducted? (d) What are the structural changes that would have to be introduced and how might they be accomplished?

After answering the questions, each group had to explain the logic behind their selections and answers. With this activity the game run ended. During the debriefing participants were asked to comment on both the mechanics of the game and the issues the game focused upon.

Experiences of the game run

From the beginning of the YUBIEC game run the working atmosphere was very good. The interest that participants showed at the end of the Issues Definition Workshop was carried over and expanded upon. Among participants motivation was very high, mainly due to the way the game started and the fact that many of the participants' expectations were fulfilled.

Once the activities of the game were explained to the participants, it became clear that active participation was demanded of them. This was different compared to most of the Workshop run, and especially different from the usual activities of previously attended seminars and workshops. Most importantly, participants realized that with gaming the engagement started there, where it usually ended at such seminars and workshops.

Participants accepted the given roles with a lot of curiosity and carried them to the end of the game run. Although most of the participants played their real-life roles, they brought many new elements to them. The tasks given were carried out with a lot of enthusiasm and new insights into their role. In performing the required activities, participants were obviously in a good mood and the atmosphere was pleasant, spontaneous and stimulative. The dialogues were very often followed by witty and humorous remarks.

At the debriefing, participants emphasized that during the game they enjoyed the opportunity to introduce their ideas and perspectives and that regardless of their differences, everyone had the same chance of expressing themselves. Because the perspectives were recognized as role perspectives and not personal ones, willingness to take other people's ideas into consideration was at a much higher level than usual. An environment in which there was an apparent interest for differing approaches to the issues of Yugoslav export competitiveness had been created.

In addition to an appreciation for other perspectives, the game established an atmosphere for both individual and group action, self-assessment, evaluation and selection of ideas, and decision making. The existence of the roles' underlying objectives and the originality of their approach was evident when every group presented very different schematics, none of which resembled the one provided by the IDC at the beginning of the game.

Integrating "intellectual" and "manual" tasks during the game run, such as determining impacts among elements of the problem environment by preparing flags on coloured pins and placing pins on the matrix, and doing that within a group and in front of all the groups, resulted in refreshing, lively and motivated problem solving and decision making.

Confrontations within and among the groups resulted in a stimulating exchange of arguments and additional clarifications of the views taken. Through these in-depth discussions, a shared image of the issues and possible ways of approaching them were developed by the end of the game run.

The participants felt that the most evident benefit was the given opportunity for taking part in the decision-making process. One of the participants emphasized that in 10 years of his expert work he was never given such an opportunity to express his perspectives and at the same time to be taken seriously and constructively by other participants.

At the end, all of the participants had a favourable opinion of the usefulness of the YUBIEC game in approaching the issues of Yugoslav export competitiveness. As an example of these opinions we have decided to quote one of the participants in the game whose corporation just contracted a similar game from the IDC.

I think that this was a very useful game, which in the beginning appeared very naïve, but as it developed it became more and more serious. Gaming taught us and showed us a lot. At moments everything seemed so clear and understandable; later we would realize that we had gotten ourselves into unsolvable difficulties. It would be very good if such games could be used in different problem-solving contexts to help open discussion of the problems.

Post-game Activities

Post-game activities lasted for about a week, during which we analyzed all of the results and experiences accumulated during the Issues Definition Workshop, YUBIEC game design process, and the game run itself. The key elements that were transformed into the project outline and project organization proposal were the joint proposals for possible solutions to key elements of Yugoslav export industry competitiveness. These proposals came directly out of the YUBIEC game run.

The final output of this gaming/simulation exercise was the following: (a) the project goal and objectives; (b) the phases of the project with main elements of the project content; and (c) the main elements of the project organizational structure. Due to the structure of this paper and the confidential nature of the materials, the output could not be elaborated upon further.

References

Duke, R. D. 1974. *Gaming, the futures language.* Beverly Bills, CA.: Sage Publications.
Greenblat, C. S. and Duke, R. D. 1981. *Principles and Practices of Gaming-Simulation.* Beverly Hills, CA.: Sage Publications.

An expert system that simulates group decision making in a stochastic environment and exhibits learning

Marian V. Sackson
Pace University, New York, USA

ABSTRACT: The objective of the expert system model development was to provide an experimental environment to test its decision-making capabilities as it simulated group decision makers. The expert system acquired knowledge dynamically as it reacted to a stochastically changing environment during three simulated years of competing in a business game environment. The experiment intended to provide a preliminary understanding of the methods by which an expert system analyses strategic scenarios and develops operating decisions as well as contributes to the growing interest in the applicability of expert systems to business decision making.
KEYWORDS: Expert system, decision making, business game, simulation, experiential learning.
ADDRESS: Information Systems Department, Pace University, Bedford Road, Pleasantville, NY 10570.

Introduction

This paper describes an experiment conducted to test an expert system model developed to simulate group decision makers in a strategy and policy-making environment. The experiment is limited in scope and is a first step to building a comprehensive expert system model. The intent of the model is to provide an experimental environment, based on a business game, to test the use of an expert system as a tool to help group decision makers learn more about the decision-making process that is used to evaluate strategic scenarios. This paper will discuss expert systems and their relevance to group decision making in a dynamically changing business environment. Then it will explore the issues of group decision makers that are frequently encountered as they evaluate strategic scenarios. Finally the business game used, the decision behaviour of the expert system, and results of the experiment will be examined.

Decision Making and Expert Systems

Decision making is fundamental to human existence. We are continually faced with problems that need solutions. How we solve these problems, the relationships between intelligence and problem solving, and the characteristics of the decision process are subjects of continual research. Decision making involves many different kinds of knowledge and planning behaviour. Past experiences may play a role in the decision process. Decision making entails congitive activity and is psychological in nature. Decisions can be made by an individual or a group.

The experiment uses the taxonomy of decision types developed by Simon (1960) that views classes of decisions on a continuum with structured and unstructured decisions on the extremities. Structured decisions are repetitive and routine, and a definite procedure can be developed for handling them. Unstructured decisions are novel and consequential, with no prescribed method for handling them, and the structure is often elusive or complex.

The background for the expert system model was taken from the research in human problem solving by Newell and Simon (1972). They developed a computer program, *The Logic Theorist*, to simulate decision-making behaviour during problem solving in specific task environments. Their early achievement has spawned further interest in the development of computer programs that became known as "expert systems" which are part of artificial intelligence.

The field of artificial intelligence (AI) evolved as a result of an interest in making machines smarter and a desire to understand intelligent behaviour. AI research has progressed along a number of tracts, but one of the most far ranging has concerned the developments of programmes for solving problems that are difficult enough to require the specific knowledge of human expertise for their solution. These programmes are called "expert systems" or "knowledge-based systems". Hayes-Roth (1984) stated that human expertise consists of knowledge about a particular domain, understanding the domain problems, and skill at solving some of these problems. Human experts generally possess large amounts of knowledge that consists largely of heuristics which enable the expert to make educated guesses when necessary, recognize promising approaches to problems and deal effectively with erroneous or incomplete information.

Research Effort

Group decisions, as made by real companies, are complex, varied, and numerous. Likewise, the need to measure the longitudinal impact of competitive decision making and "learning", necessitates observing a sequence of decisions over time. Lastly, stochastic parameters, external econometric values, and industry performance reports are part of any real organization environment. For these reasons, the experimenters chose a simulated model of companies whose managers would be making specific operating decisions in a competitive environment. The chosen simulation was the Business Strategy and Policy Game (BUSPOG) of Eldredge and Bates (1984). BUS-

MANAGEMENT DECISIONS

The simulated competition of BUSPOG requires that the management team for each company make the following set of operating decisions once each quarter of a year. The particular decisions are:

Marketing
1. Selling price for each of the three markets.
2. Advertising budget for each of the three markets.
3. Salespersons hired or discharged in each of the three markets.
4. Product research and development budget for the company.

Production
1. Scheduled production work week for the company.
2. Change in the production labour force for the company.
3. Allocation of finished product to the three markets.
4. Process research and development budget for the company.
5. Raw materials ordered for the company.
6. Plant investment budget for the company.

Personnel*
1. Sales salaries for the company.
2. Sales training budget for the company.
3. Production wage rate for the company.
4. Production training budget for the company.
5. Profit sharing for the company.

Finance
1. Bonds sold or redeemed for the company.
2. Bank loan requested for the company.
3. Dividends paid by the company.
4. Stock issues by the company.
5. Long-term savings account deposit or withdrawal for the company.

*As our four person student teams had a President role, the duties of under Personnel were allocated between marketing, production, and finance as follows:

1. and 2. went to Marketing. 3. and 4. went to Production. 5. went to Finance.

FIGURE 1   Operating decisions made each quarter

POG is a management exercise involving three levels of managerial activities: functional (marketing, production, personnel, and finance), co-ordinative, and organizational. It deals with a hypothetical refrigeration industry made up of as many as seven companies that compete within two market types; consumer refrigerators and an industrial market for refrigerator units. The management teams are all given a 4-year history of operating decisions that were made by one of the games operating companies. Then, the management team of each company makes a set of operating decisions for each quarter of the year as seen in Fig. 1.

The decisions of all the competing company groups are processed through the game computer and the estimated results of competition for the quarter are reported to each company in the form of computer printouts. A sample of these reports is illustrated in Fig. 2. BUSPOG incorporates a conceptualized

ECONOMY AND STOCK MARKET

Gross National Product 1055 ($billions)			Personal Consumption Expenditure 90 ($billions)			
No Household Formations 1504 (1000,S)			Raw Materials Cost .37 ($/LB)			
Company	Stock Price	Earnings	Dividends	Shares	Bond Interest	Loan Interest
Alaska Deep Freeze	9.74	0.27	0.21	6000	6.0	6.0
International Cooling	8.90	0.19	0.22	6000	7.2	7.2
Breezly Bruin	9.65	0.22	0.23	6000	6.0	6.0
Mr Deep Freeze	8.25	0.09	0.22	6000	6.9	6.9
Icecon Incorporated	9.26	0.18	0.27	6000	6.9	6.9
Expertcorp	7.09	0.27	0.0	6000	6.0	6.0

INDUSTRY ESTIMATES

Company	Market Prices			Average Advertising	Sales Commission	Profit Share
	1	2	3			
Alaska Deep Freeze	279	280	268	213	3.3	2.3
International Cooling	257	273	254	254	3.6	2.5
Breezly Bruin	261	252	253	304	3.7	0.0
Mr Deep Freeze	256	262	254	325	3.5	2.3
Icecon Incorporated	291	293	287	263	3.1	2.5
Expertcorp	284	293	287	263	3.4	2.5

Company	Sales Salary	Production Wages	Production Workforce	Production Output	Sales Force	Unit Sales
Alaska Deep Freeze	1033	4.31	1514	71802	39	48780
International Cool	1107	4.08	1805	81935	40	50516
Breezly Bruin	1032	4.14	1202	58546	39	50516
Mr Deep Freeze	1321	5.03	1928	84799	41	52987
Icecon Incorporated	1209	4.31	2061	86124	40	45239
Expertcorp	1030	4.26	1302	62843	40	44858

FIGURE 2 Quarterly reports of a company performance

view of the hypothetical relationship of how such an industry behaves relative to its environment. In reality, as well as in the game, the forms of some relationships are known with a high degree of certainty while those of others are only vaguely known. Stochastic parameters are used to simulate uncertainty. These are listed in Fig. 3.

The Management Department faculty of Pace University, acting as a panel of experts or knowledge source were given a set of 30 possible variables and relationships made while playing the game. They rated the importance of the variables, listed the interdependencies and correlations affecting these variables, estimated the likelihood that the relationships would shape different decisions, and determined the reactive global strategy. These faculty responses were compiled, recirculated to the faculty, new responses elicited for two iterations: a truncated Delphi technique. From this process, the model

STOCHASTIC PARAMETERS

1. Manufacturing productivity.
2. Raw material required per unit of finished product (changes randomly over time)
3. Finished product demand.
4. The number of salespersons who quit if their salary and/or profit sharing is less than the industry average.
5. Values appearing in the industry estimates reports.

1, 2, 3, 5 use normal distributions and the standard deviation statistic can be changed by the game administrators.
4 uses a binomial distribution.

Figure 3 Stochastic Parameters

was developed and a set of specific decision rules were incorporated into the initial expert system model.

The research questions for this experiment were:
1. How will the expert system react to a stochastically changing environment of the management game?
2. How and to what extent will the expert system modify the decision rules?
3. As the expert system stimulates group decision making over time, will its decisions be affected by experiential learning?
4. How will the expert system perform relative to other live group decision makers in the same arena?

Expert System Model

Expertease, the expert system tool used in this research, is a skeletal language as well as a microprocessor based computer program. This expert system development tool consists of two stand-alone pieces of software: 1. a rule set manager and 2. an interference engine capable of reasoning with rules built with the rule set manager. Figure 4 is a general model of a skeletal expert system.

An user has a particular problem that exists in the expert system task domain. The user activates the query system through the user interface and responds to the questions asked by the expert system in order to describe the problem to be analyzed. These responses are accepted by the controller/interpreter and "fire" the interference engine. The expert system executes the three parts: *interpreter* – arranges the input responses in a logical sequence, *scheduler* – determines the most likely set of internal conclusions, and the *consistency enforcer* – checks the logic of the interferences and conclusions. If the expert system needs more input in order to conclude its internal knowledge search, it will ask the user for more input responses and repeat the cycle. The particular expert system used for this research, Expertease, continues to request more input values until it can reach a conclusion (decision). This cyclical pattern terminates when the expert system can respond to the specific problem. In addition a user is able to elicit a statement

Marian V. Sackson

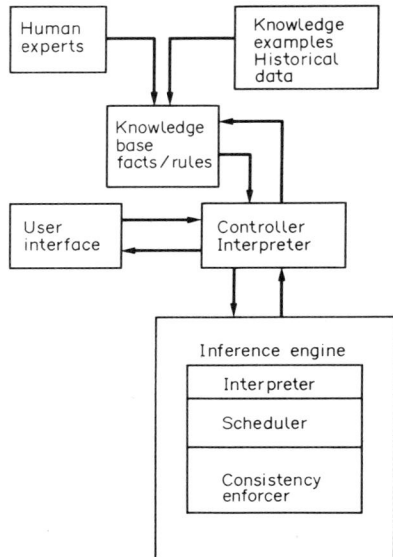

FIGURE 4   Skeletal Expert System

of the logical steps (rules) the expert system followed to arrive at its conclusion.

The expert system designer, while using Expertease, creates a matrix containing all the possible non-numeric attributes as the rows and the overall variable relationships for some decision-making process as the columns. This matrix is used as a skeletal structure for developing the numeric and non-numeric values which are "examples" of the decision-making process for a single decision. In Fig. 5. there are 16 examples (rows – the four-year quarterly history initially presented to the management teams), each row giving the values for that example of an attribute (column) that the panel of experts felt impacted on the Product Research and Development Budget decision. Each decision is a separate matrix, yet, related decisions can contain the same variables that appear in other matrices. In addition, one decision can be chained to sub-decisions that can simplify the knowledge search. The rule set manager of Expertease uses the Shannon information theoretical measure, entropy, to determine which of the variables, with the examples of previous decisions, have the "best" values (least random or lowest entropy measure) this establishing their importance in the rule generation.

The initial historical knowledge base, covering four business years (16 quarters), provides 16 examples of operating decisions in sequence. These examples determine the initial decision rules. An illustration of the variables and examples for one operating decision, the Product Research and Development budget decision, and the rule set built by the rule set manager for the initial history data related to this operating are seen in Fig. 6. The same

EXPERTEASE Example Listing, Problem:
PRODUCT RESEARCH AND DEVELOPMENT BUDGET

Data List	Company Product R&D	Company Unit Sales	(Ratio A) Industry Unit Sales/ Average Price	(Ratio B) Company Average Sales/ Company Average Price	Product Unit Decision
1	85	4979	171	172	$85
2	85	4151	142	143	85
3	85	5230	182	181	90
4	90	6045	210	210	90
5	75	3271	109	109	75
6	75	4011	135	135	75
7	75	4511	152	152	78
8	78	4177	142	142	80
9	80	3443	116	116	80
10	80	4257	144	144	80
11	80	4828	164	164	80
12	80	4481	152	152	80
13	80	3714	127	127	80
14	80	4636	159	159	85
15	85	5310	182	182	85
16	90	5718	199	199	300

(Product R&D is scaled by 1000, unit sales is scaled by 10)
All values represent last quarter values except the decision value.

FIGURE 5 Initial knowledge base

operating decision that has evolved after 21 examples have been added in sequence and the rule set for this updated version of the same decision described previously appears in Figs 7 and 8.

After each quarter during the active game playing, the company's operating decisions are added as new examples and new decision rule sets are generated through the controller. As stated previously, these new examples are considered experiential "learning" and will change the company's performance each quarter by adding to the knowledge base, thus changing the operating decisions.

Expert system validation

The paucity of attention paid to evaluation and testing expert systems left the determination of the validation method to the experimenters. The method chosen was:
1. Use a recently completed BUSPOG game consisting of 12 business quarters and replace one of the live teams with the expert system.
2. Rerun the game and use the same criteria used by the game administrators to measure the performance of the expert system in the competitive environment.
3. Compare the decisions made by the expert with the decisions of the

EXPERTEASE Rule Listing, Problem:
PRODUCT RESEARCH AND DEVELOPMENT BUDGET

```
COMPANY R&D
    <83:          COMPANY R&D
                      <77:          COMPANY UNIT SALES
                                        <4621:       $75
                                        >=4621:      $78

                      >77:          COMPANY UNIT SALES
                                        <4559:       $80
                                        >=4559:      COMPANY UNIT SALES
                                                         <4732:    $85
                                                         >=4732:   $80

    >=83:         COMPANY R&D
                      <88:          COMPANY UNIT SALES
                                        <5105:       $85

                                        >=5105:      COMPANY UNIT SALES
                                                         <5270:    $90
                                                         >=5270:   $85

                      >=88:         COMPANY UNIT SALES
                                        <5882:       $300
                                        >=5882:      $90
```

FIGURE 6 Rule set for knowledge base with 16 quarter history

human team it replaced. If the expert system performance was inferior to its competition and/or dissimilar to the team it replaced, then make the necessary adjustments to its decision behaviour and rerun the same game.

4. When the panel of experts determined the expert system performance was sufficiently modified to compete in a "live" competitive environment, the validation process terminated.

The Live Expert System Experiment

The expert system model completed three validation tests with subsequent modifications. The third set of performance measures of the expert system model had been shown to the expert panel. They concurred that the expert system was capable of making meaningful operating decisions and appeared sufficiently astute to compete with student groups in a live BUSPOG game iteration.

EXPERTEASE Example Listing, Problem:
PRODUCT RESEARCH AND DEVELOPMENT BUDGET

Data List	Company Product R&D	Company Unit Sales	(Ratio A) Industry Average Unit Sales/Average Product Price	(Ratio B) Company Unit Sales/Company Average Product Price	Product R&D Decision
1	85	4979	171	172	$85
2	85	4151	142	143	85
3	85	5230	182	181	90
4	90	6045	210	210	90
5	75	3271	109	109	75
6	75	4011	135	135	75
7	75	4511	152	152	78
8	78	4177	142	142	80
9	80	3443	116	116	80
10	80	4257	144	144	80
11	80	4828	164	164	80
12	80	4481	152	152	80
13	80	3714	127	127	80
14	80	4636	159	159	85
15	85	5310	182	182	85
16	90	5718	199	199	300
17	300	4596	177	160	305
18	125	4417	177	154	125
19	100	5088	169	193	200
20	160	4592	175	160	175
21	95	4863	174	177	97
22	250	5398	168	213	250
23	95	4709	175	170	125
24	305	5524	227	192	325
25	125	5229	229	185	205
26	175	5566	226	200	180
27	97	6079	223	223	100
28	250	7143	213	281	280
29	124	5956	222	222	250
30	300	4620	177	161	305
31	305	5627	232	211	305
32	205	6256	276	232	225
33	500	8209	261	318	600
34	180	7395	265	292	190
35	100	6993	271	258	150
36	280	7680	262	302	280
37	250	7134	268	271	375

FIGURE 7 Additional examples added to the knowledge base

The expert system model along with the other five competing student teams used the same 4-year history of decisions made by one company. This practice is common in management simulations. The changes to the pricing model, during the validation phase, enabled the expert system to select product prices that were competitive with the industry prices; but they were slightly higher. The competition gradually gained a higher percentage of the market share as they maintained lower product prices than those of the expert

EXPERTEASE Rule Listing, Problem:
PRODUCT RESEARCH AND DEVELOPMENT BUDGET

```
COMPANY R&D
< 93: COMPANY R&D
       < 83:   COMPANY R&D
               < 77:    COMPANY UNIT SALES
                        < 4621:   $75
                        > = 4621: $78
               > 77:    COMPANY UNIT SALES
                        < 4559:   $80
                        > = 4559: COMPANY UNIT SALES
                                  < 4732:   $85
                                  > = 4732: $80
       > = 83: COMPANY R&D
               < 88:    COMPANY UNIT SALES
                        < 5105:   $85
                        > = 5105: COMPANY UNIT SALES
                                  < 5270:   $90
                                  > = 5270: $85
               > = 88:  COMPANY UNIT SALES
                        < 5882:   $300
                        > = 5882: $90
> = 93:       RATIO B
              < 212:   COMPANY R&D
                       < 143:    COMPANY UNIT SALES
                                 < 4786:   $125
                                 > = 4786: COMPANY R&D
                                           < 98:    $97
                                           > = 98:  COMPANY R&D
                                                    < 113:    $200
                                                    > = 113:  $205
                       < 143:    COMPANY R&D
                                 < 238:    COMPANY R&D
                                           < 168:    $175
                                           > = 168:  $180
                                 > 238:    COMPANY R&D
                                           < 303:    $305
                                           > = 303:  CO UNIT SALES
                                                     < 5576:   $325
                                                     > = 5576: $305
              < 212:   COMPANY UNIT SALES
                       < 7064:   COMPANY R&D
                                 < 113:    COMPANY R&D
                                           < 99:    $100
                                           > = 99:  $150
                                 < 113:    CO UNIT SALES
                                           < 6106:   $250
                                           > = 6106: $225
                       < 7064:   RATIO A
                                 < 264:    COMPANY R&D
                                           < 390:   $280
                                           > = 390: $600
                                 > 264:    COMPANY R&D
                                           < 215:   $190
                                           > = 215: $375
```

FIGURE 8 Rule set with additional examples

An expert system that simulates group decision making 199

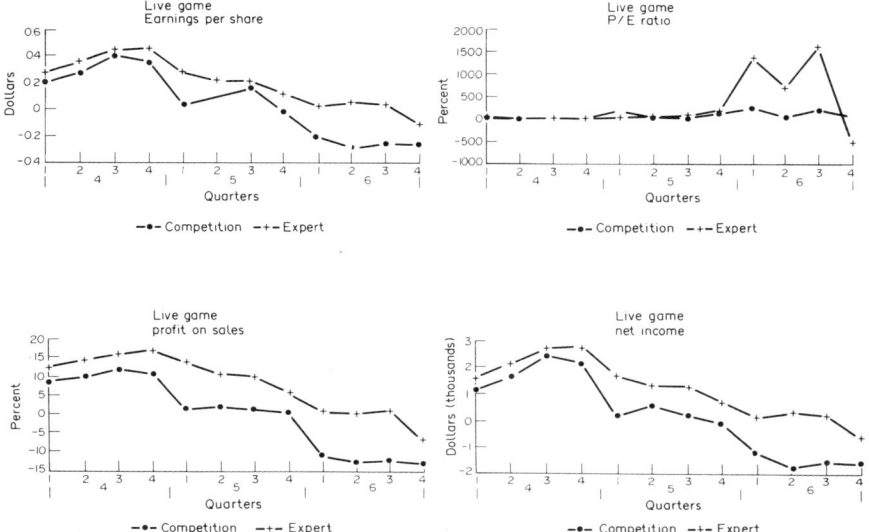

FIGURE 9 Expert System Performance Behavior

system. The expert system lowered its product prices, but consistently reacted slower than the competition.

Even though the competition continued to chip away at the expert system market share, both the net income and profit on sales ratio of the expert system were always higher than the comparable performance measures of the competition. This seemed to indicate that the expert system had a better cost model. This is in fact because the expert system's knowledge base was originally developed by the panel of experts who from real-life experience knew that a low cost structure is essential in a competitive manufacturing environment.

The expert system initially displayed a production profile similar to that of the competition. In subsequent quarters, the productivity measure for the expert system declined relative to the industry average. However, this decline did not seriously effect the competitive position of the expert system.

The financial performance measures: net income, profit on sales ratio, earnings per share, and so forth revealed that the expert system's performance measures were generally superior to those of the competition. As had been revealed from the last validation test, the cost containment model developed by the expert panel indirectly influenced the financial operating decisions. Figures 9 and 9.1 are a limited set of graphs that illustrate the comparative decision behaviour of the expert system and its competition.

Given that the investigators have made a plausible case that the expert system has captured some elements of group decision making, the conclusions are that:

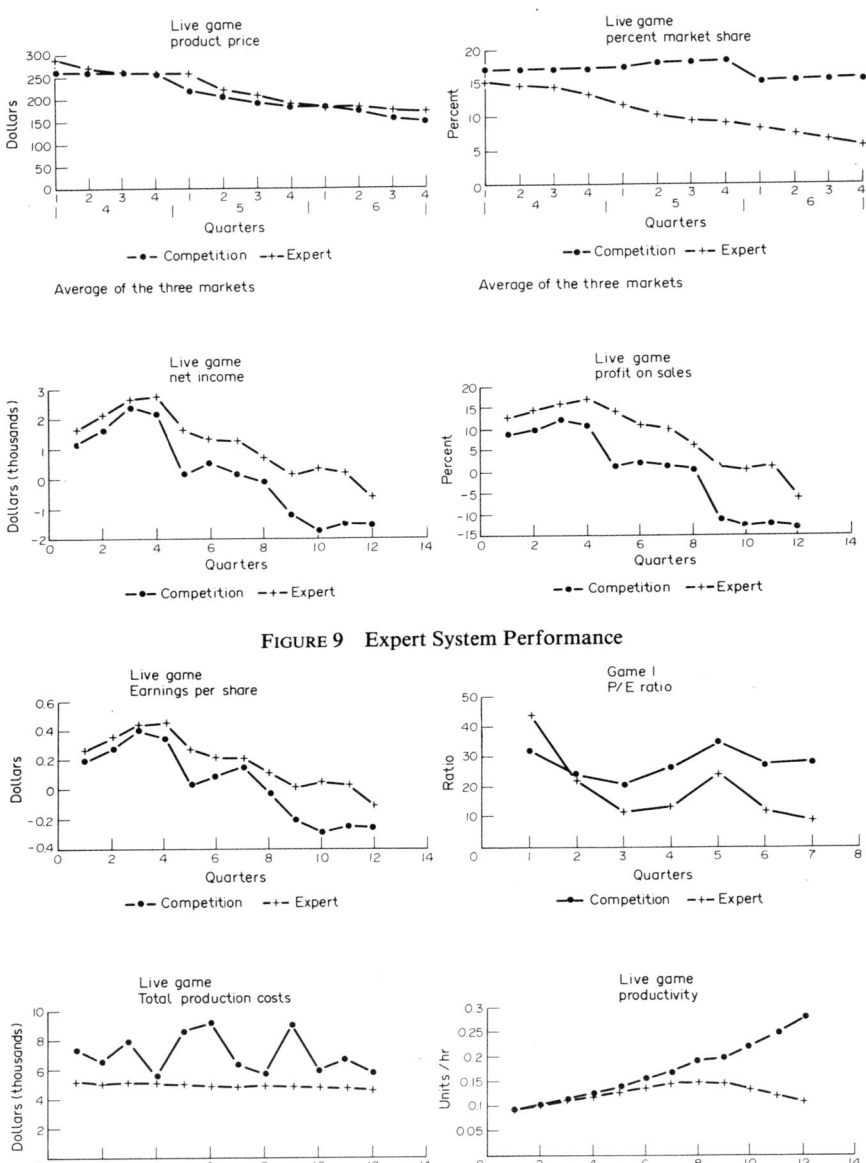

FIGURE 9 Expert System Performance

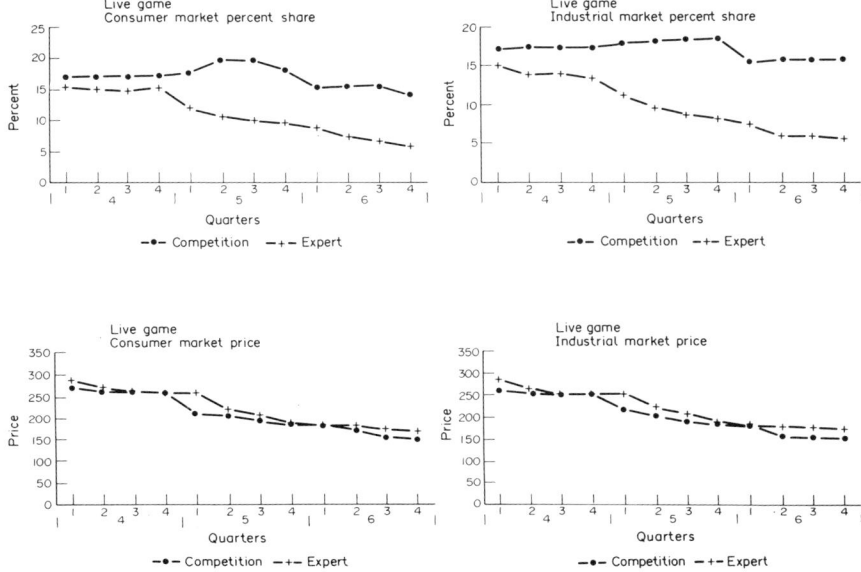

FIGURE 9.1 Expert System Performance

1. The reactive and thus conservative global strategy influenced the minimum risk taking of the expert similar to that of an actual company.
2. The structure of the expert system shell, Expertease, that uses past decisions as input to each set of operating decisions, inhibited the expert system's ability to exercise entrepreneurial decision making and instead follow decision making that reacts to the industry.
3. Since the expert system followed a rule structure inherent in its inference system, it was less able to perform sudden changes in its decision-making behaviour than those of human decision makers.
4. The changes in the decision rule structure of the expert system as more examples were added displayed a pattern of improving performance by the expert system as one would expect from human decision makers making group decisions as a result of "learning" from past performance.
5. This first step in building a comprehensive expert system model revealed that a different expert system shell that allowed more flexibility in its rule structure could improve an expert system's decision-making performance. Likewise, the next step could add the ability to modify its initial strategy as a reaction to the competition and changes in the market place.

References

Christensen, C. R. and Andrews, K. 1978. *Business Policy Text and Cases.* Homewood, Il: Irwin.

Eldredge, D. and Bates, N. 1980. *The Business Strategy and Policy Game.* Dubuque, IA: Brown.
Hayes-Roth, F. 1984. Rule-based Systems. *Communications of the ACM* **28**:9.
Newell, A. and Simon, H. 1972. *Human Problem Solving* Englewood Cliffs, NJ: Prentice-Hall.
Simon, H. 1960. *The New Science of Management Decisions.* New York, NY: Harper.
Waltz D. 1983. Artificial Intelligence: An Assessment of the State-of-the-Art. *The AI Magazine.* **13**:3.
Winston P. and Prendergast. 1984. *The AI Business.* Cambridge MA: MIT Press.

Designing and evaluating decision support systems: a group DSS for international transfer pricing

Henk G. Sol and Michael B. M. van der Ven

Delft University of Technology

ABSTRACT: A Group Decision Support System (GDSS) is presented. The aim of the GDSS is to support decisions concerning the fiscal, financial and commercial aspects of international transfer pricing. The choice for a case study approach is explained. A system for the British market has been developed in co-operation with a large, innovative, European-based pharmaceutical company. In this paper special attention is paid to the development cycle and the evaluation of the system. The development cycle starts with a descriptive empirical model. Via two conceptual models a prescriptive empirical model is designed. Evaluation of the system is based upon this prescriptive empirical model.
KEYWORDS: Decision support system; group decision support; transfer pricing; pharmaceutical industry; case study; decision process structure.
ADDRESS: Department of Information Systems, Delft University of Technology P.O. Box 356, 2600 AJ Delft, The Netherlands.

Introduction

In 1986, a research project was started to set up a group decision support system for international transfer pricing decisions. This experimental system is developed in close co-operation with a European based pharmaceutical company. It has been in use since spring 1988.

Although transfer pricing is a phenomenon of considerable importance, little is known on how decision making actually takes place within multinational companies. What is known, however, is that transfer pricing is considered an extremely complex area and that current transfer pricing research does not significantly contribute to clarifying the problem (see also Krens, 1984). Partly this is due to its multidisciplinary character, which renders it difficult to cover all management control, commercial, fiscal and financing aspects. For the other part, the cause of lack of research results lies within the multinational industry itself. As transfer pricing is considered a highly complex and sensitive area, inside as well as outside the industry, multinational companies are not very keen on disclosing their transfer pricing policies and co-operating in transfer pricing research (see Plasschaert, 1979).

The choice for a case study (sometimes also called "field study") approach is based on the following considerations (see Benbasat *et al.*, 1987):

- Little research exists which integrates all aspects of international transfer pricing;
- The project has a revolutionary character (see Yin, 1984, page 47) as not much is known about the way transfer pricing decisions are taken in practice;
- As improvement of "real-life" decision making is one of the aims, close co-operation with the industry is imperative.

It should be realized that "case study" as used above does not mean that the researcher is watching the organization from the sideline. He is actively involved in the design process and should be regarded as a member of the organization.

From the viewpoint of the co-operating company (guest company), the goals of the research project are the following:

- Increase problem understanding concerning international transfer pricing;
- Structure the decision process;
- Provide decision support tools.

These goals, which all relate to the decision *process* should contribute to a better long-term fiscal and financial policy which complies with all, sometimes even conflicting regulations. Also the financing situation should improve.

From a scientific point of view, the aims of the research project are (see also Sol and Van der Ven 1988a):

- Gain insight in the transfer pricing decision making as done within multinational companies. The existing situation should be compared to transfer pricing theory;
- Develop a conceptual model for the decision process, based on the actual situation, transfer pricing theory and decision support theory (see Bosman, 1986);
- Build a system on the basis of the conceptual model and determine the impact of the system on organizational decision making.

The system is a combination of roles of organizational members (group decision making), decision support modules and decision process structure. It is explicitly based upon transfer pricing theory and decision support literature. To integrate the judgement on various aspects, a Multi Criteria Decision Making technique is used. For more details on Multi Criteria Decision Making, see Lootsma (1986). Also characteristics of the pharmaceutical industry are included. For details on the design and the activities of the decision process, see Sol and Van der Ven (1988b).

Development of the System

The point of departure in the development of the system is the breakdown of the decision process in various stages (see also Bosman 1988 and Bosman and Sol, 1985).

We started with the description of the existing situation. For details on the object representation that is used (see Bots and Sol 1987). Objects and procedures are described in detail. The case studies which describe the actual situation are called the "descriptive empirical model". Here it means that an existing situations is described. Actual company data form an integral part of this description.

From this empirical descriptive model, the "descriptive conceptual model" is extracted. We conceptually describe how decision making actually takes place and which methods are normally used before introduction of the system. Company data or other specific details are not included. It is a general process description.

Description of the existing situation, however, is not sufficient. The aim of decision support is to aid decision makers and to improve decision making. Therefore, the descriptive conceptual model is analysed carefully. Concepts from decision theory and transfer pricing theory are used to improve the decision process. (For more details see Sol and Van der Ven, 1988b).

Improvements of the decision process are suggested. The model which describes the improved decision process is defined as "prescriptive conceptual model". In our definition "Prescriptive" refers to the normative element that is involved: The proposed decision process should, among others, be better structured, increase problem understanding and communication. Takkenberg (1983) uses the term "normative" in this case. These two conceptual models (descriptive and prescriptive) form the core of our research. In combination with computer support and the roles taken by the organizational members involved, they should lead to better company performance. The concept should be applicable to analyse and improve decision making with regard to financing problems of any subsidiary of any European based innovative pharmaceutical company. So, to a limited extent, this concept is generalizable. We would like to stress, however, that the aim of this model is improvement and not the determination of an optimal solution.

To be able to test the prescriptive conceptual model, it needs to be worked out for a specific company and filled with data. We then have a "prescriptive empirical model". The use of this empirical model to solve "real-life" problems was monitored during half a year and compared with the descriptive process model. Data gathering was done by means of structured and unstructured interviews, audio and video recordings, key stroke logging of computer use and analysis of documents.

	Descriptive	Prescriptive
Empirical	1: Analyses existing situation	4: Implementation of the improved decision process
Conceptual	2: Abstraction	3: Improvement of the decision process

FIGURE 1   The development of the system.
(numbers indicate the order of the activities)

Evaluation of the System

According to the users and top management, the three company aims as described in the first section, are reached.

The system clearly leads to high user satisfaction. This is, however, only one aspect of performance measurement. Alongside with system development, other criteria for performance measurement of the system evolved (see also Keen, 1986).

– *Criteria related to the quality of the chosen solution.*

As far as fiscal and commercial matters are concerned, the aim is to create a better long-term policy, which is acceptable to both the company and the authorities involved. From a financing perspective, the aim is that the financing situation should match the long-term financing needs at minimal cost. So the criteria which are related to company goals are:
- The distance between the effect of the chosen financing alternatives and the preferred effect. The preferred effect depends on the aspect that is considered: fiscal, commercial or financial. So this criterion should be split up in three sub criteria, each with its own weight.

- The degree of sensitivity of the chosen financing alternative for external influences like changing exchange rate, interest rate or sales figures. Also the sensitivity for internal influences like changes in production and investment figures can be considered.

With regard to the first criterion, two problems arise. The first problem is that it is not completely clear what the preferred position is as all rules that are set by the authorities are subject to interpretation. Also long-term financing needs are not known in advance. This means that the preferred solution is not a "hard" figure which can be established in a 100% objective manner. It depends to a certain extent on what the participants consider consistent in view of their long term fiscal and commercial policies. Or, in case of financing, it depends on the expectancy of the long-term financing needs.

A second problem is that the guest company prefers not to disclose what its preferred solutions are, nor what its actual solutions are. So a measure needs to be found which serves well as a performance indicator but which does not disclose actual figures.

- *Criteria related to the decision process.*

These measures include:

- Number of alternatives considered
- Total manhours spent on decision making
- Effective manhours spent on decision making
- Elapse of time before a decision is made
- Frequency of contact between participants

- *Perception of decision quality by participants and top management.*

These measurements include:

- Confidence in the decision outcome by the participants
- Confidence in the decision outcome by top management
- Perceived increase in problem understanding by participants
- Perceived increase in effective communication between participants
- Perceived insight in effects of the alternative courses of action

This scale of scores should be the basis of an overall evaluation of the system.

Conclusion

Initial conclusions with regard to the project are:

- The provision of computerized decision support tools is in itself not sufficient for improving the decision process. In addition to the tools, clear roles of the support staff and participants should be defined. Problem decomposition, use of the support tools and the roles of organizational members are integrated in the decision process structure. Process structuring plays a more dominant role than expected at the start of the project.

- Within the organization, usually sufficient hardware and software are available to provide decision support. In our case, we used two Personal Computers and the fourth generation tool Personal Wizard (see Comshare 1986). High technology is not a prerequisite for effective decision support. It might even be argued that sophisticated technology distracts the attention from another important issue: the decision process structure.

The two conclusions above relate to the project as a whole. With regard to development methodology and evaluation methods, the subject of this paper, the following can be said:

- In complex cases like ours, in which it is very difficult to establish a cause – effect relationship between target and instrument variables, the aim of the system is to aid decision makers by prescribing the steps of the decision process in detail. Improvement and not optimization is the focus of attention.
- Evaluation of a system like ours should take place by means of a broad scale of criteria for performance measurement of the system. These criteria relate to:
- The quality of the chosen solution
- The decision process
- Perception of decision quality by the participants and top management

Above, some initial conclusions of our research are drawn. It will be clear that comments on this short paper, from the multinational industry as well as from the scientific community, will be highly appreciated.

References

Benbasat, I., Goldstein, D. and Mead, M. 1987. The Case Research Strategy in Studies of Information Systems. *MIS quarterly* **11**:3.
Bosman, A. and Sol, H. G. 1985. Knowledge representation for Decision Support Systems. *In* Methlie and Sprague (1985).
Bosman, A. 1986. Ontwerpen, procesmodellen en beslissen. *Informatie* **28**:4.
Bosman, A. 1988. *Beperkt rationeel, een bron van verwarring*. Internal memorandum. Groningen: Rijksuniversiteit Groningen. (in Dutch).
Bots, P. W. G. and Sol, H. G. 1987. An environment to support problem solving. *Decision support systems* **3**:3. Amsterdam: Elsevier Science publishers.
Comshare 1986. *System W documentation: Personal Wizard*. London: Comshare Ltd.
Duffhues, P. J. W. *et al* (eds.) 1984. *Financiele leiding en organisatie*. Alphen aan den Rijn: Samson. (in Dutch).
Keen, P. G. W. 1986. Value Analysis: Justifying Decision Support Systems. *In*: Sprague and Watson (1986).
Krens, F. 1984. Overdrachtsprijzen. *In*: Duffhues *et al.* (1984). (in Dutch).
Lee, R. M., McCosh, A. M. and Migliarese, P. (eds.) 1988. *Proceedings of the IFIP conference on Organizational Decision Support Systems*. Amsterdam: Elsevier Science Publishers.
Lootsma, F. A., Meisner, J. and Schellemans, F. 1986. Multi-criteria decision analyses as an aid to the strategic planning of Energy R&D. *European Journal of Operations Research* **25**. Amsterdam: Elsevier Science Publishers.
Methlie, L. B. and Sprague, R. H. (eds.) 1985. *Knowledge representation for decision support systems*. Amsterdam: Elsevier Science Publishers.
Plasschaert, S. R. F. 1979. *Transfer pricing and Multinational Corporations: an Overview of Concepts, Mechanisms and regulations*. Westmead: Saxon House.
Singh, M. G., Salassa, D. and Hindi, K. S. (eds.) 1988. *Proceedings of the 1st IMACS/IFORS Colloquium on Decision Support Systems*. Amsterdam: Elsevier Science Publishers.
Sol, H. G. 1985. Aggregating Data for Decision Support. *Decision Support Systems* **1**.
Sol, H. G. and van der Ven, M. B. M. 1988a. A Group Decision Support System for International Transfer Pricing Decisions within the Pharmaceutical Industry. *In*: Singh, Salassa and Hindi (1988).

Sol, H. G. and van der Ven, M. B. M. 1988b. Integrating GDSS in the organization: The case of a GDSS for international transfer pricing. *In*: Lee, McCosh and Migliaresse (1988).
Sprague, R. H. and Watson, H. J. (eds.) 1986. *Decision Support Systems, putting theory into practice*. New Jersey: Prentice Hall.
Takkenberg, C. A. Th. 1983. *Planning en Methode van Onderzoek*. Ph.D. thesis. Groningen: Rijksuniversiteit Groningen. (in Dutch)
Verlage, H. C. 1975. *Transfer Pricing for Multinational Enterprises*. Rotterdam: Rotterdam University Press.
Yin, R. K. 1984. *Case study research*. Beverly Hills: Sage Publications.

Gaming as an environment for testing the effectiveness of decision support systems

P. W. G. Bots, F. D. J. van Schaik and H. G. Sol

Department of Information Systems, Delft University of Technology

ABSTRACT: Decision support systems have become quite popular during the last decade. A crucial question is: Do decision support systems lead to better decisions? Does the decision maker actually perform better when he uses a decision support system?

This paper presents a gaming approach towards testing the effectiveness of a decision support system and a so-called task structure. This approach enables the assessment of their effectiveness before they are actually introduced in a practical situation. This paper reports on the development of a management game that is intended to simulate as closely as possible an actually existing company in which the introduction is considered.

The initial results of an evaluation approach using a gaming simulator suggest that decision makers using a DSS do not perform any better than those who do not use it. Experiments show that a DSS, even though its users are very enthusiastic about it, does not cause them to make any better decisions. Decision quality does improve dramatically, though, when decision makers are guided in the way they structure their problem situation by providing them with a task structure.

KEYWORDS: Business game, decision support systems, effectiveness, management game, task structure.

AUTHOR'S ADDRESS: P.O. Box 356, 2600 AJ Delft, the Netherlands.

The value analysis of the management information systems, problem analysis and gaming

Eduard Rădăceanu

Academia de studii social-politice, Bucuresti, R. S. Romania

ABSTRACT: In this paper a complex application is presented concerning the design and/or development of the MIS used at the management development programmes, with the aim of bringing together the managers and the informaticians, as users and designers of the MIS of the enterprises. The Value Analysis (VA) approach of the MIS and Problem Analysis (PA) are also presented, as the main structural elements around the business game which can be considered the kernel of the complex application which is run as a workshop.

KEYWORDS: Business game, MIS, value analysis, problem analysis, modelling, management development.

ADDRESS: Academia de studii social-politice, Bd. Armata Poporului nr. 1-3, Bucuresti 77202, R. S. Romania.

Section Five

Methodology

Editor: **Jan H. G. Klabbers**

Methodology – behavioural and social systems design and evaluation of games/simulations, classification, taxonomy

Jan H. G. Klabbers
Utrecht University, The Netherlands

Basic ingredients for designing games/simulations are *adequate theories*, a *reflective methodology* allowing recursively re-framing of problematic situations through conversation with the stakeholders involved (a partizan approach), *tools* that are suitable for shaping virtual worlds, worlds that are susceptible for players, that tickle their imagination and support various strategies for intervention. Games/simulations deal with multiple reality. Consequently their *problem setting* should make players understand both mutual and conflicting interests, enabling them to re-frame the problem situation.

A narrowly technical perspective is not considered to be adequate, for it depends on the agreement about ends, which are regarded as fixed and clear (Schön, 1983). Social systems do not obey a narrowly technical and context-free scientific approach. In coping with complex social issues, different interests may impinge together as actors involved are being confronted with conflicting role frames. In this respect each social setting is unique. Consequently, games/simulations are most suitable when they are tailor-made for unique problem settings, for a specific audience in a definite context of use.

These epistemological considerations bring forward a dilemma, that is especially relevant for the field of gaming-simulation. Elsewhere I have mentioned that gaming and simulation belong to two different academic cultures (Klabbers, 1987). The dividing line is not the ditch where natural sciences and the social sciences meet. Berting (this volume) has amply argued that the epistemological differences cut right across the social sciences as well. It is obvious therefore, that different approaches become evident in this section on methodology of games and simulations.

Rainer Siebecke reflects on problems of the taxonomy for macro- and micro-economic gaming. In general he questions what morphology and anatomy of games are suitable in terms of the goals to be pursued. He correctly stresses that we should not search for one universal criterion for classifying games/simulations because they are a product of social life, evolving over time.

Instead of looking for a final taxonomy, Siebecke asks himself what purpose the taxonomy should serve. He aims at storing information on games in a database in order to make it available for further use for designers (authors) and game operators. Accordingly, descriptions should not be very detailed because data may be stored that are seldom used. He warns for "cemeteries of information".

This is especially important for setting up and running an information centre for economic games. Siebecke reports on experiences with the taxonomy for economic games in the context of the information centre that he and his colleagues have established in GDR.

Jan Klabbers and Barbara van der Waals elaborate on rules, which form the spine of most social systems and consequently of games/simulations as models for social systems. Reflection on rules originates from the well-known distinction between rigid-rule and free-form games. It turns out however, that here again it is important to realize that rigidity and flexibility of games/simulations depend on the position a person finds himself in. What is a frame game for a game designer, offering him much flexibility, may be a rigid rule game for the players/participants. The issue is becoming even more complicated by stating that each description of reality obeys rules embedded in the grammar of language. This notion blurs a clear distinction between descriptive and prescriptive rules, for descriptive rules on one level of abstraction become prescriptive ones on another level. Although from gaming theoretical point of view these new insights may be very worthwhile, the user-oriented taxonomy runs the risk of becoming too complex. The database that is being developed, similar to the one Siebecke has presented, may become rather complicated. Keeping in mind Siebecke's warning not to create a cemety of information, a database with several levels of aggregation seems to be appropriate, generating a hierarchy in the information retrieval for more or less advanced users. In this regard, the taxonomy can be utilized both for scientific and practical reasons.

In applying gaming methodology Kioshi Arai takes a user-oriented approach. Although community planning may be viewed as a process of conflict resolution, a game theoretical approach is beside the question. He considers community planning a learning process with scenario-development as the common tool for the problem setting. Knowledge about the community is generated in-action, the role of the scenario maker being a facilitator of the mutual learning process. Scenario-development seems to be better suited for problem setting from the point of view of the participants. It requires another type of theory of social systems than mathematical game theory. Such a theory is necessarily based on concepts that are self-referential. The way Arai describes SIMPLE (scenario-making with mutual learning process) shows that reflection and self-reference by the participants are necessary conditions for developing a community plan. We are noticing a shift from the spectator's position of the game-theoretician to the participatory perspective of the facilitator in the scenario/game development process (see also Introduction Section 1).

Charles Plummer and Tadeusz Selbirak (abstracts) discuss two examples of computer-based simulation and gaming. Plummer elaborates on the behavioural science approach, while Selbirak focuses more on the traditional management science approach.

Charles Plummer discusses relationships among theory, simulation and reality in the context of designing and applying simulation technology to theoretical and practical issues in behavioural and social systems. He views behavioural/social system simulation as a controlled representation of a real world behavioural/social system. In his definition he emphasizes the re-enactment of an existing system. He points out however that such simulations do not need to be restricted to representing historical events, they may very well be suitable for experiencing a "utopian" (ideal) or a dangerous system. Thus behavioural/social systems simulation can be used for instruction, for research purposes and for creating environments that "provide opportunities for participants to encounter experiences otherwise unavailable". Systems theory, learning theories and social psychological theory on attitude formation and change are used as a basis for the simulation design. He distinguishes ten major design, development and testing phases. In evaluating the educational effectiveness of computer-based social systems simulations, he is basing himself on a classification scheme, that is similar to the one used in Siebecke's taxonomy and the user-oriented taxonomy I have introduced. The context of use of Plummer's simulations is the educational environment.

Tadeusz Selbirak is attempting to classify different types of conflict resolution concepts. He discusses an interactive procedure based on this classification for analysing potential resolutions of two-player conflicts. His approach creates an interesting dilemma of on the one hand, academic rigour of game theory and on the other hand, the benefit of the theory in terms of real-life conflict situations. With regard to the Nash equilibrium point it is assumed that players judge the benefit of a unilateral change in their strategy not taking into account the possible reactions of the opponent. Selbirak calls this hypothetical player myopic. He generalizes the Nash equilibrium concept by departing from the classical game theory, including a "soft" approach represented by meta-game and hyper-game theories. When Selbirak concludes that the underlying concepts of the static theory used provide poor information on the question how participants of conflicts really attain stable outcomes, he questions the narrowly formal perspective of the theory. In staying within the realm of game theory he however shows how he uses a reflective methodology by incrementally generalizing on concepts. He uses a meta-game set up with the analyser/investigator acting as meta-player. From that point of view we argue that the meta-game perspective provides an adequate perspective for problem setting for the person who is carrying out the study, but not for the players. Consequently the tools are fruitful for the meta-player i.e. scientist, but not for the players as such.

Selbirak's approach is an example of knowledge driven research. The classical game theory started with a narrowly technical approach based on a formal, mathematical theory of rational behaviour. Increasingly this approach has been shown to be too restrictive.

References

Schön, D. A. 1983. *The Reflective Practitioner*. New York: Basic Books.

Problems of taxonomy for gaming in macro- and micro-economic subjects

Rainer Siebecke

Friedrich-Schiller-University, Jena

ABSTRACT: The taxonomy of games is the result of a preceding classification. To classify games they are arranged according to various features. In the German Democratic Republic, the focus of game classification has been put on macro- and micro-economic subjects. For a classification/taxonomy one should start from the following questions:
1. Which aims are pursued with the classification/taxonomy?
2. To what purpose is the classification/taxonomy to be put?
3. For which area is the classification/taxonomy to be valid and how general is the registration form for the games to be?
4. How detailed are the data to be registered?

KEYWORDS: Classification, taxonomy, information centre.
ADDRESS: Friedrich-Schiller-Universitaet Jena, Sektion Wirtschaftswissenschaften, Oekonomisches Labor, Goetheallee 1, Jena, 6900 GDR.

Introduction

If one wishes to develop the theory of games, it is indispensably necessary to provide a taxonomy of games. The taxonomy will be the result of a preceding classification. In my paper I intend to deal with the following two points:

1. Questions that have to be considered or answered before setting up a taxonomy
2. Experiences we made when classifying economic games

Questions Before Setting up a Taxonomy

Everywhere there is a multitude of individuals belonging to a basic entity their grouping or classification becomes essential for research purposes. The best example is provided by biology with its taxonomy of flora and fauna. It does not only enable us to exactly determine every plant and animal species, but this taxonomy is also an important auxiliary device for the other disciplines of biology.

How far can biology be of help for a taxonomy of games? I hold the view that we can put to use only the pattern of the taxonomy. This means
(a) that we must try to provide a hierarchical system and

(b) that we must examine the games as to how far they contain a common feature – similar to morphology and anatomy in biology – according to which they can be classified as universally as possible

Both points are closely related to each other. It should be relatively easy to cope with the first demand. The second is by far more difficult. For this reason we are of the opinion – together with Professor Klabbers – that several aspects have to be taken into consideration when one describes games. As games are systems, there are at least two classification criteria:

(a) classification according to functional features, e.g. their field of application
(b) classification according to structural features, e.g. the sphere of reality it reflects

Having said this I do not believe that there will be one universal criterion on classification. I am also sure that we will never have a final taxonomy. Contrary to plants and animals, games are a product of social life. Social progress and the resulting requirements at various levels are the cause that constantly new games are developed. This social determinism makes it necessary for a game to be described in various aspects.

I am sure that in spite of these peculiarities a taxonomy of games can be arrived at. Therefore we are faced with the problem of having to develop such a pattern that enable us to gather the features of a game. At the same time the pattern should take into account the interests of all participants. By "participants" we understand

– the authors of the game and
– the users of the taxonomy

of the various disciplines and fields of interest and from various countries.

Let us first turn to some practical questions which should be solved before we set up a taxonomy.

1. Which aim do we want to pursue?

We should be concerned with developing patterns with the help of which games can be systematized. This objective is shown in Diagram 1. This pattern represents only a very rough grid. Certainly a much finer representation of the levels could be found.

We should proceed similarly to biology where a division into two has been made. To my opinion, our most essential concern should be to include only such games in a taxonomy that serve to acquire new knowledge. All entertainment and sports-games should for the time being be left out of consideration.

2. Which purpose is the taxonomy to serve?

First: by storing the data of the games in a data file and making them available to the persons interested we also promote their follow-up use.

Problems of taxonomy for gaming 221

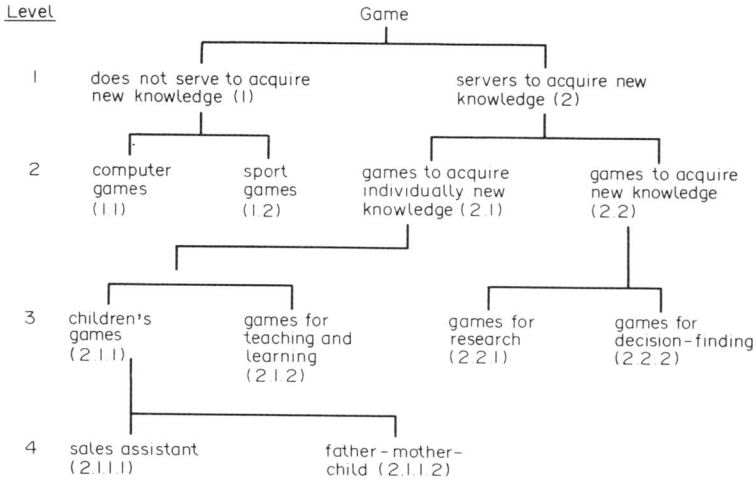

DIAGRAM 1   Classification of games according to their role in acquiring knowledge

Second: the classification/taxonomy of games is essentially necessary for research purposes on the field of games.

3. For which area is the taxonomy intended to be valid?

The answer to this question decides upon what kind of form the registration of the game will be used. It is necessary to clarify what is understood by "area". The answer that can be given points in two directions of adequate combinations (Diagram 2):

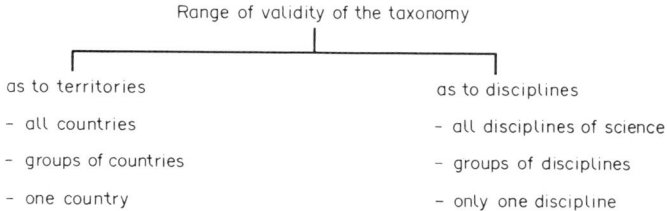

DIAGRAM 2   Range of validity of game taxonomy

4. How detailed are the data to be registered?

This question will have to be closely connected with the preceding questions. Here one should strike a happy medium. If your taxonomy is based only on few data, the game can easily be wrongly grouped. If, however, we start from detailed descriptions, we may be sure not to have forgotten anything and classified the game adequately. On the other hand, we also store data that are seldom or never required; we thus create "cemetries of information".

In the second part of my paper I would like to inform you of our experiences

Experiences When Classifying Economic Games

It is a pre-condition of setting up a taxonomy that you have a single pattern at your disposal in order to register all games according to the same criteria. When we began building up the Information Centre in 1978, our first step was to provide such a pattern – the "Nomenclature for the description of economic games". When in 1982 we also took on the task of the Information Centre for the other socialist countries, this pattern was kept in its shape and validity.

As the registration form of the Information Centre was and still is chiefly intended to promote the follow-up use of games, our approach was based on the question:

"Which data are required by a potential follow-up user or any person interested?"

Having this in mind we came to the conclusion that it is necessary to several complexes.

If you compare Professor Klabbers' taxonomy pattern with our nomenclature, you will find great accordance of the six points and their sub-groups with the individual complexes of the nomenclature.

Although we have concentrated on economic games for training purposes and further education, we tried to make the registration form flexible enough to use it for the registration of other games, too. Therefore we often included such data as
- further information (e.g. 4.3.4. of the nomenclature)
- explaining the data given more precisely (e.g. 4.3.5.)
- additional data (e.g. 5.7.)

This approach has proved useful until now. In this manner we were able to register technological, legal, paedagogical, and psychological games.

Which possibilities of classification and aspects of registration can be derived from the "nomenclature"?
- contents (6.)
  - area reflected (6.1.)
  - subject (6.2.)
- methodology
  - kind of game (1.2.)
  - field of application of the game (3.)
  - educational objectives (4.)
  - conditions of use (10.)
  - stage of development (11.)
- structure (7.)
  - participants of the game (7.1.)
  - management levels (7.3.)

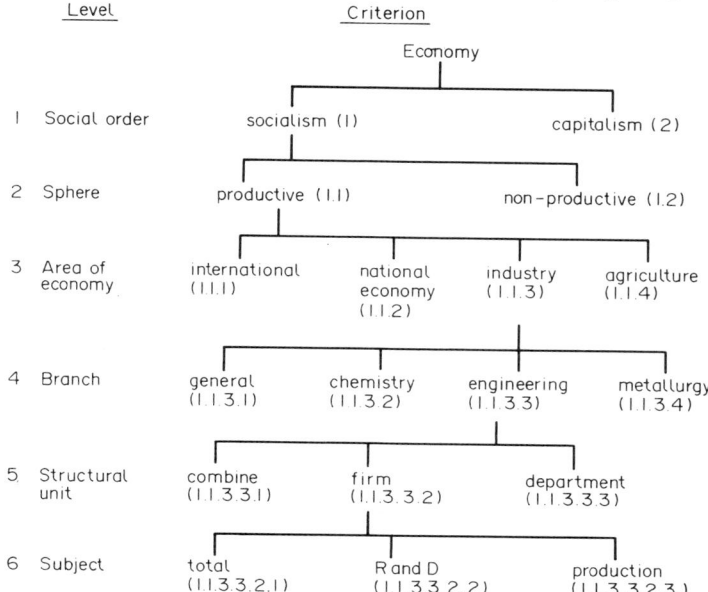

DIAGRAM 3 Classification according to the area of reality reflected

- model characteristics (7.4.)
- technical aspect (9.)

Using this grouping we are aware of the fact that not every criterion meets the requirements of a classification. And we also know that often we have just lists of criteria for the description of a game. Up to now we have found about 30 criteria for classification. Among them, there are e.g.

- the criteria of the purpose of application (teaching, research, practice)
- the criteria of teamwork of the participants (groups working isolated or in co-operation)
- the criteria of the number of management levels (one- or multiple-level game)
- the criteria of the dynamics of the management area (static, dynamic games) etc.

In our nomenclature, however, we dispensed with the strict application of these criteria for methodological reasons.

The following examples show relatively clear structures or levels in the classification pattern:

(a) grouping games according to the area reflected (Diagram 3):

At level six we stopped our classification. It may be possible to go on and break down e.g. production even further. But this would be contrary to the above-mentioned principle of clearness. In addition, our nomenclature would extend to a booklet no longer practicable. It proved valuable that the author of the game could supplement additional or more precise data.

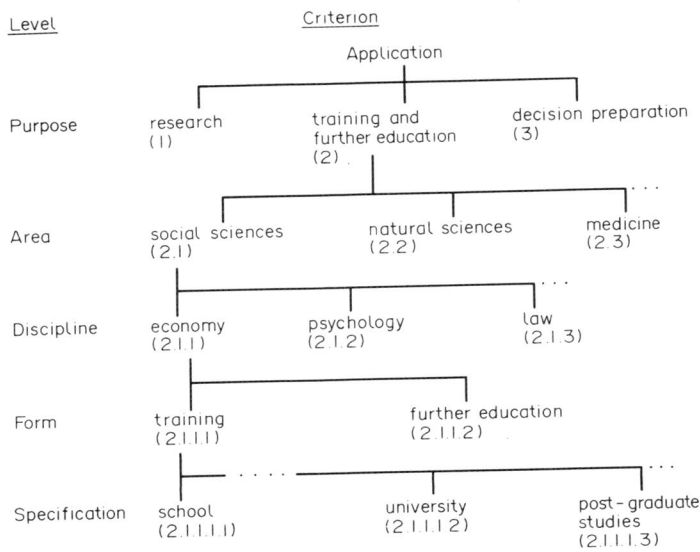

DIAGRAM 4 Classification according to the field of game application and purpose

This diagram however, reveals a serious problem of game classification:
Not every game necessarily exists at all levels. For example, we have games of level three, but also of level six.

(b) The second example shows the classification according to the field of game application (Diagram 4.). This is a classification taken from the area of methodology, mainly point three of the nomenclature.

This classification once more illustrates that our focus was and is on training processes. We did not make any further sub-division for the use of games in research and decision preparation.

These two possibilities of classification (as shown in the diagrams) indicate a second problem. "Economy", for example, in one instance appears as the head of a classification (Diagram 3) and in another it is an element of a classification (Diagram 4). It may also be arguable if levels four and five in Diagram 4 are justified.

I do not intend to go into detail with further potential classifications. Although the nomenclature has proved valuable as a registration form for games, we will make certain changes. This mainly refers to the sub-division of the non-productive sphere. The reason is that the Information Centre will extend its activities beyond economy to other disciplines too.

From rigid-rule to free-form games: observations on the role of rules

Jan H. G. Klabbers and Barbara van der Waals
Utrecht University The Netherlands

ABSTRACT: At the 1986 and 1987 ISAGA conferences the user-oriented taxonomy of games and simulations has been introduced and explained. In this paper we report on some adjustments, which concern the rule-base of games/simulations. Reflecting on the respective positions of game designer, game operator and players, it has become evident that notions like free-form, rigid-rule and frame game may have different meaning for each of them. Therefore we have made further inquiries on rules in social systems and on the rule-base of games/simulations. The results are summarized in the discussion about game components versus strata of social systems.
KEYWORDS: Rigid-rule vs. free-form, rule-base, distinct positions with regard to game-situations, game components, strata of social systems.
ADDRESS: Faculty of Social Sciences, Department of Gamma-Informatics, Utrecht University, Heidelberglaan 1, 3584 CS Utrecht, The Netherlands.

Introduction

For several years already ISAGA is paying attention to classifications and taxonomies of games and simulation during its annual international conferences. As part of this activity a "User-oriented taxonomy of games and simulations" has been introduced and discussed (Klabbers, 1987, 1988).

As we are continuing research on this topic we will report on the latest adjustments in the taxonomy. We have become aware of the need to focus on the rule-base of games and simulations, which plays such a decisive role both for game designers, game operators and players. It is obvious, that the distinction between rigid-rule and free-form or frame game is too vague.

The questions that we like to raise here are: "What do notions like "rigid-rule" and "free-form" actually mean? What do we mean by the "rule-base" of a game/simulation?"

Images of Rigid-Rule and Free-Form Gaming

Ståhl points out that a complete model of a game situation consists of both institutional – and behavioural assumptions about a game situation (Ståhl, 1983). The institutional model is a representation of the physical properties of a situation, while the behavioural model concerns the properties of the players.

In his view a game designer can at most provide and control the institutional model, while the behavioural model is provided by the players. We think that Ståhl's view is rather limited maybe because he uses mathematical game theory as a frame of reference. This type of theory applies a highly abstract representation of the institutional assumptions. In gaming-simulation however, the so-called reference system, representing the institutional assumptions, usually is described in more detail. As we have chosen to focus on gaming-simulation as the basis for our taxonomy, we have pointed out, that on the basis of the game theoretical model (Klabbers, 1987 and 1988), rules not only apply to the institutional assumptions in the game-situation, they apply to the actors as well. This complicates the issue of rules considerably, although this notion fits according to our opinion better the variety of game-situations.

We intend to complicate this issue even further by making a distinction between inside-actors, which are the players, and (remote) outside-actors like game-operators respectively game designers. For example, by means of role instructions the designer can influence the behaviour of the players directly. Indirectly he also influences the behaviour of the players via the institutional model.

We are aware that many games do not contain an institutional model like for example in pure role-play. One could argue that this type of gaming is completely free-form because the designer does not influence the way the players handle the institutional model. This view is not correct, as we know from theatre, which tends to be a rather rigid way of role-playing.

Consequently gaming/simulations which do contain an institutional model can be either rigid or free, depending on the perspective of the institutional model, the actors i.e. players, game operators and game designer.

In this paper we will present a scheme providing a synopsis of various forms of rigidity. It will be used to assess the type and measure of rigidity of games/simulations. Before elaborating on this scheme, we will start reflecting on the significance of rules of games/simulations.

The Role of Rules

Before examining rules, let us first take a glance at some synonyms. Rules are associated with regularity, measure (of length), custom, influence, government, and law. It is beyond the scope of this paper to describe these concepts in detail. What is noteworthy however, are the various positions that can be taken when considering rules. The first position is the outsider noticing patterns or regularities. The second is the insider who recognizes customs and rituals influencing his social-orientations. The third position is the boundary position usually taken by government and law-makers. The first position view rules as descriptions or interpretations of reality. The second one stresses handed-down interpretations of reality i.e. story telling, narratives etc. The third position is associated with prescriptions i.e. rules in the strict sense of the word.

Generally rules are seen as prescriptions, i.e. instructions, procedures, guidelines, laws, etc. (Duintjer, 1977, Gottlieb, 1968). Others regard rules as reality descriptions (Greimas and Courtes, 1982). Greimas and Courtes limit their definition to a particular type of descriptions i.e. descriptions of processes which are the result of the execution of rules. Thus, they focus on the regularities that result from prescriptions. Consequently they take an outsider/spectator position.

We point out that any description of reality implies excluding alternative descriptions. This not only decreases variety, it makes a description implicitly or explicitly prescriptive, for people often view their personal image of reality as the only legitimate one. Scientific descriptions tend to become formative in this regard. Game designers are well aware of the ethical implications of this notion of single versus multiple reality.

In the context of games and simulations, descriptions contain implicit prescriptions. Since gaming involves pretending, descriptions literally generate a virtual world. A designer may shape reality in a gaming/simulation in many ways. Players are invited to pretend to be in a virtual world with a high degree of realism. The syntax of a game/simulation is a powerful tool in this regard.

The distinction between pre- and descriptions often is rather fuzzy. We know that we are not very precise in the daily use of language. Although prescriptions should principally contain some sort of imperative, many times it is being neglected. For example the context-free statement "No trespassing" could mean that trespassing is not allowed, but also that trespassing does not take place.

Now that we have characterized rules both as prescriptions and descriptions, we wonder what relationship does exist between rules and rigidity. Generally speaking, rules do increase regularity and subsequently rigidity because they exclude alternative paths of actions. Prescriptions act straightforwardly in this respect, because they define what should or should not be done. The same holds for implicit prescriptions implied in descriptions. They exclude alternative ways of looking at reality. Insofar as descriptions symbolize reality they may also bring forward rigidity, because only particular actions may be open for discussion in such a (virtual) world.

Considering the three positions mentioned earlier, it is rather difficult to judge how much rigidity is generated by specific rules. We will briefly mention a few difficulties in assessing the degree of rigidity. We will not make a distinction here between injunctions and prohibitions. First of all rules can differ in their measure of detail. The more detailed they are, the more they exclude, implying that more detailed rules cause more rigidity than less detailed ones. Dependent on one's point of view, very detailed rules may cause "sclerosis" or tranquillity in the social system involved.

A second difficulty lies in the domain of a rule. How many activities are allowed by a particular rule? According to this criterion we assume that the more extensive the prohibition, the more rigidity it causes.

The combination of both difficulties (detail and domain) can however cause a dilemma. It may well be that very detailed rules may be applicable to a very limited domain and vice versa. In such a situation detailed rules call for many rules. A game designer will have to decide how important detailed rules are for a specific game. Compare for example the following instructions:

"Bake an apple pie for George"
"Cook a dinner for eight people"

The first instruction (rule) is the more detailed one. It leaves at best the freedom to choose among various recipes for apple pies. However, there are far more alternatives open for cooking a dinner. Nevertheless, cooking a dinner may be more time and energy consuming than just baking an apple pie.

Another point is that the sheer existence of a rule does not necessarily imply that anticipated processes actually will take place. A rule taking effect, presupposes both acceptance by people involved and competence to execute it.

Acceptance can have various grounds such as coercion, morals, self interest, etc. Competence involves practical skills necessary to execute the prescribed behaviour. Competence refers also to the interpretation of the rule; when should it be followed, what does it actually prescribe, etc? Here we enter the realm of the equivocality of rules and the choice between bending to or getting around them. This becomes especially tricky when dealing with multiple rules that are more or less consistent, coherent or mutually exclusive. For the game designer this is an important aspect to consider, because it influences the transparency and playfulness of a game.

These deliberations have led us to review the rule-base of the taxonomy. In the next section we will present our new point of view.

Rule-Base Versus Rigidity

Elaborating on the rule-base of the user-oriented taxonomy emphasizing various forms of rigidity, we will keep in mind the notion of gaming/simulations as social systems, (Klabbers, 1987, 1988). Regarding social systems three interlocked strata are distinguished.

- *Culture*, i.e. norms, values, beliefs, morals, interests, etc.
- *Structure*, i.e. vertical and horizontal communication and co-ordination.
- *Technology*, i.e. the whole complex of routine and non-routine procedures to handle material processes.

From rigid-rule to free-form games 229

These strata resemble Ståhl's institutional- and behavioural models. The institutional model corresponds with the technology stratum and the institutionalized part of the structure stratum. Human action in the structure- and the culture stratum constitute the behavioural model.

In the game theoretical model, mentioned elsewhere we have distinguished three components:

- *Actors* with their individual respectively collective cognition (history), their roles, tasks, social orientations, expectations, interests, etc.
- *System of rules* (of conduct); procedures, regulations, rituals, etc.
- *(Re-)constructions of reality*; states and processes of natural and artificial environments. (Klabbers, 1987, 1988).

Considering our discussion on rules, in gaming-simulations "actors" and "(re-) constructions of reality" may refer to prescriptions and descriptions. Concerning actors, the roles, tasks may be prescribed in role-instructions, while (re-)constructions of reality may be symbolized via descriptions. This means that it is not made clear in advance which rules are part of the "system of rules" and which rules are implicit in the description (syntax) of actors and (re-)construction of reality.

Therefore we will re-examine the game components and restrict ourselves to rules imposed by designers, because only in this way the impact of rules on the rigidity for all players involved will become more clear. We will not take into account rules that the players impose on themselves.

Suppose we view the game components as one dimension for indicating the meaning of rules from game designer's point of view. The other dimension will be constituted by the strata, mentioned earlier, and it will be considered the player's perspective i.e. their experience while playing a game/simulation

- *Actors*: everything that is being brought into the game by the players via the roles which are described by the designer, e.g. norms, expectations, cognitive schemata, tasks, skills, etc.
- *System of rules*; all prescriptions and descriptions imposed on the players by the designer.
- *(Re-)construction of reality*; the actual (re-)presentation of the reference system, e.g. signs and symbols, boards and tables, chairs, etc. (Van der Waals, 1988).

Gaming-simulations are highly stylized examples of social systems. The components of the game theoretical model are the constructs for the game designer to enact social systems. The strata form their empirical basis i.e. their content for the players. Another way of looking at the matrix is to regard the components as the syntax of the designer, and the strata as the semantics of the players. Both dimensions combined bring forward a matrix, the cells of which give indications about the potential role of rules.

strata game components	technology	structure	culture
(re-)construction of reality	1.1	2.1	3.1
rules	1.2	2.2	3.2
actors	1.3	2.3	3.3

DIAGRAM 1 Game components versus social systems strata

In Diagram 1 the mix of strata and game components is assessed. We will indicate the possible meaning of the nine cells presented in the diagram. Where appropriate we will use the game "THEY SHOOT MARBLES, DON'T THEY?" to illustrate our viewpoint, assuming that this game is generally known.

Technology versus (re-)construction of reality

In this cell game paraphernalia symbolizing the technology are being used. Think of the board, the marbles, the wooden dowels, the blackboard as a set of pieces. A computer simulation model could be used here as well. They constitute the potential resources of the players.

Technology versus rules

The lay out of the game space, and the set of places of pieces define (implicitly) the rules for allocating resources. In "MARBLES" the measure of the table, the relative length of the dowels, and obstacles on the shooting table describe the inherent rules of the game space. The sequence of shooting, physical guidelines for shooting are also examples of these type of rules. While using a computer simulation, rules define the constraints on the steering parameters.

Technology versus actors

This cell contains the number of game positions relative to socio-economic positions. It refers to the roles of players, i.e. the shooters, the judge, the opposition and law-makers.

Structure versus (re-)construction of reality

Here we refer to the physical communication structure like the spatial arrangements of the players. For example what is the spatial position of the judge relative to the marbleshooters, and the other groups of players?

Structure versus rules

This cell refers to all sorts of instructions guiding communication and co-ordination processes during a game. Think especially of the rules concerning the negotiation process between the marbleshooters. Who can talk to whom? Rules for beginning and finishing a game/simulation belong to this cell as well.

Structure versus actors

This cell refers to the way players handle communication and co-ordination activities during the game. They symbolize relationships between participants is social situations i.e. their collective structure.

Culture versus (re-)construction of reality

Values, morals, beliefs, etc. are represented suggestively by means of various material paraphernalia. Advertisers know quite well how to arouse values and expectations by giving consumer products a certain shape and colour. Sculptures and painters struggle with it as well. Game paraphernalia may create a virtual world like for example in science fiction.

Culture versus rules

Here we have to think of moves, and transformations of the positions that are allowed, based on customs, procedures, taboo's etc. The instruction that marble shooters should strive for coalitions for allocating resources fits into this niche.

Culture versus actors

This cell defines values, morals, beliefs, i.e. roles that players assume and that define the right of having marbles, the obligation to fight for a just distribution of marbles, and the right to be well informed by the media.

Some Observations on Rules and Rigidity

How may the diagram be helpful in assessing the degree of rigidity of gaming/simulations?

From game component's point of view a designer generally may have such freedom available in terms of defining the game space, set of rules of game positions, rules for starting and finishing a game/simulation, rules for roles etc. From his perspective a frame game may be a free-form game. It is the outsider's perspective. From the player's point of view, that same game may well be experienced as a rigid-rule game, for the resources of the players, and the set of places for resource allocation and consequently the socio-economic positions of the players may be limited in the game.

As has been mentioned earlier, the diagram reflects from vertical point of view the degree of realism or verisimilitude of game/simulations for the participants. It concerns the semantics of the game situation. Take for example an average management game and consider for example cell 1.1. (technology versus (re)construction of reality). There are only a few games, that make the actual manufacturing of products on the shopfloor explicit according to a recipe of rules. Those types of games tend to be rather mechanical with a rigid technology. An interesting example of this type of game is SWITCHER (van Linder, see this volume). In SWITCHER players are challenged to switch from a mechanic i.e. rigid-rule procedures, to an organic type of organization, i.e. free-form. Usually management games take technology and its respective structure for granted. Production is endogenously generated according to a specific algorithm. Emphasis is placed only on communication between special managers such as production manager, controller, marketing etc., stressing structural aspects of the respective social system. Production, sales etc. are viewed as black boxes. Profits and market share are calculated after each cycle. Rules are rather rigid i.e. mechanic. This is typical for most general management games. A management game primarily focusing on culture will in general just give a general outline of the technology, but it would from player's perspective focus more on rituals, intelligibility and jargon, taking technology as such for granted. TALKING HEADS is an example of this type of game (Christopher and Smith, 1987). It starts with cue cards for the respective managers, and emerges as an intensive negotiation and meaning-processing exercise.

Combining both dimensions, we expect that along the diagonal of the matrix from top-left to bottom-right games become less rigid for the players. The strata versus game-components show that the balance in designers' and players' autonomy can refer to quite different aspects. This balance can vary per cell within one gaming/simulation. For example the technology stratum can be defined by the designer, while the culture stratum is given a more or less regular disposition by the players. In that case, cell 1.1 will be a rigid-rule situation for the players, while cell 3.3. for them will be free-form situation. This presentation reflects Ståhl's idea of institutional and behaviour assumptions.

Secondly the dimension of the game components shows that simulations may not necessarily be very rigid. Players may build very well their own simulation models or design their own maps, the designer only providing a general scenario and a simple accounting system (see Kalff, section 1). Also role-plays are not always free form games. When morals, values, expectations, etc. are prescribed by a designer, the players have to stick to the chosen characters. The utmost example of this is a theatre play, in which actors assume a prescribed personality, adding their own physical being and their skills to give shape to the characters.

Numerous "multi-level multi-actor simulations", (Klabbers, 1987, 1988), can be identified with the diagram of the rule-base in mind. Assessing the impact of rules on rigidity we hope to be able to determine the relative weight of each cell in the diagram. To perform this rather difficult job adequately, the remarks we made in the previous sections about rules and rigidity may be of some help.

Furthermore we are of the opinion that the diagram provides a frame of reference to view rules within their context. We think that detailed (re-)constructions of reality tend to make rules more rigid. Material paraphernalia sustain the world that players will have to imagine according to numerous (hidden) pre- and descriptions.

The relationship between actors, rules and rigidity is difficult to understand from the outsider's point of view. Players themselves may enact more or less rigid rules. Consider the situation that players experience less freedom in handling their affairs than the rules as such presuppose. Are we going to decide that this refers to personal rigidity of the players and not to the syntax of the gaming/simulation? Or do we consider the players' narrowmindedness an issue for the game (designer) to take into account either in the specifications of a game/simulation, or in the debriefing? Maybe a game is designed just to teach players to switch between different perspectives, each one in itself presenting a closed world based on rigid rules. This is an important issue for a game designer, because the meaning of a game by its players seriously affects the message that games are trying to convey. The hidden rules behind all sorts of descriptions of the game situation resemble the well-known adage "the medium is the message".

Summary

Further reflection on the rule-base of games/simulations has enriched the taxonomy and its game-theoretical basis. The resulting diagram complicates however characterizing particular games and simulations from practical points of view. By conveying these new insights to game-designers, we hope that they become even more careful. We have become more aware of the many hidden values and representations of reality that are embedded in games. They become manifest in the situation that one plays a game that has been developed for social situations in an unfamiliar culture, whether it is a different organizational or national culture. Transfer of those type of games to other cultural settings apparently is tricky business, especially when rigid-rules are involved.

References

Christopher, E. M. and Smith, L. E. 1987. *Leadership Training through gaming*. London: Kogan Page.

Crookall, D. Greenblat, C. S. Coote, A. Klabbers, J. and Watson, D. 1987. *Simulation-Gaming in the late 1980s*. Oxford: Pergamon.
Crookall, D. Klabbers, J. Coote, A. Saunders, D. Cecchini, A. and Delle Piane, A. 1988. *Simulation-Gaming in Education and Training*. Oxford: Pergamon.
Duintjer, O. D. 1977. *Rondom Regels*. Meppel: Boom.
Gottlieb, G. 1968. *The Logic of Choice*. London: Allen and Unwin.
Greimas, A. J. and Courtès, J. 1982. *Semiotics and Language: an analytical dictionary*. Bloomington: Indiana University Press.
Klabbers, J. H. G. 1987. A user-oriented taxonomy of games and simulations. *In* Crookall *et al.* (1987).
Klabbers, J. H. G. 1988. Frame of reference underlying the user-oriented taxonomy of games and simulations. *In* Crookall *et al.* (1988).
Shubik, M. 1983. Gaming: A-state-of-the-art survey. *In* Ståhl (1983).
Ståhl, I. (ed.) 1983. *Operational Gaming, an International Approach*. Oxford: Pergamon.
Waals, B. J. van der, 1988. *Soorten en Maten van Rigiditeit in Spel/simulaties*. Master's Thesis, Faculty of Social Sciences, Utrecht University.

Simulation References

SWITCHER. van Linder, B. 1989. *In* this volume.
TALKING HEADS. 1987. *In* Christopher *et al* (1987).
THEY SHOOT MARBLES, DON'T THEY. Goodman, F. L. School of Education, University of Michigan, Ann Arbor.

A simple method of scenario-making: two Japanese cases in community planning

Kiyoshi Arai
Kinki University in Kyushu, Japan

ABSTRACT: In order to cope with planning in a situation where we cannot see what is the issue and what logic aspects are interrelated, scenario-making is a useful tool to break through the situation. This is because a scenario-making process should imply a learning process.

We tried to develop a simple method and performed two experiments. The method is a sort of free form game, similar to Delphi, and is based on a methodology to pursue a concrete and compatible scenario. People concerned with the planning process are divided into a core-group and participants. It is expected that a core-group should play a role of a controller and that participants should improve a scenario structured firstly by the core-group. Two experiments show that this method could be useful for a group of people to make clear their situation in a relatively short range of time.

KEYWORDS: Free form game; scenario; consensus; planning; learning; Delphi, participation.
ADDRESS: Faculty of Engineering, Kinki University in Kyushu, Iizuka-shi, 820 Japan.

Introduction

We are often forced to cope with planning in a situation where we cannot see what is the issue and what logic aspects are interrelated. This seems to be a sort of symptom we recently have become aware of in public planning. The reason why we cannot see what aspects and what logic should be relevant to the nature of public planning, lies in involvement of a variety of interest groups.

One origin is surely from the complex nature of social systems. There are a lot of feedbacks in the systems, and causal relations of events are very complex. A plan often leads to an unexpected result. Another origin stems from the existence of different stakes that people have in accordance with their social roles. The same things often have different meanings for different persons, and is even regarded as a different fact.

In the planning process, we should positively recognize our complex situation. It is very often that we cannot have a clear answer to our problem, and even that we cannot see a common problem to our situation. It should be due to the nature of situation itself rather than to our incompetence of managing our situation. When we try to do something in this situation, an ordinary survey method, i.e. a method of surveying the general public by an inquiry sheet, seems to be inappropriate to see what is the issue. We need more sophisticated methods that incorporate mutual learning processes between planners and interest groups and between interest groups themselves.

A real planning process partly permits feedback in order to make a reasonable decision to a problem. However, costs of feedback are sometimes very high in a real process, so some simulation technology is needed. We have tried to develop a simple simulation method for planning, which incorporates some feedbacks in a process of scenario-making.

Scenario-making with mutual learning process

Illustration of SIMPLE

This method is named SIMPLE, which is short for Scenario Improving Procedure with Learning and Evaluation. The groups are divided into a core-group and participants as indicated in Fig. 1. The core-group makes a first

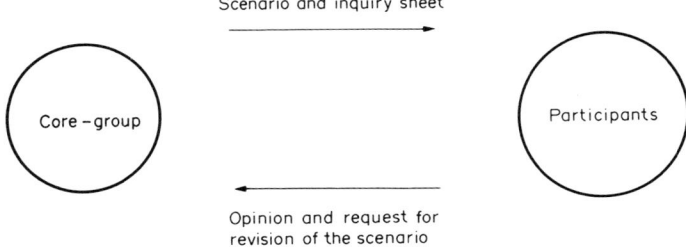

FIGURE 1   Relation between core-group and participants

draft of a scenario, and manages the process. Participants are recruited by the core-group. They consist of administrators, engineers, and planners, who are expected to represent a variety of interests. The first draft is improved through the first stage of inquiry, where the core-group revises the draft by hearing the participants in face-to-face communication. The second draft is improved through the second communication stage. This is revised again, and the final draft, the third one, is made. This process is illustrated in Fig. 2.

Improving a scenario

It may usually happen that a request for compatibility between various opinions leads to a too abstract scenario, resulting in a lack of realism, and that a request for concreteness leads to a too incompatible scenario including different interests, resulting in a lack of a mutual appreciation. We pursue

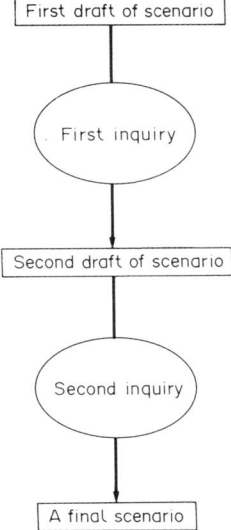

FIGURE 2 Procedure of improving a scenario

both compatibility and concreteness at the same time through the process of scenario-making.

In the stage of inquiry, we present a scenario and an inquiry booklet to participants. A scenario is a booklet which describes a realm of consideration, recognition of the situation, and prescriptions for the problems. The inquiry includes some questions and explains the meaning of the questions. Especially the second stage is described during which some opinions from some participants as well as some counter-opinions from the core-group are discarded. The purpose of this methodology is, of course, not to force participants to accept the way of thinking of the core-group, but to encourage their own creativity. Mutual learning between the core-group and participants and between participants are activated through the process.

Characteristics of SIMPLE

SIMPLE resembles Delphi because we inquire twice. But it is not Delphi since we do not ask the same questions in each stage.

To summarize, SIMPLE has four characteristics:
1. It has a scenario. Participants can share a common frame of reference.
2. A scenario is improved by a core-group, based on opinions of participants. Mutual learning is incorporated in the process.
3. A variety of interests can virtually be taken into account by participation of various participants.

4. Extensive communication is endorsed by face-to-face communication between a core-group and participants and between participants.

Two cases of applying SIMPLE

"Social Information Systems for Community planning"

This experiment was performed by financial support of National Institute for Research Advancement (NIRA), December 1982 – July 1983 (see Arai *et al.*, 1983).
1. Core-group: Arai, K., Nemoto, T., Nara, T., and two others, total five persons, co-operated by Kumata, Y., and others.
2. Participants: 42 persons, including administrators of local government, engineers of private companies, and planners.

Content of the scenario is as follows:
- Role of centre for Community Planning
- Job of the centre
- Co-operation of the centre with companies and institutions
- Foundation and management of the centre
- Realization of the centre

An inquiry was done after the scenario-making process had been finished. Then a high percentage of participants among those who answered the inquiry, 18 out of 19 persons, assessed this method to be effective (see Arai *et al.*, 1984). They, however, made suggestions to improve the process of SIMPLE as follows:
1. As for how to recruit participants and how to make them participate in the process, the suggestions are (a) to endorse the variation of participants, (b) to make participants take part in the first stage of formulating the draft, (c) to present the scenario in the presence of all participants, (d) to make an opportunity of participants' discussing each other, and (e) to have more stages for improving the scenario.
2. As for how to treat the conflict of opinions, the suggestions are (a) not to make a commonplace scenario by including all interests, (b) to make clear why some opinions are discarded, and (c) to make all opinions accessible to all participants.
3. As for how to present the scenario, the suggestions are (a) to shorten the volume of the scenario, (b) to make clear the definition of technical terms, and (c) to make much use of figures and tables to help participants' understanding the scenario.

"Community Planning of Tokyo"

This experiment was also performed by financial support of NIRA, August 1986 – April 1987 (see Kumata *et al.*, 1987).

1. Core-group: Kumata, Y., and others, total 12 persons, co-operated by Committee Members of Japan Association for Planning Administration (JAPA) and of Keizai Doyukai.
2. Participants: 66 persons, including administrators of national and local governments, engineers of private companies, and planners.

Content of the scenario is as follows:
- The image of a Humanistic Advanced Information City (HAIC)
- The image of a Humanistic Intelligent Complex
- Viewpoints for realizing the HAIC
- Redevelopment systems for the urban core
- Organizations for urban revitalization

No formal inquiry about assessment on SIMPLE was done in this experiment. But in this case, we improved some procedures through the process. For instance, we hold several symposiums to provide a close communication between participants. This was effective to help participants' understanding the scenario. In a stage of a symposium, we presented some points by video. It played an appropriate role as an introduction to the scenario.

This time the number of participants was larger than in the former case, so management of the experiment was more difficult than before. The difficulty has been reduced by co-operation of several groups, who intermediated between the core-group and participants.

Concluding remarks

This method of SIMPLE originates from a quite simple idea of improving a scenario. However, some people do not accept this simple idea. Those people are inclined to hesitate to make an incomplete draft open to many people in order to have suggestions from them. This sort of hesitation can be understood. A draft is easy to abuse because it is vulnerable to harmful groups. If the idea, that a scenario-making can be a free form game or a simulation, penetrates through people both within the group of formulating the scenario and outside the group, this method will be a powerful tool.

Two experiments, any way, show that this method could be useful for a group of people to make clear their situation in a relatively short range of time. This method should be refined to promote its potential ability.

What should be considered of the process, for the moment, is as follows:
- Mental attitude: an atmosphere free to talk about anything is necessary and emphasis should be on the process more than on final results.
- Hard and soft technology: active use should be made of audio-visual technology and face-to-face communication, e.g. a symposium, should be designed.
- Management of the process: how to formulate a core-group is important and participants should be successfully recruited.

References

Arai, K., *et al.* 1983. Social Information Systems for Community Planning: A Scenario for Making Networks between Habitants, Local Government Officers, and Planners (Machizukuri no tameno keikaku jouhou system in Japanese). *NIRA-OUTPUT(NRS-82-29)*.

Arai, K., *et al.* 1984. SIMPLE: Delphi-type Survey Method with Scenario (Scenario ni motozuku delphi-gata chousa shuhou SIMPLE no kaihatu in Japanese). *Planning Administration* **13**.

Kumata, Y. *et al.* 1987. Community Planning of Tokyo: A Humanistic Advanced Information City (Jouhou koukando toshi Tokyo no machi-zukuri in Japanese). *NIRA-OUTPUT(NS-86-8)*.

Design and evaluation of computer-based behavioural/social system simulations

Charles M. Plummer
Rochester Institute of Technology, New York, USA

ABSTRACT: Theoretical concepts, educational objectives and realistic situations are integrated in two behavioural/social system simulations. The POPULATION BOMB SIMULATION concerns population control, where attitudes, reliability/validity of information, and ethics are involved when judging others with stereotypes, labels, and obedience to authority as factors. The FUTURE SCHOOL SIMULATION applies attitude change theory in a structure encouraging "counter-attitudinal role-playing", where participants attempt to persuade others to adopt innovative proposals for improving their future school.

Relationships among theory, simulation, and reality, are discussed in the context of designing and applying simulation technology to theoretical and practical issues in behavioural and social systems. Phases and procedures for the design and evaluation of computer-based simulations are presented. We conclude with a summary discussion of the current use of simulation in higher education, and the possible convergence of relevance, problem solving, hybrid human-computer interaction, modelling, and dynamic interrelationships in simulation applications that can advance theory, improve simulation design, and solve real problems.

KEYWORDS: Population control; future school planning; simulation design, technology, theory, evaluation; computer-based simulation design and evaluation; behavioural/social system simulation.

ADDRESS:
Rochester Institute of Technology, RIT Research Corporation, 75 Highpower Road. Rochester, New York 14623.

Computer-based interactive procedure for analysing conflict resolutions within simulation and gaming

Tadeusz Selbirak

Institute of Public Administration and Management, Warsaw, Poland

ABSTRACT: An attempt to classify the different types of conflict resolution concepts from metagame and hypergame theories based upon the classical Nash equilibrium point is presented. Underlying role in these concepts play the assumed extended models of players' subjective rationality which admit their nonmyopic behaviour. Special interactive computer procedure based on the given classification developed by the author for analysing potential resolutions of the two-player conflicts is shortly reviewed. Possible applications of this procedure within simulation and gaming are indicated.

KEYWORDS: Game theory, conflict resolution, computer-based decision support system, simulation.

ADDRESS: Institute of Public Administration and Management, 02-067 Warsaw, Wawelska 56, Poland.

Section Six

Learning Environments and Communication

Editor: **David Crookall**

Learning environments and communication: an introduction

David Crookall
The University of Alabama, USA

Three papers appear in full in this section, along with the abstracts of five other papers. This introduction attempts to draw some links between all these accounts, i.e. including the contributions of which only the abstracts are published here. The purpose of this is to encourage readers not only to reflect on the broader concerns of contributors in this section and but also to make contact with contributors whose full paper does not appear here.

The eight contributions in this section cover an extremely wide array of simulation/gaming applications. A few years ago Noel et al. (1987) emphasized that

> It is difficult to imagine a more polyglot area of kindred scholarly activities than simulation. It reflects the coming together of a rather amazing variety of intellectual traditions: operations researchers, diplomatic analysts, economists, military analysts, international relations scholars, small-group and organizational sociologists, experimental social psychologists, mathematicians, computer scientists, educational psychologists, and still others. That there has been mutual enrichment is as undeniable as is the resulting terminological confusion.

This is reflected in the contributions to this section, except perhaps that the above list does not encompass all of the "intellectual traditions" represented here. We would have to add such fields as second/foreign language education and town planning.

Such traditions and fields can be conceived of as one type of environment, e.g. using a simulation/game in an international relations environment. However, environment can also refer to the smaller-world context of the kind of structure, atmosphere, technology and such like of a particular simulation/game session. The contributions in this section show variety across both interpretations of the term "environment". If I could only think of a third interpretation, it would be possible to draw a cube and then three-dimensionally categorize each paper. But this would probably lead to further terminological confusion.

It is thus that the first paper in this section is by Ken Jones, for he manages to sort out, in his usual nimble way, some of the confusion between the terms "game" and "simulation". In so doing, Ken also provides us with much food for thought on how our conceptions of these events may influence what happens in them and what learning takes place. In many ways, Ken's paper should be read as a backdrop to the other contributions in this section. His concern with what actually happens in a simulation/gaming event as an evolving environment, as opposed to the physical materials (which should properly be referred to as a simulator or game-base), is a healthy reminder to those of us who tend too easily to get caught up in the technology – sometimes to the point of forgetting the people directly concerned (the participants), the communication that takes place between them and their social realities.

One particularly good example of a simulation/game in which content and process are foremost is ME — THE SLOW LEARNER, by Don Thatcher and June Robinson. This has already become a classic exercise, and is one not only which employs fairly simple equipment (e.g., spectacles, tape, boxes), but which is extraordinarily effective in creating a truly memorable learning experience. The exercise aims to help people come to understand, both intellectually and viscerally, what it is like to have learning difficulties. In their paper (included as abstract), Don and June stress the critical importance of the affective dimensions in the learning context, in particular the emotional blocks caused by some learning environments. Thus, their simulation/game creates an environment in which participants (both teachers and learners) can learn and communicate about learning difficulties and environments.

From the relational and social aspects of learning we turn, in the next two contributions, to more physiological considerations of this process. (Perhaps this can be considered as our third interpretation of the term "environment" – an inner biological one?) Both contributions consider the mutual interdependence between simulation/gaming and aspects of the now classic left- and right-brain distinction. In the first of these (abstract only), Michael Rockler focuses on what he terms "brain-compatible" teaching strategies. With examples of actual simulation/games, Mike convincingly shows us how research into brain functioning may provide us with additional arguments for considering simulation/gaming as a particularly powerful learning methodology. The second contribution (full paper), by Peter and Gennie Raynolds, is concerned with the ways in which one particular game, the JOG YOUR RIGHT BRAIN exercise, encourages "participants to explore and share their intuitive reactions and feelings toward" a set of topics. A scoring system gives attitude, identification and self-esteem measures, and this is illustrated with data generated during the ISAGA conference session.

The above discussions on simulation environments are in many ways generic, in that they may be applied to a wide variety of situations and topic matters. The contribution (abstract only), by Alan Cudworth, continues with some of the above themes. However, Alan emphasizes a number of communication skills, and discusses these within a more specifically substantive area – that of urban planning and development. Alan also discusses the important

Learning environments and communication 247

goal of reaching a balance between theory and practice in institutional settings, and how the means by which this is reached will vary from one subject area to another.

The next type of environment we encounter in this section is the now ubiquitous computer, but many of the themes evoked above thread themselves through the subsequent papers – after all, a computer without substance or process is just a heap of plastic, copper and silicon. Much has been written recently on computerized simulation (see, e.g. the recent collections by Crookall, 1988a, 1988b; Wolfe, 1987), and the last contributions in this section are all welcome additions to the discussion.

In the first of these, Fred de Vries accomplishes two broad objectives. The first is to provide us with an outline of the multifarious configurations which computerized simulations can take.

Consideration is given to basic issues, such as type of learning task, degree of reality simplification and user interface. Fred's second objective is to provide us with a glimpse of the multiflorous nature of computerized simulation, by providing illustrations of five very different examples.

The last two abstracts in this section provide concrete illustrations of the many ways in which some of the concerns outlined by Fred are being tackled. The paper outlined in the first abstract gives a glimpse of the complex issues involved in designing computerized simulation for foreign/second language learning. The second concerns computer-based training for military purposes.

Foreign and second language learning is a lengthy and intricate task, one that requires many strands of expertise, and an integration of many factors. The introduction of a computer into the language learning environment complicates things further, mainly because the computer is not a natural language-using beast – it does not, nor can it be made to, think, act and react linguistically and communicatively like another human being. The underlying task, therefore, is to somehow get the blessed machine to give the impression of so doing. This is where simulation comes in, but this does not in itself always solve the problem. Ways have to be found of designing a computerized simulation which will allow learners to use and practice language in as natural a way as possible, ways which approximate to authentic discourse. One way is to use a simulation as a catalyst for creating a general environment for social interaction, where learners use language amongst themselves, with the computer serving as a kind of resource (e.g. similar in manner to which many business simulations work, or as a communications medium). Another way is to somehow get the computer to play the role of another language-using person, with which (or whom?) the language learner may converse in some fashion. This kind of scenario, however, immediately raises the stakes, for – as noted above – a computer is not a person. The classic example is Weizenbaum's Doctor programme, with which readers will, I am sure, be familiar. This seriously limits the kind of "interaction" (hardly social) that can take place. When we add to this limitation considerations pertaining to language *per se*, such as grammatical correctness or communicative appropriacy, as well as to the language learning process itself, such as motivation or a supportive climate, then the obstacles would seem to be almost insurmountable. Fortunately, however, humans (not computers!) are resourceful and creative beings. The contribution (abstract only here) by Doug Coleman demonstrates great ingenuity in the ways he tackles such problems; both by relying on the age-old human frailty of disbelief suspension and also by actually making use of some of the machine's capabilities, Doug goes a long way to overcoming some of its basic inherent limitations. His approach is to design computerized conversational simulations which integrate linguistically valid exchanges and learner-centred (or -sensitive) error correction procedures. The paper is a significant contribution to the field, and it also considers possible future developments.

The last abstract in this section, by Hilbert Kuiper and Jos van der Arend, is concerned with an altogether different domain, that of training in specific military skills, like aiming and firing at a target. Of interest is the breakdown of the training system into two linked modules, one of which stores and prepares the parameters and basic data for tailor-made use in the other. The whole design process is cleverly conceived and the technical specifications are impressive, but the system requires a powerful mini-computer, making its portability somewhat limited. As King and King (1988) point out, portability is a major criterion in any computerized simulation, especially if relevance to other users is considered important. It should also be mentioned that the system described could probably be applied to domains other than the military, such as environmental studies or urban planning. Such applications would be heartily welcomed, especially if such a system were made available for more commonly-available computers.

To sum up, what all these contributions illustrate is that the learning context is as important as, or more so than, the simulation materials and technology themselves. As Ken says, the materials are merely a springboard to some kind of event or learning experience, and what actually happens there depends on a whole host of factors, people and processes which make up the learning environment, on how these factors mesh together, on the communication patterns among the people and on the kinds of processes which emerge. In many ways, communication is the key which provides substance and life to these learning environments, and this appears to be especially relevant to simulation/gaming (see Crookall and Saunders, 1988). Readers concerned with the quality of learning environments and the communication patterns that come to life there, will find much that is of central interest in this section. The task that remains, for all of us, is somehow also to communicate something of this insight to our colleagues, who may neither be convinced of, nor yet even acquainted with, how effective such environments can be for learning (see Greenblat, 1987).

Rererences

Crookall, D. (ed.) 1988a. *Computerized Simulation in the Social Sciences*. (Special issue of *Social Science Computer Review*, **6**:1) Durham, NC: Duke University Press.

Crookall, D. (ed.) 1988b. *Simulation/Games and the New Technologies: Applications and Problems*. (Special issue of *Simulation/Games for Learning*, **18**:1) Loughborough: SAGSET.

Crookall, D. and Saunders, D. (eds.). 1988. *Communication and Simulation: From Two Fields to One Theme*. Clevedon, Avon and Philadelphia, PA: Multilingual Matters.

Greenblat, C. S. 1987. Communicating about simulation design: It's not only [sic] pedagogy. In Crookall, D., Greenblat, C. S., Coote, A., Klabbers, J. and Watson, D. R. (eds.) *Simulation-Gaming in the Late 1980s*. Oxford & New York: Pergamon.

King, A. R. and King, B. F. 1988. The redesign of PROJECT SIMULATION for microcomputer-assisted instruction in psychology and research methodology. In Crookall (1988a).

Noel, R. C., Crookall, D., Wilkenfeld, J. and Schapira, L. 1987. Network gaming: A vehicle for international communication. In Crookall, D., Greenblat, C. S., Coote, A., Klabbers, J. and Watson, D. R. (eds.) *Simulation-Gaming in the Late 1980s*. Oxford & New York: Pergamon.

Wolfe, J. (ed.) 1987. *A Practical Guide to Business Gaming*. (Special issue of *Simulation and Games*, **18**:2.) Beverly Hills, CA & London: Sage Publications.

Some dangers when using interactive events to improve competence

Ken Jones

Freelane author, UK

ABSTRACT: In order to use interactive events successfully to tackle complexity, uncertainty and a conflict in values it is important that the events themseves should not be a muddle of techniques with the participants not being sure what they are supposed to do or who they are supposed to be. If some participants behave like players trying to win a game, while others behave like professionals in a simulation, or actors in an informal drama, then the event can compound complexity, increase uncertainty and distort the clash of values.
 The session included SHOULD OUR STUDENTS BE PROFESSIONALS? – an unpublished simulation involving a controversy between Traditionalists, who believe in instructional methods, and Professionalists, who believe that the students should be treated as professionals by means of interactive events. It provided an opportunity to look at the dangers as well as the advantages of using games and simulations.
KEYWORDS: Interactive events; competence; ambivalents; gaming; simulations; professional behaviour; terminology.
ADDRESS: 4 Ashdown Lodge, 1c Chepstow Villas, London W11 3EE, England.

Introduction

One of the problems of ISAGA is that most Americans tend to label all interactive events as games, the British tend to label them as exercises, and the Dutch often call them simulations if they are in the field of science, and games if they are in the humanities. There is also a general tendency to use the labels interchangeably.

Consequently, any conference of ISAGA is fraught with complexity, uncertainty and a conflict in values which is often hidden beneath the flux of terminology.

A practical consequence of the interchangeable labels occurs in the classroom where many interactive events are of no consistent technique but are ambivalents – a confusion of incompatible behaviours and motives in which some participants are behaving as professionals, others as gamesters, or students, or actors, or authors. Ambivalent events are usually the result of inconsistent signals from the facilitator or from the materials. As with Pavlov's dogs, inconsistent signals tend to produce madness rather than competence.

The main reason for the muddled terminology is that most people categorize interactive events by non-essential characteristics – usually the aims, the materials or the subject matter. But interactive events are events: things happen. Thoughts, feelings and behaviour are their essence. To say "I have designed a game" is really to say "I have designed the materials, rules, instructions and so forth". Complexity and uncertainty can be reduced by categorizing according to what happens – what did the participants actually do, think, feel? Such categorization also makes it easier to use such events to explore values, including personal values and educational values.

The inside view

Educationalists usually see the classroom from the front, with themselves having the responsibility of teaching. It is a habit of mind which makes it difficult to appreciate the situation as seen from the back of the class where the key question is often "What are we supposed to do?" This question becomes particularly important when running interactive events in which the teacher becomes facilitator. It is not just a matter of what to do, but also what is the role of the person who is doing it - what are their powers, duties, responsibilities?

If the activity is a game, then the answer is "You are supposed to be players who try to win." The essential characteristic of games is that they must involve players. It is misleading to try to define games as competitive events with rules. War, business, law and examinations are all competitive events with rules, but they are not games, nor are the participants players. As argued elsewhere (Jones, 1988a) a game is a magic kingdom where the rules themselves require no outside justification. Hitting balls into small holes or moving pieces around a board under restrictive conditions embrace their own ethics and rationale. In chess no one says "Pawns are weak pieces so it is unfair to capture them". In Monopoly a player will not say "It is unfair to knock down houses in order to build hotels." The only ethics are those concerning fair play and sporting conduct, and these particular ethics are implied in the rules of the game.

If the interactive event is a simulation, then the participants are supposed to do whatever is appropriate according to their roles – journalists, bankers, diplomats – including accepting professional real-world ethics. Even if the roles are those of criminals there is the ethic of honour among thieves. Part of professional behaviour is the acceptance of the imaginary consequences which could occur after the event. Unlike games which have no hypothetical after-events, simulations exist hypothetically both before and after. In a game a successful bluff within the rules would require no further justification, but in the debriefing of a business simulation the question could be asked "Did you consider that your bluff was likely to damage the reputation of your company?" The reply "Who cares? I won, didn't I?" is to speak like a player, not like a professional. It is to confuse two incompatible techniques.

In an exercise the role and the attitude of mind is usually that of problem solver. If the event of a treasure hunt the problem solver is hardly likely to say "As a professional treasure hunter I need far more information before I set out on what might be a wild goose chase." To do so would move the event in the direction of a simulation. In an exercise the answer to "What are we supposed to do and who are we supposed to be?" might be "You are problem solvers in an exercise with the job of trying to find a solution".

If the event is informal drama then the participants have a duty to behave like actors, and also authors. Yet acting and authorship are sometimes incompatible in the sense that the time required to invent some plausible fact to sustain the role-play impedes the fluency of speech. Hesitations, stumblings and so forth are commonplace, particularly when another participant has asked a factual question which was unanticipated. Successful authorship involves the participants inventing facts which are apt, or witty, or dramatic, or profound – and this is to step into the field of art. For a detailed examination of such events see Jones (1987) and (1988a).

Ambivalents

Reporting on the running of PRISONER'S DILEMMA as part of a research project, Liebrand (1983) writes:

One of the most significant aspects of this study, however, did not show up in data analysis. It is the extreme seriousness with which the subjects take the problems. Comments such as "If you defect on the rest of us you're going to live with it for the rest of your life" were not at all uncommon. Nor was it unusual for people to wish to leave the experimental building by the back door, to claim that they did not wish to see the "sons of bitches" who double-crossed them, to become extremely angry at other subjects, or to become tearful.

Although these consequences are extreme, the symptoms are similar to well over a dozen other ambivalents (or suspected ambivalents) described in Jones (1988a) which occurred in Britain, the United States, The Netherlands, Italy, Bulgaria and Egypt.

The usual symptoms are:
1. The facilitators are perplexed by what they regard as inappropriate emotions and attitudes, and can offer no adequate explanation.
2. If the facilitators blame anyone then they tend to blame the participants, not themselves, and if the participants are aware that something has gone wrong then they tend to blame each other, not the facilitator.
3. In writing an account of the event the facilitator uses labels more or less interchangeably.
4. The interpersonal damage does not show up in the research data.

Reading between the lines of such events, it seems likely that the facilitators provided an ambivalent briefing, and the participants adopted different attitudes according to their perception of what type of event they supposed it to be.

Those most at risk in the aftermath of an ambivalent event are the people who treated it as a game and tried to win. Thinking in the role of players and gamesters they may have felt that bluffing and deceit are part of harmless gaming fun, and that the most interesting strategies are high risk ones. Almost equally vulnerable to subsequent attack after an ambivalent event are the problem solvers who saw it as an intellectual puzzle with nothing personal implied. Opposing them are the "professionals" who adopted real world ethics, including the personal ethics of honour, of conscientiousness, and duty to the community. Afterwards, in the personal exchanges or in the silent avoidances, the gamesters and the problem solvers usually do not know what has hit them or why. From their point of view it was a game, or an exercise, and they cannot see why their critics are making such a fuss over something so trivial, and the facilitator often shares their point of view.

Even if an ambivalent contains no personal clashes, the behaviour of gamesters is usually less competent than if they had taken the role of professionals – see "Why gamesters die in space" (Jones, 1988b). Ambivalents are doorways to incompetence.

The implicit criticism of facilitators in the above diagnosis is not an attempt to lay the blame solely on their shoulders. After all, designers and publishers of the materials also use labels interchangeably, as do many authors. Another problem is the over-indulgence in the use of the word "game" (player, play, gaming). As argued in Jones (1988a, 1988b), American terminology favours "game" because the educational system in the United States is similar to games – both are highly competitive, the result is more important than the process, and the scoring methods are precise, objective, and used more or less continuously. Thus, in the United States the word "game" has acquired many desirable educational connotations. This coalescing of American education and games is made easier because of the fact that some notable American games include frequent breaks for discussions of strategy and have a high element of planning with teams compiling their secret books of dozens of "plays". To import the word "game" into European culture, and certainly into third world culture, as a label for all interactive events is just asking for trouble.

Classification by participant behaviour results in games becoming a depleted category and ambivalents becoming a large one.

Damage containment

Since interactive events guarantee a high degree of participant autonomy there is no sure method of avoiding ambivalents. Even with the clearest and most consistent briefing some participants can be perverse. However, the easiest and most obvious way to tackle the problem is to discuss it with the participants before the event and to concentrate on categorization by thoughts and behaviour rather than by materials and aims. This will enable the participants to undertake a certain amount of policing themselves within the event, thus isolating and containing any saboteurs.

Another useful hint is to take a close look at the materials to see if they are methodologically consistent. Often they are not. So-called "simulations" can include role cards which require play-acting – "You are an angry politician", "You are a tough manager". Games can be sabotaged if the instructions state that the job of the participants is to imitate real world behaviour. Exercises can become ambivalents if the participants are given functional roles, with some participants treating it as an intellectual problem while others take on professional power.

For example, if instead of saying "In this exercise you are problem solvers with the job of making paper hats" the instructions state "In this event you are production workers with the job of making paper hats" then some groups may assume it is an exercise and not question the basis of the task. However, other groups might treat it as a simulation and decide to form a trade union, or go on strike. The facilitator might say to the simulation group "Oh dear, by striking you have ruined the exercise" or "Well done, by going on strike you have introduced some real-life factors". The first remark could antagonize the simulation group on the grounds that production workers have rights, and the second remark could annoy the exercise groups who might take the view that the strikers had changed the ground rules.

Ambivalents usually have the effect of reducing competence, compounding complexity, increasing uncertainty and distorting any clash of values.

Explorations of issues and values

Interactive events which are not ambivalents can result in a useful exploration of issues and values because (a) in designing the materials it is not difficult to set up a situation which provokes exploration as distinct from "learning the facts" (see Jones, 1985), and (b) the participants have roles – player, problem solver – and can think and behave in a way which does not compromise their real-life positions. This is unlike the traditional classroom discussion where people speak for themselves and are thus vulnerable. A consequence of such vulnerability is that they often seek safe ground, engage in platitudes, and their words can conceal rather than reveal. In interactive events participants can have the duty to state arguments which are not identical with their own personal beliefs and which may be completely contrary to them. However, statements of contrary belief are nothing new and are often valuable: they were formalized by the Catholic Church in the office of Devil's Advocate.

The design of 'Should our Students be Professionals?'

This unpublished simulation by the author is aimed at provoking a discussion of educational issues and values. Since most participants tend to favour the use of interactive events it follows that there is a relatively high degree of devil's advocacy among those who have the roles of the traditionalists.

Some dangers when using interactive events 255

In the first part of this simulation everyone is in one of five groups: school inspectors, traditionalist teachers, professionalist teachers, traditionalist students, professionalist students. Each group has its own confidential memo which mentions such issues as teacher training, examinations, standards. Everyone also has the research report by Barzini and Hilal (fictitious) outlining the classroom differences between the traditionalist and the professionalist approach – the arrangements of the furniture, the use of case studies, simulations, role-play, and so forth.

In the second part of the event a complete re-grouping occurs. Each participant takes on the function of leader of their group which then splits up and a series of parallel top level meetings take place.

Since this was the ISAGA conference, there was the possibility that the participants might include some of those referred to by Drew Mackie (1986) as "British gaming scene's most renowned sceptics who have often been responsible for subverting the intentions of simulations at SAGSET and other conferences." To help fend off possible sabotage the programme notes for the session remarked:

It might be interesting to see if conference-goers behave like:
(a) players trying to win a game
(b) actors engaged in drama
(c) authors inventing stories
(d) professionals involved in a simulation

Thus, the event had an element of referring to itself – a simulation about interactive events which indirectly invited participants to look at educational values, sort out complexities, tackle uncertainties, and polish up their competence in the process. Although unpublished the theme is similar to WE'RE NOT GOING TO USE SIMULATIONS which is a complete simulation published in the appendix to *Simulations in Language Teaching* (Jones, 1982).

Session Review

The session lasted for 2 hours, and the eight participants were Dutch, Danish and Italian. Fears that there might be play-acting or authorship were unfounded; everyone behaved with professional intent.

The groups were told that they had half an hour for private discussion before the joint meeting started, and that they could discuss their policy anywhere. Three groups then went in search of more private surroundings. Unfortunately, the two Professionalist Teachers became so involved in their discussion and creation of ideas that they neglected to look at their watches and arrived for the joint meeting 20 minutes late.

At one of the two joint meetings a rather subtle correction occurred when the inspector summarized the position of the others:

Inspector (Dutch): Now you, the Traditionalist teacher, think that the ..

Traditionalist teacher (Danish): Feel that.

Inspector: Feel that the . . .

The tone was polite, friendly and helpful on all sides. Compare this with the more direct dialogue of 15-year-old secondary school pupils in a deprived area of London:

Professionalist teacher: All right then, what is the first thing you do when you walk into a classroom?

Traditionalist teacher: I tell them to get out their workbooks.

Professionalist teacher: There you are, you tell them to get out their workbooks. . . .

My conclusion is that the ISAGA conference-goers and the secondary school children were doing the same thing – talking about themselves and arranging their environment to suit their own favoured modes of communication.

Before the session began I warned the conference-goers that one of the questions for the debriefing would be their hidden thoughts. However, it turned out to be difficult to get them to reveal any differences between what they had said in public and what they had thought and felt in private. They seemed to have an aversion to the word "I". An intimation of hidden depths occurred on the following day when I was chatting to the only male participant. I learned that he had felt patronized and somewhat put down when the Inspector had welcomed him as a student. Was this, I asked, a gender problem. No, he replied, that did not come into it. He had felt the inspector's special welcome to him as a student was double edged, and that although it was polite and friendly he thought it was indirectly intended to emphasize the superior position of the inspectors and teachers. I then asked why he did not mention this during the debriefing – did he forget, or was he just being polite. He smiled and said maybe it was because he was Dutch. Relating this conversation to a Danish member of the group she exploded slightly and said that students could not expect equality with teachers. This led to an interesting conversation which was later joined by the male ("We've just been talking about you") participant.

Although the above paragraph refers to "the" debriefing, the example illustrates the point that there can be a succession of unofficial debriefings after any simulation, like ripples on a pond. Even the "official" debriefing can be the subject of comment in the peripheral debriefings. These reverberations are part and parcel of the nature of the technique. Thus, "the" debriefing is not the final chapter, and second thoughts recollected in tranquillity (by the facilitator as well as by the participants) are often more useful than the immediate conclusions of the official debriefing.

Rererences

Jones, K. 1982. *Simulations in Language Teaching*. Cambridge: Cambridge University Press.
Jones, K. 1985. *Designing Your Own Simulations*. London: Methuen.

Jones, K. 1987. *Simulations: A Handbook for Teachers and Trainers*. (2nd edition) London: Kogan Page; New York: Nichols Publishing.

Jones, K. 1988a. *Interactive Learning Events: A Guide for Facilitators*. London: Kogan Page; New York: Nichols Publishing.

Jones, K. 1988b. Why gamesters die in space. In Crookall, D., Coote, A., Saunders, D., Klabbers, J.H.G. Cecchini, A. and Delle Piane, A. (eds.) *Simulation-Gaming in Education and Training*. Oxford: Pergamon.

Liebrand, W.B.G. 1983. A classification of social dilemma games. *Simulation & Games*. **14**:2.

Mackie, D. 1986. Simple games for complex situations. In Craig, D. and Martin, A. (eds.) *Gaming and Simulation for Capability*. Loughborough: SAGSET.

ME — THE SLOW LEARNER and some of its implications

Donald Thatcher and June Robinson
Solent Simulations, UK

ABSTRACT: The use of simulation ME-THE SLOW LEARNER to promote empathy with learning difficulty or handicap is described briefly together with a more detailed description of the different stages of the debriefing.

There is a consideration of the Learning System as it is related to the simulation and to the development of learning handicap together with a brief consideration of the System of the Learner.

Aspects of the growing data bank of responses to the simulation are discussed particularly as they relate to the origin and development of learning problems or handicaps. Some of the major learning difficulties identified by past participants in ME-THE SLOW LEARNER are explored.

A brief consideration of the complex way in which learning difficulties or handicaps develop is considered and some theoretical perspectives are identified. Finally, the importance of emotion in learning contexts is explored.

Full details of the simulation are to be found in Thatcher, D. C. and J. Robinson, 1987, *ME-THE SLOW LEARNER-A Manual*. Fareham, Hants.: Solent Simulations.

KEYWORDS: Learning; the Learning System; learning block; learning handicap; learning inhibition; empathy; debriefing.

ADDRESS: Solent Simulations, 80, Miller Drive, Fareham, Hants., PO16 7LL UK.

Simulation/gaming: brain-compatible teaching strategy for the improvement of competence

Michael J. Rockler
Rutgers University, USA

ABSTRACT: This paper demonstrates the relationship between research on the brain and the process of simulation/gaming as a teaching strategy. It provides advocates and users of the process with evidence that simulation/gaming facilities learning because it is a brain-compatible teaching strategy. CULTURE CONTACT, PARTICIPATIVE DECISION MAKING, and BUILDING THE FUTURE'S COMMUNITY are examples of games that can be used to demonstrate this relationship.

KEYWORDS: Brain compatible learning; teaching strategy; brain research; learning theory; proster theory.

ADDRESS: Rutgers University, 311 North Fifth Street, Camden, NJ 08102, USA.

The "JOG Your Right Brain Exercise" at ISAGA 88

Peter A. Raynolds[1], Gennie H. Raynolds[2]
[1]*Northern Arizona University, USA,* [2]*Projective Awareness Research Center, USA*

ABSTRACT: JOG Your Right Brain (JOG) demonstrates the projective differential (PD) procedure, which may be utilized in various educational, consulting, developmental and research efforts for individuals, groups and organizations. PD responses are rapidly made choices from a set of pairings of abstract visual images created by the presenters. The choices are in response to the question "which image from each pairing seems somehow to be more like ----", with a topic filling the blank. JOG allows participants to explore and share their intuitive reactions and feelings toward four topics which are tailor-made for the session. Topics always include the participants themselves and may also include organizations, plans, projects - even games and simulations. Individual results are scored and interpreted during the session on a two-part form which permits both the administrator and participants to retain copies. The scores provide nonverbal measures of attitude and identification toward the topics as well as of self-esteem. A PD study conducted in Japan and America showed the measures to have intercultural validity. The JOG exercise provides participants with new insights into their perceptions of the specific topics covered and also develops new abilities in using intuitive processes as well as new appreciations of how their thoughts and feelings blend together. Results from JOG at ISAGA 88 are discussed.
KEYWORDS: Projective; nonverbal; right-brain; intuitive; visual; attitude; identification; self-esteem; cross-cultural; congruence.
ADDRESS: College of Business Administration, Box 15066, Northern Arizona University, Flagstaff AZ 86011, USA.

Introduction

The purpose of the JOG YOUR RIGHT BRAIN (JOG) workshop was to provide a demonstration of the nonverbal projective differential (PD) procedure which is contained within the JOG exercise. The intention was to illustrate some of the ways the PD is relevant to concerns of the International Simulation and Gaming Association (ISAGA) in general, and also to the theme of this year's conference held in Utrecht, The Netherlands, August, 1988.

ISAGA is concerned with the use of simulations and games in management education, research and practice. The theme of ISAGA 88 focused upon our rapidly changing world and the need for an "improvement of competence in dealing with complexity, uncertainty and value conflicts". Simulations and games can address management issues such as these in a manner which embodies some of the ambiguities actually faced by practicing managers. This fact makes simulations and games useful in managerial education in ways that more traditional, non-experiential and strictly cognitive approaches cannot address. That is, in simulations and games, there is a greater resemblance between learning and professional practice contexts. This qualitatively different approach to what is learned translates into "better" learning.

Indeed, complexity, uncertainty and value conflicts in actual practice can reach levels requiring different strategies in managerial behaviour than those usually taught in management courses. When adequate information is unavailable for rational decision making, and when organizational environments are no longer relatively homogeneous, stable and placid, then the logical, analytical and deductive skills that we emphasize in training managers (in the United States, at least) are insufficient. It is futile for management to search for a finite set of relatively simple causal linkages between the variables they control and their desired outcomes. It is impossible to maximize *or even optimize* allocations of resources, because conditions are worse than uncertain, they are turbulent (see, e.g. Emery & Trist, 1965; McWhinney, 1968; Thompson, 1970; Raynolds, 1972; Coffey *et al.*, 1975). As should be apparent, the conduct of intercultural management amplifies these problems of complexity, uncertainty and value conflicts.

Under dynamic, disturbed and even chaotic environmental circumstances, which do seem to be increasingly present nowadays, competent managers are ones who not only have high tolerances for ambiguity, but also, it has been recently suggested, have a high capacity for the effective use of intuitive perceptions (see, e.g., Fineberg & Levenstein, 1982; Mintzberg, 1976; Agor, 1986, 1984; Taggert & Robey, 1981). To be utilized effectively in organizational settings, such seemingly nonrational perceptions require that managers achieve demanding levels of self-knowledge (including their value-orientations), combined with interpersonal openness and congruence between thoughts, feelings and verbal expression. They make keen discernments between relevant insights and mere wishful (or fearful) thinking. As participants in managerial decision making, they qualify themselves as reliable, rather than fickle, reporters and explorers of their own and others' hunches and "gut feelings" in addition to being expert analysts. They have developed a sort of phenomenological "expertise". Purely "rational" conclusions, while necessary, are insufficient in and of themselves for effective management in highly unstable organizational environments.

In our view, important ways of enhancing competence include an increased emphasis upon developing intuitive, interpersonal and communications skills among individual managers, and an improved integration of qualitative and quantitative considerations in collective planning and decision making. Alone, or in conjunction with other strategic planning, organizational development and managerial training progamme, the PD procedure and the JOG exercise provide flexible tools which can be used for advancing these aims. This paper begins with some background on the PD itself and then provides an overview of the JOG exercise. The final section is a summary of the JOG demonstration at ISAGA 88, as it relates to the conference theme (and, also, to thinking about the planning of future ISAGA meetings).

PD Background

The JOG YOUR RIGHT BRAIN exercise is built upon PD responses which comprise its unique centrepiece. A PD response is similar to a semantic differential response (Osgood et al., 1957), except subjects choose which of two abstract visual images is more like a topic, rather than which of two adjectives is more like it. A topic is defined as the object (usually denoted by a word or phrase) being rated on a semantic or PD instrument. The PD task requires subjects to make a simple forced choice rather than indicate where the topic falls on a continuum, such as the semantic differential's seven interval scale. A PD item consists of a pairing of abstract visual images projected onto a screen. Since the PD choice is nonverbal and is a selection from abstract visual alternatives, we consider it to reflect what Betty Edwards (1979) refers to as an "R-mode" mental process. On the other hand, the semantic differential, being verbal, evokes an "L-mode" response. A simplified PD exercise consisting of one item and not involving 35 mm slides, TAP YOUR RIGHT BRAIN, can be found in Raynolds and Raynolds (1986). It provides an illustration of a PD pairing.

Research has shown that most respondents readily make the seemingly nonrational PD choices. Furthermore, it is common for 60-90% of subjects to make the same choices. This is the PD response phenomenon (McInnis, 1981; Raynolds, 1970; Raynolds et al., 1981; Sakamoto, 1980).

The PD response is a pictorial, nonverbal, analogical and holistic choice occurring in less than one second. Consequently, it avoids some of the effects of verbal response biases due to respondents' aware or unaware experiences of pressure toward making socially desirable, cognitively consistent and/or self-serving responses. Binary representations of PD choices can be analysed in ways similar to semantic differential data. Reliable and valid general purpose nonverbal attitude scales have been constructed producing coefficient alpha reliabilities of .7 to .8 from 4 to 12 PD items. Such scales have been used to predict voting intentions and consumer brand choices in, as yet, unpublished studies. PD ratings of the topics GOOD and STRONG by American and Japanese college students were highly correlated, indicating promise for the intercultural validity of PD measures. A 12 item PDF stimulus set is being employed in measuring emotional reactions to TV ads along 18 dimensions. When verbal and nonverbal PD data are combined, predictions of behavioural intentions are significantly improved. But, data from the nonverbal and verbal sources sometimes disagree. This suggests respondents are undecided, highly ego-involved, under internal stress and/or in conflict with respect to the topics (Cummings, 1971; D'Antoni, 1973; Raynolds et al., 1981; Raynolds et al., 1983; Sakamoto, 1980; Raynolds et al. 1988).

Overview of the JOG Exercise

The exercise was developed to introduce people to the PD response phenomenon, and to do so in a manner that would be fun. We have found most persons must experience the PD to understand it. Just describing it doesn't work. Over the past several years, we have also found the JOG exercise to have a place in the classroom, in organizational decision making and in organizational development, while also providing useful research data!

During a JOG session, participants are taken on an experiential journey through some of the lingages between their nonverbal and verbal perceptions of the topics employed. This "jogging" highlights the two channel nature of all complex mental functioning and constitutes a major learning that participants can take home. The procedure tends to trigger creative and intuitive processes which often take some time, perhaps several days, to reach culmination. Additionally, JOG provides new insights into the topics themselves. The exercise consists of several steps which are outlined in Table 1 and discussed, in detail, in Raynolds and Raynolds, 1989.

JOG employs four topics, two of which are "anchors", and two of which are "working topics" for a particular session. The anchors are usually MYSELF (THE WAY I REALLY AM) and THE IMAGE I PREFER (OR LIKE BETTER). These anchors allow us to compute scores reflecting attitudes toward and identification with the two working topics, as well as a nonverbal measure of self-esteem. JOG employs only five PD images, taken from Raynolds and Raynolds (1982), and presented in all ten of their possible paired combinations. Each image appears in four of the ten pairings, and may be chosen as being like one of the four topics from zero to four times. Participants can thus identify the single image that is most like each of the four topics. These images are the ones which are "somehow" most salient, relevant, important – most connected with the topics for individual participants. When a topic involves shared experiences among participants, there often is a group consensus.

Part of the JOG exercise is set aside for participants to jot down a name and/or brief description of each of the five images. These verbal associations to the single images often connect in unusual, but important, ways with the topics. A flood of new insights and energy is released. A sort of "alogical" amplification occurs when participants are asked to connect the visual image epitomizing a topic (i.e. the image chosen most often as being like the topic) with ideas and reasons as to why this was the case. The evocativeness of the images seems to open up new ways of expressing one's thinking and feeling about the topics. It is fruitful for members of an organization, group or

TABLE 1 Outline of JOG Your Right Brain Exercise Steps

Step 1 The projective differential technique was administered on four topics: I - MY PERSONAL PROJECT, II - ISAGA 88, III - MYSELF (THE WAY I REALLY AM), IV - THE IMAGE I PREFER (OR LIKE BETTER). All ten pairings of five PD images were shown in the same order for each of the four topics.

Step 2 The five images were shown for about a minute each, one at a time. Participants named and/or described each one, as though finding it displayed as a piece of art on a wall somewhere.

Step 3 Scoring: First, participants counted (tallied) the number of times each of the five images was chosen (in Step $\neq 1$) for each of the four topics. Next they counted the number of times (out of the ten slides in Step $\neq 1$) that the same image was chosen for topics I & II (Same Choice Score) and repeated the procedure for each of the other combinations of topics ("I - III", "I - IV" . . .).

Step 4 Interpretation: The *First Stage* consisted of tentatively interpreting attitude, identification and self-esteem measures from the Same Choice Scores. These were tentative hypotheses for respondents to "try out":

Attitudes were defined as the participants' orientations toward topic I and II in terms of favourableness or unfavourableness, posititiveness or negativeness. Attitude:

Toward Topic I = Same Choice Score (IV - I)
Toward Topic II = Same Choice Score (IV - II)

Identification was defined as the degree to which topics I and II were viewed as being similar or dissimilar to oneself. Identification:

With Topic I = Same Choice Score (III - I)
With Topic II = Same Choice Score (III - II)

Self-Esteem was defined as the degree to which participants felt good about themselves at the time of exercise (moods will affect results):

Self-Esteem Score = Same Choice Score (III - IV)

A score of "5" (50%) was neutral. Scores over "5" were positive and scores under "5" were negative. Scores closer to "0" or to "10" indicated greater extremity.

The *Second Stage* consisted of testing these tentative hypotheses by finding the image(s) which epitomized Topics I, II and III (i.e. were chosen most often in Step $\neq 1$ and tallied in Step $\neq 3$) and then looking at the names given to these images (in Step $\neq 2$).

Step 5 Participants completed the exercise by writing verbal associations to two target words or phrases: (1) the TOPIC being studied further and (2) the name given to the plates(s) epitomizing it. Flashes of insight during the associations were registered as "AHAs", and conclusions for action were also noted on the form. This was a "private" portion of the exercise, with no data being collected.

family to compare and discuss the names given to images epitomizing their units.

ISAGA 88 Session Review

The PD demonstration was held on Wednesday afternoon in the third of four workshop sessions during the conference. The following four JOG topics were used: MY PERSONAL PROJECT, ISAGA 88, MYSELF (THE WAY I REALLY AM), and THE IMAGE I PREFER OR LIKE BETTER.

TABLE 2 JOG Scores from ISAGA 88 (Percent)*

	PD (Nonverbal) R-Mode	Verbal L-Mode	Disparity
Attitudes toward			
PERSONAL PROJECT	45%	75%	30%
ISAGA 88	55	60	5
Self-esteem	73	70	3
Identification with			
PERSONAL PROJECT	50	NA**	NA
ISAGA 88	60	NA	NA

Note: Scores over 50% are positive, under 50% are negative.
**No L-Mode data were obtained for indentification scores.

Between steps 1 and 2, a verbal (L-mode) attitude score was additionally obtained toward the first two topics, as well as a rating of self-esteem. These verbal scores were on a 0 to 10 scale, with "0" meaning completely negative, "5" neutral and "10" completely positive. This scale allowed differences, which we refer to as "disparities", to be directly computed between verbal (L-mode) and nonverbal (R-mode) scores.

Of the 16 participants, 12 returned copies of their response forms. Tallies of these responses yielded the mean results, shown in Table 2 as percentages of the maximum possible scores. With respect to MY PERSONAL PROJECT, the nonverbal mean attitude was slightly negative at 45%. This is in contrast with the positive verbal attitude of 75%. The 30% disparity between the two scores suggests participants were experiencing difficulties with their projects. But these difficulties were not reflected in the verbally expressed attitudes. Such disparities have been associated in prior PD research with excessive ego-involvement, stress or conflict with respect to a topic. It should be noted that, although neutral, the mean nonverbal identification score of 50% toward MY PERSONAL PROJECT was higher than the attitude score. To us, this suggests participants were committed to expending additional efforts in order to overcome the obstacles they were experiencing (a "healthy" sign).

Nonverbal (R-mode) and verbal (L-mode) SELF-ESTEEM scores were both positive, 73% and 70%, respectively. The nonverbal score was, in fact, the highest we have seen at any professional conference, perhaps reflecting something about the stature of those who attended this international conference in general, and those who attended the JOG session in particular. Mean disparity (3%) was low, suggesting an absence of over-involvement, stress or conflict. Of course, there is no way to judge this sample's representativeness of all persons attending ISAGA 88.

Regarding ISAGA 88, the nonverbal and verbal attitude scores were slightly positive at 55% and 60% respectively. There was also a more positive nonverbal mean identification score of 60%. These are consistent with results

TABLE 3 JOG Image Selection Tallies (Percent of possible choices)

PD Image	ISAGA 88	THE IMAGE I PREFER (OR LIKE BETTER)	Difference
"a"	45.8	50.0	-4.2
"b"	56.3	68.8	-12.5
"c"	50.0	43.7	6.3
"d"	58.3	39.6	18.7
"e"	39.6	47.9	-8.3

TABLE 4 Names and descriptions of Images "d" and "b"

Image "d" (too much in ISAGA 88)	Image "b" (would have liked more in ISAGA 88)
Control, static concentration; the watching angel; antennae; bear hug; extracted tooth; brain tumour; masked figure in the fog; swimmer with snorkel in Hanauma Bay (coral); bat, cold; bats in hell; greed; he-man, transformer.	Warmth, freedom, play; sheep's head from front; the ram; animal watching you; bee in a snapdragon; fox; running attacking bull, the steer . . . attacks matador, the bullfight; animal head, raging bull; red world; brown study bird; Christ in Spordobo (?); in a way, compassion.

from several other professional conferences. Again, the mean attitude disparity (5%) was small, between nonverbal and verbal results.

We now turn to another way of utilizing the JOG exercise. It provides an exploration of the JOG results to gain new insights into one of the working topics, in this case ISAGA 88. These are preliminary ideas emerging from the data, and could potentially be of assistance in planning and decision making. We begin by finding which images epitomized ISAGA 88 and the topic THE IMAGE I PREFER (OR LIKE BETTER). Then, the names given to these images are studied in a search for possible underlying themes or issues in the minds of participants. These notions can become hypotheses later tested by less projective, more direct methods.

An examination of the number of times each of the five JOG images was chosen as being like ISAGA 88 discloses that image "d" was most like ISAGA 88 (chosen in 58.3% of its appearances). But, this same image, "d" was also chosen the *least* number of times as THE IMAGE I PREFER (OR LIKE BETTER), chosen in only 39.6% of the possible selections. See Table 3 for these results. Given the difference of 18.7%, we conclude there was, somehow, a bit "too much" of image "d" in ISAGA 88.

On the other hand,, image "b" was most chosen as THE IMAGE I PREFER (OR LIKE BETTER), 68.8%, while it was *less* often chosen as being somehow like ISAGA 88, 56.3%, for a difference of 12.5%. Thus, we conclude that participants would possible have liked more of the climate suggested by image "b" in ISAGA 88.

All of the names and descriptions of these two PD images are listed in Table 4. They suggested questions to us about the perceptions and feelings of

ISAGA 88 participants: For example, was the conference possibly seen and felt to be a bit too controlled, static, negative and/or self-centred (rather than playful, supportive, fertile, active and compassionately confrontative)? What might be done to encourage more warmth and playfulness to increase the rapport among conference attendees coming together, many for the first time, from widely different cultures? These are questions we offer to the programme committee for consideration in planning future ISAGA conferences.

Reactions to the session: Although there was a shortage of time for full discussion during the JOG session, later conversations with attendees disclosed the results to be interesting, even surprising. New insights were gained into nonverbal (intuitive) processes as well as into the topics, especially the personal projects. Some expressed dismay at first and then appreciation for seeing how they reacted to the high levels of ambiguity and uncertainty embodied in the exercise. A post-conference JOG session was held for several persons who were unable to attend at the scheduled time. The JOG data from that session are not included here, since ISAGA 88 was not a topic. Finally, a number showed interest in using JOG in their classroom, research or consulting activities.

To summarize the results of the session in terms of the conference theme of "improving competence" participants seemed to be competent in dealing with ISAGA 88 and self-concepts, but not in dealing with their PERSONAL PROJECTS. This was evidenced by the low disparity scores on the former topics but high disparity scores of the latter. We believe that bringing high disparities to the attention of participants serves as an important first step in "improving competence", because it provides an impetus to examine and explore perceptions and feelings which are at odds with verbal communication. As mentioned earlier in this paper, organizational effectiveness can be seriously hampered if there is a lack of congruence between thoughts, feelings and expressions. The JOG exercise addressed this issue. Additionally, the exercise provided participants with new insights into their own intuitive processes and methods of dealing with ambiguity and uncertainty.

References

Agor, W. H. 1986. *The logic of intuitive decision making: a research-based approach for top management*. New York: Qurom Books.
Agor, W. H. 1984. *Intuitive management: integrating left and right brain management skills*. Englewood Cliffs, NJ.: Prentice-Hall.
Coffey, R. E., Athos, A. G. and Raynolds, P. A. 1975. *Behavior in organizations: a multidimensional view*. Englewood Cliffs, NJ.: Prentice-Hall.
Cummings, T. G. 1971. A methodology for reconstructing and studying social systems linkage processes. (Doctoral dissertation University of California, LA). *Dissertation Abstracts International* 31.
D'Antoni, J. S. 1973. Content-oriented and process-oriented value systems: an examination of the valueing mechanism as it relates to measures of congruence and cognitive style. (Doctoral dissertation, University of Southern California). *Dissertation Abstracts International* 33:8.

Edwards, B. 1979. *Drawing on the right side of the brain*. Los Angeles: J. B. Tarcher.
Emery, F. E. and Trist, E. 1965. The causal texture of organizational environments. *Human Relations* **18**:1.
Feinberg, M. R. and Levenstein, A. 1982. How do you known when to rely on your intuition? *Wall street journal*, June 21, p. 16.
Hai, D. M. (ed.) 1986. *Organizational behavior: experiences and cases*. St. Paul: West Publishing Co.
McInnis, N. 1981 Rorschach revised: new window to the unconscious? *Brain/mind bulletin* **6**:12.
McWhinney, W. H. 1968. Organizational form, decision modalities and the environment. *Human Relations* **21**:3.
Mintzberg, H. 1976. Planning on the left side and managing on the right side. *Harvard Business Review* **54**:4.
Osgood, C. E., Suci, G. J. and Tannenbaum, P. H. 1957. *The measurement of meaning*. Urbana, Ill.: The University of Illinois Press.
Raynolds, P. A. 1972. Developing managerial capabilities for coping with turbulent environments. *Working Paper No. 13*. Research Institute for Business and Economics, Graduate School of Business Administration, University of Southern California.
Raynolds, P. A. 1970. The projective-differential: a general purpose inkblot technique for studying denotable objects. (Doctoral dissertation University of California, LA). *Dissertation Abstracts International* **31**.
Raynolds, P. A. and Raynolds G. H. 1989. Jog your right brain: fun in the classroom (and research too!), *Organizational Behavior Teaching Review* **13** (in press).
Raynolds, P. A. and Raynolds, G. H. 1986. Tap your right brain: a multipurpose classroom exercise. *In* Hai (1986).
Raynolds, P. A. and Raynolds, G. H. 1982. *Projective differential images*. Flagstaff, AZ: Projective Awareness Center.
Raynolds, P. A., Sakamoto, S. and Raynolds, G. H. 1981. Consistent responses by groups of subjects to projective differential items. *Perceptual and motor skills* **53**.
Raynolds, P. A., Sakamoto, S. and Raynolds, G. H. 1988. Consistent responses to projective differential items by American and Japanese college students. *Perceptual and motor skills* **66**.
Raynolds, P. A., Saxe, R. and Sakamoto, S. 1983. Toward projective differential scales. *Working paper series*. 83-1. Flagstaff, AZ.: Northern Arizona University.
Sakamoto, S. 1980. Contingency severity and individual performance in a probabilistic game setting. *Human Relations* **33**:10.
Taggert, W. and Robey, D. 1981. Minds and managers. *The academy of management review* **6**:2.
Thompson, J. D. 1967. *Organizations in action*. New York: McGraw-Hill.

Thurston-Parkin – a case study in team communication

Alan L. Cudworth
Trent Polytechnic, Nottingham, England

ABSTRACT: THURSTON-PARKIN, a communication simulation/game, was devised to enable students to develop effective communication skills. The skills developed in this exercize include negotiating verbally, drawing, writing and presentational skills.

The role-playing simulation/game follows the planning and development of the village of Thurston-Parkin situated somewhere in England and provides a model which could be adapted to suit other disciplines.

Participants undertake the role of landowners, developers, architects, planners and council officials, and the purchase, planning and development of the sites in the village can be simulated.

Although participants play individual roles, the overall success of each team can be seen and discussed. The debriefing, an important aspect of the learning process, explored both the individual and the team contribution to the development process with encouragement given to participants to comment and evaluate their own and group performances. It also explored the emotions and motivations of the participants whilst playing the game and evaluated the effect of their decisions on the environment, considering what kind of environment had, in fact, been created.

The debriefing concluded with an evaluation of the simulation/game as a learning experience and the value of the model as a framework for developing other innovations.

KEYWORDS: Simulation/game; role-play; environmental studies; town planning.
ADDRESS: Department of Surveying, Trent Polytechnic, Burton Street, Nottingham, NG1 4BU, England.

Student learning invoked by simulations embedded in a learning environment

Fred de Vries

Centre for Educational Technology, Open University of the Netherlands

ABSTRACT: Simulation programmes running on stand-alone microcomputers can be used for educational purposes. A simulation will have different configurations' depending on the educational purpose for which it is designed. The advantages of simulation for learning activities are discussed. From a designer's point of view that are several aspects to take into account when developing a simulation programme for education, such as: what kind of learning task is involved? To what extent is reality simplified in the simulation? What level of fidelity is applied? What is the nature of the system's response? What kind of user interface is applied? To what extent are help facilities offered? What is the place of the computer programme in the whole learning environment. In the workshop five educational simulation programmes were analysed according to the above questions.
KEYWORDS: Education; simulations; computer-assisted instruction; design.
ADDRESS: Open University of the Netherlands, Centre for Educational Technology, P.O. Box 2960, 6401 DL Heerlen, The Netherlands.

Introduction

Sometimes simulations play a part in courses developed at the Open University of the Netherlands (Ou). The premises underlying open distance education (freedom of time, place and pace of study) make it necessary to find alternatives for situations in which traditional universities arrange group meetings and practical sessions. This is where simulations come in. In some Ou courses simulations play an important role. For several reasons an educational simulation at the Ou usually takes the form of a computer programme which runs on a personal computer. At the Ou:
- the computer-programme allows students more freedom to choose their own study hours.
- the computer-programme makes it possible for a student to work on his or her own, especially in simulations where otherwise fellow students play a part.

Following the situation at the Dutch Ou, in this workshop only educational simulations that use a stand-alone microcomputer are discussed. As a consequence, only part of the domain of simulations is covered. In the workshop the different configurations of an educational simulation are introduced.

The use of Computer Simulation Programmes in Education

Simulations are used in different contexts, for instance management, health care and natural sciences. Simulations are typically presented according to the goals and objectives of the course they are a part of (Miller, 1984). Several types of simulations can be distinguished. Several definitions can be given all stressing various aspects of simulations. What you expect a simulation to be depends on the way you use it.

For educational purposes the primary characteristics of simulation programmes are (Gredler, 1986):
a. The programme is a realistic setting in which the student is presented with a problem. (The level of realism can be varied.)
b. The student makes inquiries, makes decisions, and takes action.
c. The student receives information about the ways in which the situation evolves and changes in response to his or her actions.

Thus there are innumerable possibilities for designing an educational simulation.

A fairly common situation is one in which a mathematical model of a system you want to simulate (often developed for research purposes) is available. To have students gain insight into the relations of several components of the system, you let them work with a simulation. For this purpose the model is put into a computer where the students can change input parameters and observe the effects of these changes on the output. Students are hopefully learning about the model by manipulating it and seeing what happens.

There are also simulations in which there is no mathematical representation of reality. When using such a programme you follow a path with several branches. In a simulated situation the student may only intervene by answering specific questions asked by the programme. The student is informed as to whether the answer is right or wrong, and he will follow a predetermined path in the programme. Gredler (1986) argues that such simulations tend to act as drill and practice exercises.

When using a CAI environment it is easy to add an additional feature such as a tutorial, to help students gain knowledge of the basic concepts used in the simulation part. Otherwise you can help students to learn about the model by guiding them through the simulation under study.

As you can see there are several types of simulation that can be used. Whether the use of a chosen simulation programme works well, depends on at least one aspect: Does the simulation succeed in the teaching role you have reserved for it? If not, there will be a gap in the relation with the other learning materials used in a course. Alessi and Trollip (1985) distinguish four phases of teaching with CAI. They state that a simulation may be used for any of the four phases separately or for any combination of them. The phases are:
a. presenting the student with information;

b. guiding the student in acquiring the information or skills
c. providing practice to enhance retention and fluency; and
d. assessing learning.

Choices must be made on the phases you want to take care of in the computer's simulation. The rest should be taken care of outside the programme. For example, if you use a simulation that only contains a model, in which the students can change parameters, it is likely to be applicable to provide practice (c.). The presenting of the basic concepts (a.) and the guiding of the student can be taken care of by other learning resources such as written materials and the teacher.

As you can see there is no one accepted view on educational use of simulations with microcomputers. I do not plan to discuss what the precise definition of an educational simulation should be. It is clear that a simulation has some characteristics that make it useful in education. Worth mentioning are the following:

Exploration Compared to practical work in a laboratory the consequences of the student's actions are not really serious in a simulation: nothing is spoiled or broken. In a simulation students will explore more of the available options.

Transfer of learning An educational simulation gives some practice in the subject. So one should expect that what is learned in a simulation usually transfers well to the real situation (Alessi & Trollip, 1985).

Efficiency A simulation compared with the system it represents can be more efficient in different ways. For example, a laboratory procedure of a chemistry course can in its simulated version take less time by compressing the time scale. The efficiency can also be enhanced by eliminating actions not really necessary for the learning process.

Consistency Simulations make it possible to present the same problem with controllable characteristics for all students. Compared with real life experiments, simulations usually work out exactly as you planned.

Aspects to take into account in the development phase

From an educational point of view, the following topics have to be taken into account when developing an educational computer simulation.

Learning task

In general one has the idea that simulations are perfect to induce discovery learning. In the simplified world of a simulation, the student solves problems, learns procedures, comes to understand the characteristics of phenomena and how to control them, or learns what actions to take in different situations (Alessi & Trollip, 1985).

Whether discovery learning in a simulation is really enhanced, depends on the task the student is given and the learning environment.
- If given too much to discover, the student will not be able to integrate it systematically in his/her current knowledge. The student must have a basic knowledge of the variables in the simulated system to be able to understand the proceses.
- Working with a simulation package in itself does not induce planned learning. It is the environment (task, feedback, explanations, etc.) that set the learning task.

The educational use of a computer simulation programme can be divided in four categories (adapted from Min, 1987):
- free discovery learning Students are free to do their own experiments with the simulation programme. They think up their own problems and try to solve them.
- task oriented learning The student has several tasks to perform, which can consist of exercises on paper or can be integrated in the programme.
- coached learning While working on the simulation, the programme can generate an assignment appropriate to the learning history of the student.
- problem solving Here cases are used. In a case, the system reacts in a particular way. The student has to analyse the situation and try to intervene in the system to allow it to return to a desired state.

Simplification

When developing a computerized simulation, there are two reasons to simplify the system. The first consideration is that you are not able to describe or programme it more precisely. The second is that you might simplify it for instructional reasons. The simulation can, in its ultimate form, be too complicated to serve learning tasks. In a course you can increase the complexity of the simulation, by gradually adding new theoretical concepts in the course of the simulation. Simplification can also be seen as a disadvantage, as Winer and Vazquez-Abad (1981) do. They state that students should be able to "discover" the system's behaviour by exploration and manipulation. Educational simulations are either oversimplified descriptions of complex situations or provide only restricted manipulation. Whether and how much you simplify depends on your goals for the simulation within your course.

Fidelity

When a simulation is simplified you also decrease the fidelity of a simulation. Fidelity concerns the level of realism, both on the visualization and the time frame of a simulation. Simplifying does not necessarily mean the effectiveness of a simulation also decreases. Trollip (1987) argues that fidelity should be as low as possible. Only the aspects absolutely necessary to maintain the relation with the modelled system (real world) should be in the simulation. On the other hand there are applications where a more realistic simulation is useful. When you want to enhance transfer of learning to the real environment in which skills should be practised, then it is appropriate to use a simulation programme with high fidelity. In general the level of fidelity is a function of the learning objectives and the practical possibilities.

Reactions of the system

In a simulation you want the student to experience the influence of the interventions with the simulated environment and if necessary correct their behaviour. In education, complete feedback is then desirable (Joyce and Weil, 1984). Two forms of feedback can be distinguished in simulation programmes: natural and artificial feedback (Alessi & Trollip, 1985).

Natural feedback in a simulation is similar to what would occur in reality. Artificial feedback is usually corrective. For example when in a chemistry experiment the heat is too high, the experiment fails. Natural feedback could result in presenting wrong results. Artificial feedback could be the message on the screen such as: The temperature is too high, try again.

A computer programme offers the possibility of giving immediate feedback. When giving natural feedback, it is appropriate to give it at a time that is concurrent with reality. This can be immediately after the action of the student, but sometimes it is more realistic to delay the natural feedback. For example in a medical simulation the effects of giving a medicine to a patient are shown after some time. The same distinction can be made for artificial feedback. The actions of the student can immediately be corrected or an evaluative feedback after an exercise can be given. Whether it is appropriate to do so depends on the level on which you want students to explore their own strategies.

The balance between artificial and natural feedback, delayed or not should be decided on the characteristics of the course and the students. In a simulation with merely natural feedback, one should be careful of the distracting effects that may occur. Natural feedback can on the other hand be useful to add elements of gaming. A lot of artificial feedback can make students feel uncomfortable. They feel a strict eye is being kept on them. It may be more appropriate to give detailed evaluative feedback on the activities of the student afterr the completion of an exercise.

Another more elegant way of giving feedback is to make a shift when the student is working on the programme. First the student must get to know the ins and outs of the simulation. While doing that he or she is led by artificial direct feedback. After some time the feedback can change to a delayed natural one, so as to confront the student with the consequences of his or her actions.

User-interface

On user-interfaces of educational computer programmes one can hold a whole workshop. In the context of this workshop a few things must be kept in mind. Make the user interface as transparent and easy to operate as possible. Do not let the student lose too much time with learning how to operate the interface. Some options are especially important in a simulation. When students have made a mistake in their own opinion, it should be possible to end and restart the simulation.

Help facilities

Help can be offered on both the operation of the computer programme and on its contents, the simulation. If a more complicated user interface cannot be avoided, there should be help available on the operation of the programme. It is a good habit to make this information always easily accessible. A special function key or a reserved field on the screen when using a mouse is a good possibility.

When variables can be changed during the simulation, additional facts on the parameters should also be easily accessible, so the student knows what he/she is doing. A more general solution is to have background information available on the programme itself or via reference to other learning materials in the course. This may be information which should be known by the students before starting with the simulation. The background information should also be easily accessible.

Learning environment

So far I have discussed some aspects of simulations, which put the computer programme in the centre. Of course it is only a part of a course. The integration of the programme with accompanying learning materials such as basic study materials, in which the learning task is set and the concepts necessary to work with the simulation programme are taught, is important. Booklets with exercises structure the study activities using the computer simulation and a teacher can give assistance. And last but not least, it can be useful to let students work in pairs on a simulation. They learn from each other when discussing the actions to be taken.

Session Review

The workshop participants worked in small groups with at least two of the five available programmes. This was useful to get a good impression of the major differences between the programmes. Short descriptions of the programmes are now given.

HUMAN is a simulation based on a mathematical physiological model of a human being. The programme offers the possibility of intervening in the model by directly changing variables of the model. The educational context lies outside the computer programme in an accompanying exercise-book. The values of the different variables appear in digits on the computer screen.

A second medical programme is CHIPS, which stands for Computerized Highly Interactive Patient Simulation. One can ask questions to a patient and do some investigations. The programme gives textual feedback on this. The student has to formulate his own diagnosis, on which the programme reacts.

WASMEER is a simulation embedded in a series of tutorials of a ground water pollution problem. While working through the tutorials, more and more complications are introduced and the handling of the simulation shifts gradually simple demonstration into free usage of the full model. The results of the calculations on the mathematical model are presented in a map of the area. The groundwater flow and the resulting pollution are shown graphically.

POPDYN offers several models of population dynamics where in contrast with the programmes already mentioned the mathematical models themselves are under study. After changing the variables the programme gives a graphical representation of the results.

The fifth programme DESTILLATIE is an Interactive Videoprogramme on destillation. Experienced process operators lack theoretical knowledge on the destillation process. To a certain extent the programme allows decision making. The results are shown in a realistic way by video. When making a mistake, the student can immediately refresh his theoretical knowledge.

The five programmes were deliberately chosen for their differences in the aspects mentioned in the previous paragraph: learning activities, level of simplification and fidelity, reactions of the system, user-interface, help facilities and learning environment. In the group of participants these differences where noticed as features of educational simulation programmes, serving different educational goals. Of course everyone has their preferences. But in general one can say that depending on the role a simulation programme is given in an educational setting, its sophistication is important. For example, using HUMAN in a classroom situation where a tutor is present and an exercise book is given, the programme can be an effective educational tool. A programme that is meant to be used individually such as WASMEER, must have a structure that guides the student and a totally transparent user interface.

In the group discussion we did not get complete cut-and-dried answers to all the questions raised. In general the conclusion is that a simulation can provide a unique learning experience. In what form the simulation should be offered best, depends on the educational goals and the educational environment.

References

Alessi, S. M. and Trollip, S. R. 1985. *Computer-based instruction: methods and development*: Englewood Cliffs, NJ: Prentice Hall.
Joyce, B. C. and Weil, M. 1984. *Strategieën voor onderwijzen: theorie in praktijk* Apeldaorm: Van Walraven.
Gredler, M. B. 1986. A Taxonomy of Computer Simulations. *Educational Technology*, 26:4.
Miller, M. D. 1984. The use of simulations in training programs: A review. *Educational Technology*, 24:11.
Min, R. M. 1987. *Computersimulatie als leermiddel : een inleiding in methoden & technieken* Schoonhoren: Academic Service.
Trollip, R. 1987. *Workshop on Computer-Based Simulations* EARDHE-workshop, 23 April 1987, Utrecht, The Netherlands.
Winer, L. R. and Vazquez-Abad, J. 1981. Towards a theoretical framework for educational simulations *Simulations/Games for Learning*, 11:3.

SIMULATION REFERENCES

CHIPS. Gerritsma, J.G.M. *et al.*, State University at Utrecht, The Netherlands, 1987.
DESTILLATIE. Van Asselt, E. *et al.*, MediaVision, 1988.
HUMAN. Coleman and Randall, 1986.
POPDYN. Huisman, W. H. T. *et al.*, Open university of the Netherlands, 1988.
WASMEER. Huisman, W. H. T. *et al.*, Open university of the Netherlands, 1986.

On modelling in CALL conversational simulations

D. Wells Coleman

The University of Toledo, USA

ABSTRACT: Conversational simulations for computer-assisted language learning (CALL CS's) involve the computer on one side of a dialogue in the target language (TL) - the language to be learned - and one or more language learners on the other. The benefits of CALL CS's derive from their focus on communication in the TL, rather than on explicit "rules of grammar". They do this by dealing with TL structures without "lecturing" about them and thus help learners overcome errors without explicit correction. This is possible because their conversational form permits the use of modelling (Underwood, 1984; Terrell, 1982). A CALL CS designed by the author (RELATIONS) uses modelling to respond to several types of learner input, maintaining the focus on communication in a conversational context.

KEYWORDS: Simulation; CALL; modelling; parsing; ESL/EFL

ADDRESS: Dept. of English Language and Literature, The University of Toledo, Toledo, OH 43606-3390, USA.

Interactive simulation in a computer-based training environment

Hilbert Kuiper and Jos van der Arend
Physics and Electronics Laboratory TNO, The Hague, The Netherlands

ABSTRACT: Advances in computer technology have brought us the potential of real-time interactive Park Task Training (PTT) with commercially available hardware, Park Task Training permits, for example, the learning of certain time critical procedures for operator training with a graphic workstation. In this Computer Based Training environment interactive simulation and model building (student model, tutorial model, etc.) play an important role. In a training environment two systems can be distinguished.

The first one is called the Instructional Design System. With this system the instructional designer specifies the training situations which have to be created, including the simulation environment. The result of the design process is a library with a collection of possible environments. The instructional designer has computer supported tools (e.g. colour touch screen, graphics tablet, graphics editor) available to specify these environments.

The second system is the Tutoring System. The (real-time) situations specified in the design process are presented to the student, who has to respond adequately to the situation. The reactions of the student responding to the simulation are measured, compared with the desired actions ("tutor's solution") and judged. The collected data of the student is evaluated and the model of the student is updated with the received information. The simulated tutor defines, depending on the results and the history of the student, what will be the next exercise material to be presented. This Student Model (and the related teaching strategy) is an application of Artificial Intelligence in the computer-based training world, called ICAI (Intelligent Computer Assisted Instruction).

KEYWORDS: Interactive simulation; CBT; ICAI; real-time simulation.
ADDRESS: Physics and Electronics Laboratory TNO, Oude Waalsdorperweg 63, PO Box 96864, 2509 JG The Hague, The Netherlands.

Section Seven

Special Topics

Editor: **Willem J. Scheper**

Special topics: an introduction

Willem J. Scheper
Utrecht University, The Netherlands

We are living in a world characterized by an acceleration in the rate of change. Everything seems to be in a state of flux and transformation, solutions to problems that are found today can be outdated tomorrow. Technological changes, demographic changes and an increase in the fierceness of competition (among other factors) all are related to this phenomenon affecting communities from the global level down to the local level. Under these circumstances, decision making becomes a very difficult job to perform. Decision makers find themselves entangled in an almost inextricable cobweb of problems. For example, it seems impossible to stop our ecosystems from deteriorating or to control the costs of health care, while at the same time maintaining the level of quality. Relations between industries in different countries as well as relations between different governments are becoming more and more important, at the same time complicating decision making. However, in this field of international relations it is of utmost importance to deal competently with problems that - if not properly dealt with - might have destructive consequences for the global community.

Gaming-simulations have proved to be a powerful combination of methods and ideas in dealing with complex issues involving value conflicts between various parties such as the ones just described. In this part of the proceedings attention is given to several simulation-games which address those problems.

In the first section simulation-games concerning environmental planning are presented. Four simulation-games are described showing different ways of handling and conceptualizing problems related to the field of environmental planning. For a detailed introduction to these simulation-games, the reader is invited to read the excellent review of the workshop on environmental planning, written by Bert de Vries who was chairman of this workshop during ISAGA 88.

Following the papers on environmental planning, two papers are included that are dealing with health care. *Bronkhorst, Truin, Klabbers and Plasschaert* focus on the dental health care system. They present a simulation-game dealing with the imbalance between supply and demand of dental health care services. The main goal of the model is to provide government and local dental health planners as well as insurance companies and the dental profession with a tool to assist them in programme and policy planning. *Ten Brummeler and van Dijkum* describe a simulation-game directed at the quality of treatment and the related costs of physiotherapy. They devote special attention to the use of a tutorial in introducing their simulation-game to participants.

Both simulation-games have in common several actors or parties with conflicting interests ranging from material to value conflicts. Both games show suitable ways of acquainting people with problems related to the managing of complex health care systems.

As countries do have different interests, potential conflict is inherent in their (multi-)lateral relationships. Because of the enormous consequences of a conflict (armed or unarmed), it is in everyone's interest to keep the probability of a conflict as low as possible. It is in this realm of international relations that diplomats are operating. While representing and defending the interests of their countries, they encounter in their daily work a wide variety of issues such as trade, security, development, culture and human rights. In his paper on diplomatic games, *Meerts* presents a game designed to improve the skills and insights of diplomats in dealing with complex multilateral negotiation processes, the uncertainties involved in these processes and the value conflicts that distort negotiation processes. His comments on how to improve the personal involvement of participants in a game are of interest to anyone designing games.

The three last papers in this volume are devoted to gambling. Related to almost all simulation-games discussed in this book are concepts like uncertainty, information, strategy, policy, conflict and competition. These concepts form the basis of gambling theory. Using the core-concept, i.e. risk taking, as their starting point, *Saunders and Turner*, while reflecting on the linkages between gambling and gaming, arrive at the central theme of ISAGA 88: competence. In their interesting contribution they distinguish five levels of competence: intellectual, social, emotional, manual and personal. They argue that in speaking of competence regarding gaming-simulation these five levels of competence need to be considered.

Cudworth presents the simulation-game SURVIVAL which focuses on the procedures adopted in traditional tendering operations for building work. More generally, it directs the attention at risk taking and gambling practices many firms perform. In doing so SURVIVAL draws upon the field of gaming-simulation as well as on gambling theory, illustrating some comments made by Saunders and Turner on the linkages between both fields.

Finally, *Brown* in his contribution (included as an abstract) discusses gambling-addictions. He views the phenomenon of excitement and arousal as an important stimulus for gambling, thus stressing the emotional level of (in)competence proposed by Saunders and Turner. A model outlining a possible development towards gambling-addictions is presented. This model accentuates the role live-television plays in early-childhood as the main stimulus of a child.

Environmental planning: workshop review

Bert de Vries

State University Groningen, The Netherlands

ABSTRACT: A brief description of the four simulation-games presented is given. Some of the purposes are discussed, with emphasis on the use of simulation games as a means to increase environmental awareness. Finally, the relation between game and decision maker, the availability of data and the cultural context for the four games are compared.
KEYWORDS: Simulation-games; environmental planning.
ADDRESS: Centre for Energy and Environmental Studies, State University at Groningen, PO Box 72, 9700 AB Groningen, The Netherlands.

Introduction

During the workshop on Environmental Planning, four games have been shown and discussed. The first was Future Voltage (FUVO), a simulation-game about electric power planning and developed by De Vries and Benders at the Centre for Energy and Environmental Studies (IVEM) of the State University Groningen, The Netherlands. The second was a simulation-game on Biosphere and Underdevelopment (B&U). It has been designed by Schapira of the Universidad Nacional de Cordoba, Argentina, within the UNESCO Man and Biosphere programme. The third demonstration concerned a simulation-model on integrated rural development, called IRDEM and developed by Rijsbergen and Baarse at the International Federation of Institutes for Advanced Studies (IFIAS) in Maastricht, The Netherlands. Finally, a simple but interesting game on the nuclear accident in Chernobyl USSR, was shown by Bourlès, France.

The workshop has been attended by 6–13 people. Due to good computer support, the participants have had the opportunity to exercise themselves with the models/games.

In this review, I briefly describe the four games. Then, some of the characteristics of environmental planning and the use of simulation-games in this context are discussed. Finally, some key issues in each of the presentations are discussed.

Brief Description of the Presented Simulation-Games

The first (FUVO) and the third (IRDEM) of the presented simulation-games are discussed in detail elsewhere in these Conference Proceedings. FUVO is an interactive simulation-model (written in Turbo-Pascal), which asks the players to decide on power plant investments and energy-environmental parameters in 3-year steps. Based on the outcome of their decisions in terms of reliability, cost and environment indicators, players make their own strategy through a 20 to 30 year period. IRDEM is an acronym for Integrated Rural Development Model (using Lotus 1-2-3 spreadsheet). It allows the users to explore and evaluate rural development strategies and is implemented on the basis of experience in the Lake Basin area in Kenya. Major variables are population, income and water and land use characteristics.

B&U is a model which attempts to establish an on-going discussion with participants about development strategies which explicitly incorporate environmental aspects. The model is designed by university staff in such a way that (public and private) decision makers can analyse hypothetical future situations and communicate about it with a high "realidad de la simulacion". To this purpose, the regional economy of the Municipality of Alta Gracia, Argentina, in an input-output framework. Important objectives c.q. decision variables are the rate of (un)employment, the rate of return on sector investments, economic output and wage levels and the level of basic services (water, electricity, sanitation etc.). As part of model construction, interviews and role-playing exercises have been organized.

Chernobyl is a norms-oriented game about the way regional governments behave in a situation of catastrophe. A cloud of radio-active particles is emitted from the central area no. 1 and is carried by the wind (random – a dice is thrown) to one of the six surrounding area's on the game-board.

Using a pack of cards and a dice and a set of rules, the players can decide about how their decontamination teams are going to be used. The psychological trade-off is between avoidance of health hazards for the population in your area on the one hand and a concerted effort to stop the radio-active emissions on the reactor site on the other hand. Negotiations – and a possibility to blame other people – are an important element, too. The set-up of the game illustrates aspects of the well-known theory of social-dilemmas.

Environmental Planning and the Use of Simulation-models

Simulation-games have first entered the policy domain in the fields of military operations and business management. Since then, many situations have been explored with this tool. Simulation-games about environmental planning have only rather recently been introduced (e.g. STRATEGEM) and are often developed out of more detailed operational models (e.g. Future Voltage or FUVO). I suspect that one of the reasons is, that the major thrust in modelling has been oriented towards socio-economic developments.

Only recently, policy makers are acknowledging that environmental issues really matter: they demand their share of [scarce] capital and labour resources, they affect people's actual and perceived well-being and they may, if neglected, undermine the very basis for long-term development. As a result of this, energy and environmental management became more than just economic engineering. Biosphere resources and their sustainable management are becoming an integral part of techno-economic development.

As in other fields, simulation-games serve a spectrum of purposes:
1. educational, to learn about [the dynamics of] resource management,
2. decision support, to help policy makers in understanding the possible and exploring the feasible,
3. a communication device, to learn about each other's "mental maps" and to invite people to participate in societal valuation processes, and
4. a heuristic device, to uncover problem area's and create or explore strategies for the future. For example, STRATEGEM was designed for decision support, but is now widely used as an educational tool. FISHBANKS is meant to be an educational as well as communicating device. ECCO has been used for decision support, whereas PLANSPIEL-E focuses on learn-how-to-operate. The "policy exercises" as developed at IIASA serve mostly communication and heuristic purposes (Toth, 1986).

The simulation-games in the present workshop differ as to their purposes, and consequently in their relationship with participants. FUVO is mainly an educational tool; it has however been successfully used among policy makers and students to communicate about basic assumptions and concepts. IRDEM is specifically designed to assist policy-makers in long-term resource development issues, although it may serve other purposes as well. B&U explicitly aims at interaction with decision-makers, whereas the Chernobyl game mainly serves communication purposes.

It seems to me, however, that many simulation-games on environmental issues have one other purpose: to raise the awareness of players (students, managers, policy makers) with regard to the importance and the complexity of resource dynamics. As such, these games are a natural extension of the recent development towards a more integral approach of development, as is exemplified a.o. in the UN report *Our common future* (1987) and the IIASA report on *Sustainable development of the biosphere* (1986). There is a "message" in most of these games: ecosystems should be exploited in a sustainable way, taking into account surprise events, nonlinear behaviour, long-term feedbacks etc., if farmers and industrialists are to avoid the gradual deterioration of their very support system. Recent research on trends in forest quantity and quality, soil productivity, desertification and species extinction have greatly enhanced our awareness of the need for sustainable resource management.

Discussion

Interaction Between Game and Decision Maker

Both games B&U and IRDEM are explicitly aimed at providing decision makers with a tool to get a better understanding of [complex] reality and to discover policy variables and evaluation criteria. One implication is that a good relationships with a group of decision makers has to be established for a period of several years. Such a simulation-game may gradually gain a high "reality content" in terms of techno-economic structure and develop into a major tool for future policy making (Toth, 1986). However, they are not transferred to other situations without a major effort in terms of data and user experience. Chernobyl, on the other hand, is of a more abstract character and claims to illustrate psychological processes. FUVO is in-between: once a rather large and country-specific data-set is implemented, players can in a few hours learn some general principles (construction delay, feedback etc.). Using FUVO for policy exploration and evaluation would however also require additional effort.

Data Availability

One of the problems modellers are confronted with is the amount and type of data required. Are they available, are they reliable, will underlying relationships remain etc. Both B&U and IRDEM are using the simulation-game exercise as a tool to get and gradually improve the database.

This is an interesting strategy, its aim being to involve policy makers actively in the acquisition of knowledge about "their" system and thus ensure that the model inputs and outputs are really meaningful to them. In FUVO this aspect has been introduced with help of so-called scenario variables: GDP-growth, oil-price etc. can be set in advance to create a policy-environment which if not realistic is at least deliberately chosen.

Cultural Context

The four games presented were quite different – so is their cultural context. For instance, Schapira's B&U-game faces a situation in which speculation in the financial sector drains the economy and the regional administration is fond of building new things instead of maintaining old things or is solely concentrated on the "dream of a multinational corporation" within its region. The B&U-game intends to help visualize these cultural aspects – quite a task in the realm of social change.

Rural development in Kenya is faced with many constraints which are not explicitly included, yet play a major role in the implementation of development projects. For example, private roof catchments for rain could relieve water supply problems but the access to capital is a major constraint. Similarly, migration out of the region seems almost inevitable to maintain a decent income – but where have these people to go? And how about (un)employment rate as an indicator when the roots of the income problem is related to socio-cultural aspects? Thus, it seems that resource-accounting oriented simulation-games like IRDEM and ECCO rightly emphasize the economic, technical and energetic feasibility of development strategies but have to get a follow-up to explore the implementation aspects.

FUVO is basically oriented towards policy exploration within a technical energy-economy-environment framework. It does not address matters of values and implementation explicitly. However, as experience has shown, it induces discussion about issues like risk acceptance, import dependence, the quest for ever more economic output etc. Such discussions may be structured by the game operator.

Finally, the Chernobyl game represents implicitly the apparent bureaucratic decision-making processes in the USSR. Solidarity and hypocrisy are two key elements in the set-up of the game. One may wonder, however, if France would not be faced with similar psychological phenomena, were a major nuclear accident to happen in one of France's forty-or-so nuclear power stations.

References

Brundtland, Gro Harlem. 1987. *Our Common Future.* Oxford: Oxford University Press.
Clark, W. C. *et al.* 1986. *Sustainable development of the biosphere.* Cambridge: Cambridge University Press.
Lienhard, H., Steiger, F. and Weber, K. 1975. *Planspiel Elektrizitätswirtschaft.* Bern and Stuttgart: Verlag Paul Haupt.
Rijsberman, F. and Baarse, G. 1989. Simulation of integrated rural development with IRDEM. *In* this volume.
Toth, F. 1986. Practicing the future: implementing "the policy exercise concept". *Research Report IIASA WP*-86-23. Laxenburg: IIASA.
Vries, B. de and Benders R. 1987. Future voltage user manual. *Research Report IVEM-22.* Groningen: IVEM.

Simulation References

B & U. Schapira, L. *et al.* 1989. *In* this volume.
CHERNOBYL. Bourlès, C. 1989. *In* this volume.
ECCO. Slesser, M. and King, J. 1988. Edinburgh: Centre for Human Ecology.
FISHBANKS Ltd. Meadows, D. 1988. Durham, NH: University of New Hampshire Press.

FUVO. Vries, B. de and Benders, R. 1987. *In* de Vries and Benders (1987).
PLANSPIEL-E. Lienhard, H., Steiger, F. and Weber, K. *In* Lienhard, Steiger, and Weber (1975).
STRATEGEM. Meadows, D. Biesiot, W. and Geerts, M. 1987. *Research Report IVEM-13*. Groningen: IVEM.

Learning about electric power planning: a gaming approach

Bert de Vries

State University Groningen, The Netherlands

ABSTRACT: This paper discusses the development of electric power modelling from supply- and technology-oriented to more integral, culture-oriented approaches. This development is illustrated within the framework of individualist-historicist/rationalist-structuralist dichotomies. Next, the simulation model Future Voltage on electric power planning is described in terms of performance criteria, levels of variables and input-output formats. Experiences with Future Voltage in university curriculae are briefly discussed. The model has proven to be an effective learning tool. Extension towards a utility-oriented management game is the next phase, to be carried out in a joint Dutch-Indian research project.
KEYWORDS: Electric power planning, energy modelling, environmental modelling.
ADDRESS: Centre for Energy and Environmental Studies, State University at Groningen, PO Box 72, 9700 AB Groningen, The Netherlands.

Introduction

Simulation modelling has become an ever more important tool in planning electric power systems and articulating and implementing electric power policy within a broader socio-economic context. The modelling domain gradually expanded from power engineering and micro-economic accounting relations into the wider and more complex areas of environmental engineering, macro-economics and ecology. At the same time, due to e.g. construction times for nuclear power plants of 10–12 years and the recognition of the long-term character of acid rain and greenhouse effect, the longer term got more attention. Klabbers proposed a taxonomy of forecasting and planning methods on the basis of time horizon degree of accuracy of system description (Klabbers, 1985). Within the context of this taxonomy, one can trace the development of electric power models in The Netherlands. I suspect developments to be rather similar in other countries (Baumgartner a.o. 1987).

Initially the well-defined, short-term models dominated decision making e.g. the simple econometric models to forecast demand or the detailed operational models to optimize plant operation. Longer term forecasting was scarce and hardly used explicit forecasting methods (see e.g. STT 1971). During the seventies, broader and longer term issues like OPEC-dependence and nuclear waste disposal came on the political agenda and both managers and scientists were forced to respond. At first, the scenario method as a means to deal with the more complex and less well-defined aspects of (sociopolitical) systems rose to the foreground (cf. LSEO 1976, Gordon, 1978, Economische Zaken, 1979). In an attempt to make the scenarios more specific and amenable for policy-formulation, more extensive simulation models with emphasis on the medium to long term were developed e.g. SCELEC (de Vries *et al.*, 1985, Dijk *et al.*, 1987) and SELPE (Boonekamp, 1982). Emphasis, however, remained on the supply-side (production planning). Only minor efforts were put into modelling the demand side, one obvious reason being the lack of reliable data.

An interesting exception to the lack of interest for demand side dynamics was the system dynamics approach of Ford (1983). He was invited to participate in the dutch debate on nuclear power and his approach played a role in the formulation of the environmental, so-called CE-scenario. It also initiated, in combination with an interactive version of the large simulation model SCELEC, a research project at IVEM towards the development of an interactive simulation model for a micro-computer (Benders, 1985). This model has been developed successfully into an interactive simulation game with the name Future Voltage (FUVO). Structural relations with the "environment" (economy, ecology) are added to provide an adequate decision-making context. In the terminology of Klabbers, this development represents an initial move from the causal, technology-oriented stratum to the culture-oriented norms stratum. When the policy-making process continued, the emerging confrontation between these two strata and the corresponding methods has initiated a move towards interactive simulation-games as the appropriate tool at the decision-making, structure-oriented stratum.

A recent conference on modelling made it clear, however, that this approach has its own drawbacks and opponents and is considered as a complementary rather than a substitute method (PTW 1988). During a demonstration of FUVO some modellers and managers showed scepticism, expressing the familiar complaint that much more detail had to be added to make it a realistic tool for decision making. Participants with an educational or policy-making interest responded more positively. The demonstration also revealed the importance of the participants' willingness to have an open discussion if a simulation-gaming approach is to be successful in policy formulation and implementation.

Demonstration of Energy Planning

To place these developments in perspective, I first digress a bit further into the rise of simulation-gaming in the field of energy and environmental policy. The 19th ISAGA Conference dealt with the role of simulation-gaming with respect to the improvement of competence in dealing with complexity, uncertainty and value-conflicts. In his opening address to the Conference, Berting proposed a "four cultures" framework for evaluation. These four cultures arise from the dichotomies individualist – structuralist and rationalist ("Nature") – historicist ("Culture")[1]. As such he discusses simulation-gaming within the social science context: "As gaming and simulation also deal with social and cultural life . . . it is no wonder that they have to deal with the same types of fundamental theoretical oppositions as the "traditional" social sciences" (Berting, 1989).

In Berting's classification, gaming is characterized as historicist-individualist approach whereas simulation and scenarios are considered adequate tools within the rationalist-structuralist approach. He suggests that at least four different conceptions of game or play can be distinguished: an outcome-oriented decision support (rationalist-individualist), a process-oriented device for communication and self-reflection (historicist-individualist), a case-study oriented instrument to learn about social life (historicist-individualist) and a functionalist conception as in theatre (historicist-structuralist). Each conception has its own "culture", its own conception of man. Berting's analysis can put some light on the extent to which (simulation-)gaming can provide a communication device between a rather coherent goal- and technology-oriented group of decision makers on the one hand and the often incompatible, less well-articulated and value-oriented claims from societal groups. Let me illustrate this briefly with respect to electric power planning.

Most west-European countries have experienced over the past decades a discussion of fluctuating intensity about the necessity and acceptability of nuclear power as a source of electricity. The initial response from electric power, research and government circles belonged to the rationalist-structuralist "culture" (see e.g. Uitham *et al.*, 1977). Formal models in a scenario context provided the required expertise to decide about the future – at least, it was claimed and hoped. Faced with growing anti-nuclear opposition and surprise events like the Iranian revolution, most governments attempted to settle the most controversial issues with "information campaigns". Whereas this was successful for some parts of the debate, it made even more clear that other parts of the debate were deeply rooted in the participants' values. As a result, electric power policy became dominated by "political games" and formal models were discredited at an alarming rate[2]. In countries like the US, Sweden, Austria and the Netherlands, the turning point – and the realization that *political* decision making was required – was in the years 1980–1984.

Scientists partly prepared, partly followed this transition from a rationalist to a more historicist culture. The homo-economicus type of "aggregate consumer" fell apart into a collection of individuals with conflicting values, seldom well-informed, often behaving irrationally and always caught in some type of social dilemma – either consciously or unconsciously. The social aspects of "the energy problem" were more and more recognized and analysed. Formal models got back their proper role of supporting intelligent, case-study oriented analysis. Metaphors emerged to enrich and inspire the all too technical discussions within the modelling and planning departments (see e.g. De Geus in PWT 1988).

In short, we regained our freedom: even if we feel ourselves actors in a local and global drama, there is no such thing as a predictable, predetermined future. This changing perspective on modelling reality has to some extent reformulated environmental problems, along with the insight that complexity (nonlinear behaviour, surprise, delays) has to be an essential part of ecosystem modelling. Within this context and the advent of cheap and powerful microcomputers, simulation-gaming enters the arena of energy and environmental policy-making, providing a process-oriented outlook with emphasis on learning. It emphasizes our possibility and responsibility to improve communication, to stress participation and reflection and to explore and structure our social relationships. This new outlook is now beginning to help us in exploring strategies for a (more) sustainable management of the biosphere (see e.g. STRATEGEM).

Future Voltage as a Decision-making Tool

General Set-up

Before examining actual experience with FUVO as an operational and educational tool, I will briefly describe its set-up, give some details on the gaming context and state goals and performance criteria. The simulation game is set up from the management perspective of a central utility. Their basic goal is to design a strategy for the future which satisfies a number of – partly incompatible – *performance criteria*. These are:

- system reliability, expressed as the ratio between installed and required capacity (this so-called reserve factor has some predetermined, system-dependent optimum value),
- electricity cost, in money units per kWhe generated and transmitted,
- environmental impact, about which the decision maker is informed in terms of annual emissions of air pollutants and generation of solid waste, and
- strategic aspects, expressed in the remaining reserve of an indigenous high-quality fuel resource and the ratio of oil-fired respectfully nuclear and total capacity.

The criteria are covered in the formal goal of utilities "to provide electricity at the highest possible reliability and the lowest possible, socially acceptable cost". The four criteria are shown to the players in the format of five indicators – see fig. 1.

The model distinguishes three *levels of variables*. First, there are system parameters of a techno-economic nature which are set at the beginning of the

Game and status indicators   Country [   ]

Indicators (the lower the better)

Period	1	2	3	4	5	6	7	8	9	10
Economic	[ ]	[ ]	[ ]	[ ]	[ ]	[ ]	[ ]	[ ]	[ ]	[ ]
Capacity planning	[ ]	[ ]	[ ]	[ ]	[ ]	[ ]	[ ]	[ ]	[ ]	[ ]
Environment	[ ]	[ ]	[ ]	[ ]	[ ]	[ ]	[ ]	[ ]	[ ]	[ ]
Cum ind resource used (%)	[ ]	[ ]	[ ]	[ ]	[ ]	[ ]	[ ]	[ ]	[ ]	[ ]
Strategic risk (oil/nuclear)	[ ]	[ ]	[ ]	[ ]	[ ]	[ ]	[ ]	[ ]	[ ]	[ ]

Status

	1	2	3	4	5	6	7	8	9	10
Gross national product	[ ]	[ ]	[ ]	[ ]	[ ]	[ ]	[ ]	[ ]	[ ]	[ ]
Electricity demand	[ ]	[ ]	[ ]	[ ]	[ ]	[ ]	[ ]	[ ]	[ ]	[ ]
Import percentage	[ ]	[ ]	[ ]	[ ]	[ ]	[ ]	[ ]	[ ]	[ ]	[ ]
Electricity - GNP ratio	[ ]	[ ]	[ ]	[ ]	[ ]	[ ]	[ ]	[ ]	[ ]	[ ]

FIGURE 1

game session. Among them is the set of power plants for the specified country as of the first period of operation (each period covers 3 years). Secondly, there are a number of scenario variables which can be set – partly graphically – before the session starts. They create the game "environment" and can or cannot be the object of explicit discussion before the session starts. Most important among them are the oil price path, the Gross National Product (GNP) growth path and the electricity-GNP elasticity development in the future. The third type of variables are the decision variables, which are presented in the format of an input screen or paper – see Fig. 2. They consist

Inputform   Period [ ]   Country [   ]

New investments

Cogeneration	(5 MWe)	[ ]
Windpower	(5 MWe)	[ ]
Hydropower	(25 MWe)	[ ]
Nuclearpower	(900 MWe)	[ ]
Coal fired power	(600 MWe)	[ ]
Oil/gas fired power	(300 MWe)	[ ]

Thermal efficiency (in %, ≤ 60%) and emission reduction (in %, ≤ 95%)

	Efficiency	NOx/SO2-em.red.
Coal fired power	[ ]	[ ]
Oil/gas fired power	[ ]	[ ]

Fuel quality (in % sulphur) and import (in % of oil/gas requirement)

Sulphur % coal	[ ]
Sulphur % oil	[ ]
Desired import %	[ ]

FIGURE 2

of new investments in power plants, efficiency and emission coefficients and fuel quality and origin.

Learning about electric power planning 297

To help the players in formulating their strategy, they can ask for additional screen or print information. For instance, they can view the expected peak demand growth (constant electricity cost) and the forecasted installed capacity (including under-construction and out-of-operation units). They can also ask for more detailed information on the economic and environmental status from the first period onwards. Once they made their decisions, the computer presents them with the outcome and they can enter the next cycle of evaluation and decisions. More details about the model structure and the computer presentation have been given elsewhere (de Vries *et al.* 1987).

Experiences

Over the last 2 years we have gathered experience in the use of Future Voltage as interactive simulation model, both in an educational and an analysis context. Here, I will mainly focus on its use in university courses on energy and environment (see for an analysis application e.g. Stikker, 1988).

Future Voltage is used as a regular part of the IVEM-curriculum on energy and environment. The students (8-10) are asked to present a 25-year Electric Power Plan for The Netherlands. The allotted time is 2 to 3 days. They constitute the Board of Directors, each of them has a role (Director, Public Relations, Research and Planning etc.) and access to information which is relevant for him/her in view of this role. How to communicate and organize themselves and how the simulation model is to be used in the design of their final strategy is part of the learning experience.

The main conclusion has been, that Future Voltage provides teachers in energy and environmental issues with an excellent tool. Students got – after an initial period of confusion – completely involved, engaged in fierce debates of continuously improving quality and were forced to be explicit about issues and trade-offs which most probably would not have been noticed let alone experience without a game-setting. Of course, there is still room for improvement (see also the workshop review).

One important point is the complexity of electric power planning, in the sense that it has a techno-economic kernel which most people are unfamiliar with, that it covers both short and long term, that the decision indicators are of rather different nature etc. Consequently, it has turned out that even for a group of graduate students the game leader should start with an introductory lecture on the system elements and structure, to ensure that there is real understanding and policy-formulation instead of just trying to hit the target. It has confronted us with the need to choose between either a more simplified "educational version" and a more elaborate "management version". The educational version has to focus more on some basic lessons, like the impact of long construction times (delays) and the price-demand interactions (feedback). The management version, on the other hand, should evolve into a game with more elaborate roles and rules, and at the same time provide an even better model representation of the real-life system. Only then, management issues will arise in such a way as to be relevant and challenging for utility managers. Some extensions e.g. a more detailed model of the demand-side and transmission-distribution relations and options, are required to this purpose.

Prospects for the Future

In my view electric power planning will remain an important public policy issue in the decades to come. There is no doubt, that both the developed and the developing countries face some serious challenges in relation to energy and environment issues. European governments, in the face of large-scale forest degradation, aim for air pollution reduction agreements. At the same time, however, the Chernobyl-accident has made the nuclear option less attractive and less feasible. Developing countries should give high priority to electricity – an infrastructural good and a basic need. At the same time, they are confronted with the possibly catastrophical consequences of ever-increasing global carbon-dioxide emissions and the large risks involved in nuclear fuel and waste handling and storage. Due to shortage of capital and skilled labour, efficiency improvements – often the most attractive option – are introduced at a slow pace if at all.

Therefore, it is most welcome if not urgent that students in the engineering, economic and social sciences have the opportunity to learn about this important topic. The next step will be, that Future Voltage is extended into a management tool to be used by south-east asian utilities. The project, financed by the Dutch Ministry of Foreign Affairs, will be a joint research project of IVEM, Groningen, and the Systems Research Institute, Pune, India.

Notes

1. This is not to say that I fully support Berting's framework. For example, the proposed dichotomy "Nature" vs. "Culture" is rather misleading in assessing environmentalists like Arne Naess and his deep ecology. Naess rejects explicitly a narrow rationalist approach of "Nature", emphasizing instead the need for and value of a wholistic, participatory approach. See e.g. A. Jones in *The Ecologist* 1983:141.
2. This became most evident during the last phase of the public debate on energy policy in The Netherlands around 1982. Politicians felt no need to respond to the outcome of this 3-year, twenty-five million guilders debate. The newly appointed Energy-Director at the Ministry of Economic Affairs was a juridical expert on negotiations between governments and environmental citizen groups.

References

Baumgartner, T. and Midttun, A. (eds.). 1987. *The politics of energy forecasting*. Oxford: Clarendon Press.

Benders, R. 1985. Future Voltage – een interactief computersimulatiespel. *Research report IVEM-17*. Groningen: IVEM.
Berting, J. 1989. Structures, actors and choices. *In* this volume.
Boonekamp, P. 1982. Beschrijving van SELPE. *Research report ESC-17*. Petten: ECN.
Dijk, D. and Kok, M. 1987. A comparison of different modelling approaches to strategic energy planning. *European Journal of Operational Research* 29:1.
Economische Zaken, 1978. *Nota energiebeleid*. Den Haag: Staatsuitgeverij.
Ford, A. and Youngblood, A. 1983. Simulating the spiral of impossibility in the electric utility industry. *Energy Policy* 1.
Gordon, J. 1978. Lange termijnscenario's voor de olie- en steenkoolvoorziening van Nederland. *Energiespectrum* 2.
Klabbers, J. H. G. 1985. Instruments for planning and policy formation. *Simulation & Games* 16:2.
LSEO. 1976. *Interim rapport landelijke stuurgroep energie-onderzoek*. Den Haag: Staatsuitgeverij.
Meadows, D., Biesiot, W. and Geerts, M. 1986. STRATAGEM. *Research report IVEM-13*. Groningen: IVEM.
PWT. 1988. *Preliminary proceedings of the conference on "de werkelijkheid van het model"* Den Haag: Stichting Publieksvoorlichting over Wetenschap en Techniek.
Stikker, A. 1988. The Taiwan 2000 study-experiences and impressions. *Futures* 20: 4.
SST. 1971. Electrical energy needs and environmental problems, now and in the future. *Research report SST-7*. The Hague: Stichting Toekomstbeeld der Techniek.
Uitham, K., Vries, B. de and Zijlstra, G. 1977. Kernenergie in Nederland. *Acta Politica* 12:2.
Vries, B. de and Dijk, D. 1985. Fuel and investment options for electric power generation for The Netherlands to 2000. *Energy Policy* 3.
Vries, B. and Benders, R. 1987. FUTURE VOLTAGE user manual. *Research report IVEM-22*. Groningen: IVEM.

Biosphere and underdevelopment

Leopoldo Schapira

Universidad Nacional de Cordoba, Argentina

ABSTRACT: BIOSPHERE AND UNDERDEVELOPMENT is a simulation of the problem of subregional development in Argentina, incorporating the environmental dimension into the planning of the Government's action.
This simulation is a part of the programme PARTICIPATIVE GENERATION OF REGIONAL PLANNING PROJECTS IN THE MUNICIPALITY OF ALTA GRACIA AND PARAVACHASCA AREA - PROVINCE OF CORDOBA - ARGENTINA which is carried out with UNESCO's funds.
Through this simulation the Mayor and the City Council of this small town should learn how to reconcile the transformation needs implicit in economic and social development with the conservation of the environment as regards their natural resources and their historical, social and cultural patrimony. The democratic authorities should recover the practice of actively exercising the municipal faculties, which was lost after long periods of authoritarian and administrative centralism.
The economic stagnation suffered by Argentina does not stop the change process in the socio-spatial system: it slows the evolution of problems generated by growth, but it accelerates the deterioration of the investments already carried out and the demand for solutions of social conflict.
The simulation is based on an econometric - environmental impact model which is operated by groups of participants who represent private and public sectors in the region.
KEYWORDS: Environmental planning, participation, reality of the simulation, variable scenarios.
ADDRESS: Avda Olmos 15, Piso Primero B, 5000 Cordoba, Argentina.

Introduction

BIOSPHERE AND UNDERDEVELOPMENT is a simulation of the problem of underdevelopment in Argentina, incorporating the environmental dimension into the planning of Local Government's actions.

BIOSPHERE AND UNDERDEVELOPMENT is an important step in the fulfilment of the objectives of PARTICIPATIVE GENERATION OF REGIONAL PLANNING PROJECTS IN THE MUNICIPALITY OF ALTA GRACIA AND PARAVACHASCA AREA, PROVINCE OF COR-DOBA - ARGENTINA which is carried out with UNESCO's funds - Programme of Man and Biosphere, Project 11.

This simulation is based on an econometric - environmental impact model which has been designed to be operated by members of small towns' City Councils. During the exercise they represent the internal private and public sectors of the region.

The simulation has the purpose of generating a constant dialogue between the university team and local authorities. This is achieved by getting the former involved in the simulation, not as observers, but keeping the role of the "outside world" for them. They represent the source of uncertainty in every little region, the external private and public investors who can contribute to local development or who can break it, destroying the existing environmental system.

The simulator generates situations tending to show the most significant relationships between economic activity and biosphere impact, in order to induce changes in the authority's behaviour which, with a certain degree of automatism and the blind faith that in Argentina natural resources are inexhaustible.

During ISAGA 88, the presentation of BIOSPHERE AND UNDERDEVELOPMENT was divided into two aspects: the role of the simulation in the ALTA GRACIA PROJECT, on the one side, and on the other, a general description of the simulation and the use of the computer model.

The Simulation in the Local Authorities Improving of Competence Project

The objective of the UNESCO's project is to train the authorities of the municipality of Alta Gracia as regards the generation and use of initiative in economic and environmental planning. The public officials should learn how to reconcile the transformation needs implicit in the economic and social development with the preservation of their natural resources and the historical, social and cultural patrimony of all the Paravachasca area.

As a consequence of long periods of authoritarian and administrative centralism the practice of actively exercising the municipal faculties of planning, investment and control was being lost.

The officials who are responsible for land planning are unaware of the characteristics of the socio-spatial system either as regards its spontaneous processes or the budgetary and normative resources. They are unable to perceive the social problems whose detection requires a certain previous technical training and do not manage the possibilities of other financial sources external to the local budget.

The economic stagnation suffered by Argentina does not stop the change processes in the socio-spatial system. It may diminish the rhythm of development of problems derived from growth, but it adds to those whose origins can be found in slovenliness, in the deterioration of the investments already carried out, in the paralysis of markets, etc.

Thus, it is essential to make use of the resources of current legislation and to generate initiatives at the local level, not only as a way to induce development in the short run, but also as a means to help solve the current acute social problems.

But to do that, it is fundamental to introduce changes in decision-making behaviour. The disillusion after the hopeless beginning of democracy, has produced in politicians a feeling of impotency and a paralysis in governmental action.

In this context, the development of this project should serve as a means to put into evidence what is going wrong in decision-making groups.

To fulfil the project objective it is necessary, on the one hand, to take more advantage of the reduced instruction period the authorities may get, and on the other, to assure the application of the acquired knowledge in the "a posteriori" governmental practice.

The simulation of real situations is the main mechanism of research and of communication between the university team and the authorities of Alta Gracia. A simulation is a powerful pedagogic tool able to make the training period more profitable. Besides this, to assure continuity between the acquired experience and the practice that follows, BIOSPHERE AND UNDERDEVELOPMENT has been designed with a sufficient level of complexity and with such a degree of generality that it allows the incorporation of variables of proved validity, so much so that the database used in its practice will be the origin of the final local information system.

As a counterpart, it is hoped that the game will awaken in the authorities a sort of "thirst" for the particular type of information contained in the database.

At the same time BIOSPHERE AND UNDERDEVELOPMENT, should be used to create an appropriate environment for the manifestation of the target group's decision-making attitudes, which are tied to culture and to the critical circumstances existing in Argentina.

Description of the Simulation

Basis for the general structure

To explain the structure of BIOSPHERE AND UNDERDEVELOPMENT, those aspects which have been devoted to represent the social-spatial system and the others which are expected to generate a framework for the revision of decision-making behaviour, it is necessary to carry out a dissection of the components of the act of simulation.

In the Social Sciences, the development of a simulation exercise is an attempt to recreate a part of social reality which, despite the artificial nature of this practice, must seem to its participants as real as possible. Or at least it must be accepted as such by the persons involved in this activity.

In the Oxford, Crookall and Saunders' essay (1987) the existence of a "reality of the simulation" is practically admitted. That is to say that in the development of a simulation exercise a particular reality is built, which is characteristic of that exercise.

From this idea it can be affirmed that the reality of the simulation is generated from a simulator which is a system of definitions and rules whose

Biosphere and underdevelopment

consideration and fulfilment during the simulation aim at representing a portion of the social reality: "the reality to be simulated".

The validity of the simulator, that is its consistency with the real world, depends on several factors derived from the fundamental fact that the simulator is an interpretation made by the modeller of a portion of the reality and is expressed in the particular language of the systems. What is left to the cultural understanding and to the participants' creativity (soft parts) can only be evaluated once the simulation has taken place.

The reality of the simulation is made up of:

(a) The elements of the "reality to be simulated" explicitly included in the simulator. These are the rigid and invariant aspects to which the participants' behaviour must be adjusted in the simulation. It is the "hard" part of the representation of the social reality.

(b) The elements of the "reality to be simulated" implicit or explicitly provided by the participants of a simulation when they up-date the simulator. Individual attitudes or group arrangements are implicit or explicitly generated as regards the aspects of the reality which they intend to simulate. These elements complete the simulator in those aspects in which it is not exhaustive either intentionally to give room for the evolution of exogenous variables, or due to omission (since what is lacking is the designer's cultural understanding of what is obvious). These could be the discovery, construction and assumption of a role accepted as "veridical" or "real" by the participants, the acknowledgement of a recurrent pattern, the elaboration of objectives and strategies, etc.

(c) The elements of the "reality of each particular actor" that "go beyond the social reality to be simulated". This is what is "veridical" in the participants' projections. They deposit in the individual and group behaviour the result of their previous experiences fixed in their psychosocial frame. It can be the competitiveness, individualism, the handling of myths, the team working habits, fears, the appreciation of what is shown in his performance during the exercise, etc.

"The reality of the simulation" thus appears as an attempt to "filling and supplementing" the simulator with the purpose of developing a shared perception of the "whole reality" which does not lack anything in order to function completely. When this collective attempt fails, the act of simulation fails as well. The "reality of the simulation" would be the result of a tendency to complete the reality, projecting into the scope of the situation, psychological and sociological aspects which reveal the culture of the group.

The attitude of building a "whole reality" would be analogous with the perceptual phenomena of good shape, background continuity, etc. or, on the other hand, with the psychological tendencies of completion of the ambiguous.

In a way, a simulation exercise shows a conflict between "the reality to be represented" and the tendencies that constitute the "reality of the simulation".

Components of the reality of BIOSPHERE AND UNDERDEVELOPMENT

According to the UNESCO's project objectives, it was particularly necessary in BIOSPHERE AND UNDERDEVELOPMENT to control this phenomenon in order to elucidate, from the aspects related to the learning of the socio-spatial system, those linked with the mobilization of beliefs, of myths, and mainly of socially negative practices which occur when a decision-making process is going on.

In our simulation the representation of the socio-spatial system (simulator) has been covered both by a computer model and by the role-playing of participants dealing with the subject of the simulation. Following Jan Klabbers' (1982) structural analysis based on the "complexity and uncertainty" components of social reality, the simulator's computer model was assigned to represent the most predictable aspects of the socio-spatial system, that is to say, the techno-economic input – output level, while the less deterministic economic and political behaviour of the social actor was left to the role-playing.

On the other side, the aim to check and measure the occurrence of cultural attitudes existing before the social interaction which are beyond the particular subject of the simulation, was planned to be achieved by the permanent introduction of changes in the scenarios, which are directed to reproduce the performance of Argentina's socio-economic framework, which is strongly conditioning individual's behaviour.

Interactions within the structure

Thus, the built-up structure composed of 1. a computer model, 2. rules for role-playing and 3. the generation of a set of variable scenarios, supposes the existence of a scheme of interactions.

1. University team – Scenario Computer Model. This relationship is established through a series of computer screens. Operators are allowed to change the Scenario Model variables.
2. University team – Local Authorities. This relationship is developed both through the Scenario Model output and by assuming the "external world" role during the social interaction period.
    The simulation is designed to allow the participation of four big sectors: extraregional private agents, extraregional public agents, intraregional private agents and municipal agents. The role-playing situations

take place from the representation of the behaviour of these social actors. The university team represents external private and public investors.

The extraregional public group is made up of public officials of the federal hierarchy. They represent the provincial and national ministries without discrimination. They are the "outsiders" of the public administration.

3. Local Authorities' role-playing. It is a social interaction period. They represent local private and public sectors.

The intraregional private groups can be subdivided into managers and trade unionists. Each of these group develops two tasks. The first one acting in defence of their sectorial interests: the managers decide on their investments and the workers negotiate their employment and salary levels. The second one consists of acting as members of the City Council, having to analyse and decide about the proposals of the Public Sector.

Generally speaking, the decision-making process and its impact refer to the allocation of resources in the private sector with the objective of bettering return on investment, to the labour struggle to improve employment levels and the average salary of state and private workers, to public efforts to provide marginal groups with their basic needs to complete the Social Basic Capital, to assist the private sector with the infrastructure for production and to improve and conserve the environment.

4. Local authorities – Computer Model. This relationship is established through a series of computer screens. Participants are allowed to decide each round's public and private investments.

5. Scenario Computer Model – Each Round Computer Model. The first one produces a hypothetical state at the beginning of each round situation. The second one processes the "true" evolution of the situation after each round's role-playing period. They share the value of some variables.

6. Finally, Local Authorities – University Team and Scenario Computer Model. This situation is produced when the need to change Capital – Product Relationship linked with Environmental Thresholds appears.

The sequence of interactions

The simulation takes place through a succession of rounds. Each of these rounds represents a variable temporal period. The environmental consequences of the decisions are projected through 10 and 20 years.

Each round-sequence is determined by the order of appearance of the Computer "screens" (Graphs 1 and 2). At first, these "screens" are divided into two big groups: those reserved to teachers (Scenario Model screens) and

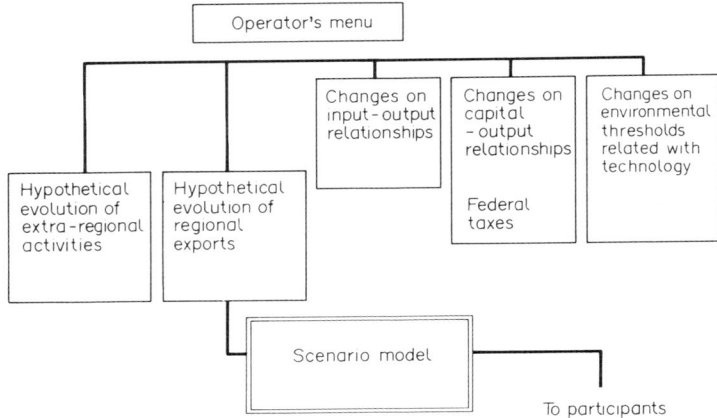

GRAPH 1 Sequence of each round scenario building

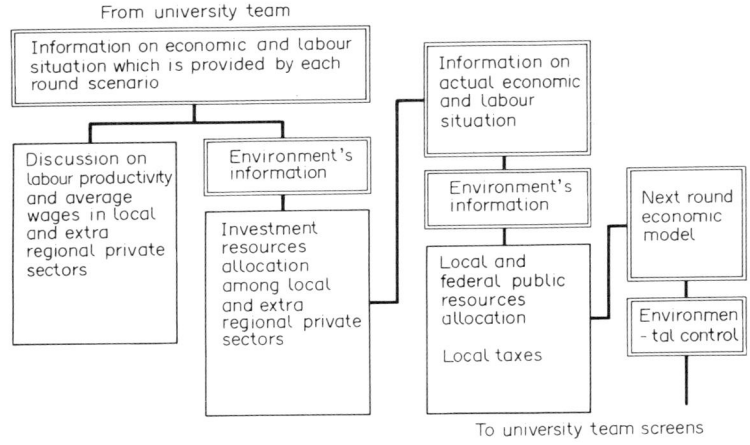

GRAPH 2 Sequence of participants interaction

those that can be used interactively by teachers and participants (Each Round's Model Screens).

The Scenario Model Screens aim at introducing the information about the environmental impact threshold of the economic activity as regards the technology on one side, and on the other, the inputs needed to generate the hypothetical situation in which the economy is supposed to develop if the external factors act as expected.

The economic system of the region greatly depends on the external economy. Its basic economy is made up of the agriculture and cattle sectors, mining, tourism and small industries which use mining and agricultural inputs.

Any possibility of growth in these sectors depends on the evolution of extraregional demand. The instability of the internal and external markets as a consequence of the fluctuations of the monetary policies (real interest rates, quotation of foreign currency), changes in the orientations of big dominant firms (among which the multinationals appear), norms for importation and exportation, international price fluctuations, etc., generally prevent the fulfilment of long-term plans and at the same time, they obstruct the continuity of local investment.

Thus, the research-teaching team varies the extraregional situation round by round, and they do not fulfil the short-term perspectives that appeared in the previous round, with the intention of representing the consequences of the situation of economic dependence (internal and external colonialism) to which the small regions of underdeveloped countries are subjected.

It is within this scenario that the participants act. The influence of cultural attitudes produced by that context is analysed during the debriefing.

In the following screens working conditions are discussed, private investments are decided, and public measures of resource allocation are taken.

Session Review

During and after the presentation of BIOSPHERE AND UNDERDEVELOPMENT, a series of comments were generated which will help to improve future versions of this simulation.

- The need was recognized to introduce into the performance of the simulation more elements to support the interaction among participants. The "abstract" Computer Model Screens should be complemented by "tangible" things such as "money", maps, etc.
- It is a general concern, among social simulation modellers, to pay attention to the way participants manage to define social needs. This is very important when playing games in "Third World" Countries. In the case of this simulation a sugggested improvement was to provide a set of concrete investment projects, giving the participants the opportunity to establish priorities by themselves.
- There was general agreement and interest in exploring the incorporation of artificial intelligence advances into the field of simulation, mainly to represent non-deterministic phenomena as environmental impacts.
- This incorporation would provide a new kind of interaction computer – participants: which is the new dialogue: social actors – expert system.

The team in charge of UNESCO's project has started working trying to add these suggested modifications to BIOSPHERE AND UNDERDEVELOPMENT.

References

Crookall, D., Oxford, R. and Saunders, D. 1987. Towards a Reconceptualization of Simulation from Representation to Reality. *Simulation and Games for Learning* **17**:4.

Klabbers, J. 1982. Investigacion sobre futuros y elaboracion de politicas en el sector publico. *Cuadernos de Politica Social* numero 34.

Simulation of integrated rural development with IRDEM

Frank R. Rijsberman and Gerrit Baarse
IFIAS and RESOURCE ANALYSIS respectively, Maastricht, The Netherlands, both formerly of DELFT HYDRAULICS.

ABSTRACT: A simulation model was developed during 1986–1987 to support evaluation of rural development strategies. The main contribution of the model is that it is an attempt to quantify the effects of projects in different economic sectors (mainly water and agriculture) in a unified framework.
  The input data describe the land and water resources of the planning area, the population and the production facilities, particularly for cropping, livestock and fishing activities. Based on both a scenario of autonomous developments such as population growth and urbanization rates and a rural development strategy consisting of a mix of projects, the model simulates the development of a series of indicators over time. Seventeen types of projects can be implemented at any time in the planning period, dealing with public or private water supply, land reclamation, resettlement, agricultural production, and provision of educational and medical facilities. The model tracks indicators dealing with quality and quantity of food available, quantity and quality of water available, land use patterns, income, employment, and availability of public services.
  The model is currently implemented as a LOTUS-123 spreadsheet. The structure of the water and agriculture sectors are based on experience in the Lake Basin area in Kenya. The model has been used for a workshop for the staff of the Lake Basin Development Authority, and was well received.
KEYWORDS: Simulation modelling, policy evaluation, rural development, water, agriculture, environment, economics.
ADDRESS: International Federation of Institutes for Advanced Study, Witmakerstraat 10, 6211 JB Maastricht, the Netherlands (phone: 043-250465), and RESOURCE ANALYSIS, A.O. Minderbroeders 12, 6211 HM Maastricht.

Introduction

Since McNamara, then President of the World Bank, placed major emphasis in development aid on helping the rural poor and rural development in the early 1970s a large number of rural development programmes have been undertaken. This shifted attention from execution of relatively large, self-contained projects to the development of rural development programmes which often consist of a large number of small projects in different economic sectors – what can be called a programmatic or process approach (e.g. Euroconsult, 1986). Many of these programmes have not been very successful, Lai and Haiden (1984) conclude, for instance, that:

> Rural-development schemes in Africa that could be described as successful unambiguously are remarkably hard to find. On the whole, outcomes range from the resounding failure, . . . , from

which there were virtually no returns despite large expenditures, to the expensive success which looks impressive and may have given great benefit to those fortunate people affected by it, but at a high cost (p.534).

It would appear that this is one area where there is a lack of competence in dealing with complex systems. One cause of this incompetence is no doubt the data required to evaluate the impact of a rural development strategy consisting of a mix of many projects in a number of economic sectors, which have not normally been available. In many respects the development and evaluation of rural development strategies has been a qualitative process, probably more art than science.

In the wake of the microcomputer revolution numerous authorities in developing countries are currently involved in the development of various types of databases. Areas for which databases are developed include, for example, environmental planning and natural resources management. The quantitative information in these databases can allow planners and managers to assess their natural resources and the condition of the natural system, and evaluate the consequences of their actions. This will only be true, however, if two conditions are met:

- the data in the database have to be reliable, complete and updated regularly – no minor task; and
- the planners and managers have to be able to use the often massive amounts of data for their specific needs – by no means a characteristic all databases automatically have.

A persistent problem with "realistic" computer models used for planning has been that they often require large amounts of data which were hard or impossible to come by. With some of the database projects the inverse situation occurs: there are large amounts of data, but no structured tools to bring them to bear on the problem.

In the framework of a project to develop a combined landuse and water resources database, and eventually a socio-economic database for the Lake Basin Development Authority (LBDA) the need for a quantitative tool to use the data in the databases for environmental planning and natural resources management was keenly felt.[1] Planning and management across economic sectors to develop a specific region, with emphasis on environmentally sustainable development, will be indicated as "integrated rural development" in the remainder of this paper. A relatively simple interactive simulation model was developed to address the small scale rural development problems that exist in a considerable part of the Kenyan Lake Basin.

Typical problems are:

- insufficient domestic water supply;
- poor public health conditions;
- soil erosion and soil degradation (fertility depletion);

- low agricultural production;
- low income (lack of primary income sources); and
- lack of public service facilities (educational, medical, transportation and marketing system, credit institutions).

This paper describes the model that was developed to create a tool that would take advantage of the data becoming available through the database projects to help evaluate rural development strategies. The paper discusses the use of simulation models for natural resources management. The paper advocates a decision support approach to modelling and outlines some of the consequences of this approach for modelling that deal particularly with the interaction with decision makers. The structure of the model is outlined and its basic assumptions and drawbacks are mentioned.

Illustrative results for an analysis of a small region in the Lake Basin area, the Kanjira sub-region, are presented. The data used are realistic estimates put together in Kenya. A workshop was organized in Kenya during which the model was used by LBDA staff. The main comments received during this workshop are presented in the section on conclusions and recommendations.

Use of Simulations for Decision Support

The authors favour a decision support approach to simulation modelling, based on their experience with development of policy analysis approaches to natural resources management issues (e.g. Baarse and Rijsberman, 1986; Baarse and Rijsberman, 1987; and Rijsberman, 1987). A decision support system approach is, for instance, discussed by Grigg (1985). The term describes the role of information, computers and modelling for planners: that is, their purpose is to provide information for management. Without being a comprehensive definition, the key concept is that the decision support system provides decision information from a computer, data and modelling base, to help solve unstructured problems. Unstructured problems are those that require judgement and intuition in addition to quantitative information for solving. Decision makers – not planners and modellers – provide the qualitative information and therefore have to be part of the decision support system in an interactive use situation. In other words, the system does not provide "answers" to problems, but information to improve decisions.

Modelling for Decision Support Systems is characterized by Horsey (1986) through:

- incorporation of the decision makers in the analysis, to account for intuition and judgement as well as quantitative data for unstructured problems;
- adaptibility and flexibility in response to changes in decision or data structures;
- focus on ease of use in interactive mode for those less than familiar with computers; and

- information management emphasizing creative use of formats, contexts and media for use of information in the analysis.

The modelling approach used here resembles a Forrester-type dynamic simulation as used before for water supply system dynamics (Grigg and Bryson, 1975), later implemented on a spreadsheet for enhanced interaction during simulations in workshop situations (Rijsberman, 1987). The dynamic simulation is based on flows of goods and stock, with feedback provided through information about the system at any time.

The IRDEM Model

For the analysis of rural development options, a specific model was developed, the so-called Integrated Rural Development Model (IRDEM). The model is based on the following main principles. Given the available land and water resources of a defined area, a description is given of the development of population, land use, production activities, etc., over time. Status and developments of the rural system are expressed in a number of socio-economic and land use indicators, that are considered relevant for planning and evaluation in rural development. Desired developments of various kinds are brought about by human actions which are referred to as "projects", while an integrated mix of projects is referred to as a "strategy". The model aims to facilitate the identification, comparison and evaluation of such projects and strategies.

The overall structure of the model is given in Fig.1.

IRDEM should be regarded as an integral planning tool which provides an overview of the overall situation in the modelled area with respect to land use, food production, income and water availability and the development of these quantities over time. The results of more detailed analyses to be carried out with more specific models can be inputs to IRDEM.

A version of IRDEM was developed and made operational during the Kenyan database project. It is based on a spreadsheet approach and is programmed in LOTUS-123. The model is user-friendly and easily understood and modified. A full description of the model structure takes up too much space in the context of these conference proceedings, but is available from the authors. An understanding of the level of detail in which the various processes are modelled can be obtained from the Tables 1 through 4 that list key elements of the model, that is, input data, projects and scenario variables considered, and model output.

The Kanjira Cast Study Area

Description of the Kanjira sub-location

In selecting an appropriate example study area, the following criteria were used:

Simulation of integrated rural development with IRDEM 313

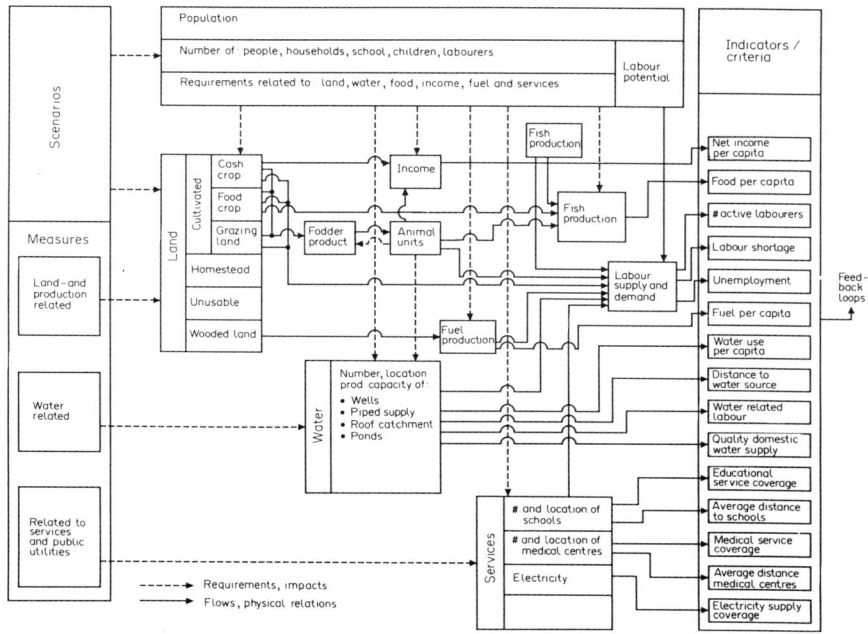

FIGURE 1   IRDEM structure

- a relatively densely populated area of limited size (sub-location);
- the existence of at least a number of the above problems;
- not too far from Kisumu, so that the area could be easily visited.

Based on the impressions of a field trip, which was attended by all members of the LBDA project staff, the Kanjira sub-location was adopted as a suitable area. A second field yielded the input data not already in the database.

Kanjira sub-location is a relatively small and densily populated area. Problems relate to domestic water supply, land resources, agricultural production, income situation and the availability of public services and facilities.

Water Supply. The major source for domestic water supply in the Kanjira sub-location is Lake Victoria. The average distance to the Lake for the people living in Kanjira is about 5 kilometres (ranging from about 1 to 10 kilometres). Other important water sources at present are the seasonal ponds and

TABLE 1 *Population and land input data for Kanjira illustrative analysis*

a. Population-related input variables	VAR	VAL	DIM
Number of people (t = 0):	NP	7500	–
Average number of people in "household":	NPH	6	–
Potential labour rate:	PLR	50	%
Fraction of people of school going age:	FPSA	40	%
(Net) population growth rate:	PGR	4	%/yr
Food (protein) requirements per capita:	PRC	20	kg/yr
Number of households with electrical supply:	NHES	5	–
Water requirements per capita:			
• distance class 1:	WRC1	25	l/d
• distance class 2:	WRC2	15	l/d
• distance class 3:	WRC3	10	l/d
Water carrying capacity per manyear:	WCMY	150	m3.km/y
Weights for water quality index computation:	W1	10	–
	W2	7	–
	W3	3	–
	W4	1	–
Number of schools (t = 0):	NSC	8	–
Number of medical centres (t = 0):	NMC	3	–
Average capacity of school (number of people):	NPSC	280	–
Average number of people served by medical centre:	NPMC	1500	–

b. Land-related input variables			
Total area:	TA	1900	ha
Total agricultural area (t = 0):	TAA	1500	ha
Area wood land (t = 0):	AWL	200	ha
Area not useable (t = 0):	ANU	100	ha
Fraction fallow land (of crop land):	FFL	0	%
Urbanization fraction agricultural land:	URAL	0.4	–
Urbanization fraction wood land:	URWL	0.3	–
Urbanization fraction unuseable land:	URUL	0.3	–
Desired area cash crop:	DAC	500	ha

streams which only hold water during the rainy seasons and some period after. The people are divided into groups and every group has rights of use to a pond. As can be inferred from the 1:50,000 topographical map, in total at least (some) twenty of these seasonal ponds and streams exist. The groundwater is too deep in most places and saline in others. As a result, little use is made of groundwater in the present situation.

From some houses with corrugated iron roofs, rain water is collected for domestic water supply. Although the roof catchment seems to be a potentially feasible supply source, the use of roof catchments is limited to a few homes only. The main reason given for not using roof catchment is the lack of finance to purchase the storage tanks, gutters and collecting panels.

The water use per capita per day varies with seasons; during the rainy seasons, when the ponds are full the average use per capita per day is estimated at 20 litres. Alternatively, an amount of 10 litres per day is used in the dry season. Estimates given for the water use of animals are: 30 litres per day for cattle and 5 litres per day for sheep and goats.

TABLE 2 *Water and production input data for Kanjira illustrative analysis*

c. Water-related input variables	VAR	VAL	DIM
Number of "households" with wells:	NHWE	10	–
Number of "households" with piped supply:	NHPS	0	–
Number of "households" with roof catchments:	NHRC	200	–
Number of public wells:	NPWE	0	–
Number of public piped supply points:	NPPS	0	–
Number of public roof catchments:	NPRC	5	–
Number of ponds or small streams (seasonal):	NPS	20	–
Seasonal pond use fraction (part of year):	SPUF	0.66	–
Average water yield public wells:	WPWE	0	l/d
Average water yield public piped supply:	WPPS	0	l/d
Average water yield public roof catchment:	WPRC	800	l/d

d. Production-related input variables			
Average food crop yield:	FCY	1.5	t/ha
Average cash crop yield:	CCY	2000	Ksh/ha
Average fodder yield food crop land:	FYFL	3	t/ha
Average fodder yield cash crop land:	FYCL	2.5	t/ha
Average fodder yield grazing land:	FYGL	1.5	t/ha
Fuel wood production wood land:	WPWL	10	m3/ha.y
Organic fuel production crop land:	OFPC	0.5	m3/ha.y
Number of livestock (livestock units):	NLU	1200	lu
Livestock growth rate:	LGR	10	%/y
Livestock slaughter rate:	LSR	5	%/y
Livestock marketing rate:	LMR	5	%/y
Average meat weight per livestock unit:	LMW	375	kg/lu
Average money value per livestock unit:	LMV	3000	Ksh/lu
Fodder required per livestock unit:	FRLU	3	t/y
Water requirement per livestock unit:	WRLU	40	l/d
Annual milk production per livestock unit:	AMIP	300	kg/y
Annual fish production:	AFP	50	t/y
Nutritional value (protein) food grain:	PFG	100	kg/t
Nutritional value (protein) milk:	PMI	30	kg/t
Nutritional value (protein) meat:	PME	180	kg/t
Nutritional value (protein) fish:	PFI	190	kg/t
Cost of inputs food crop land per year:	CFC	150	Ksh/ha
Cost of inputs cash crop land per year:	CCC	200	Ksh/ha
Labour inputs food crop land per year:	LFC	0.8	my/ha
Labour inputs cash crop land per year:	LCC	1.0	my/ha
Labour inputs livestock per year:	LIL	0.1	my/lu
Labour inputs fisheries per year:	LIF	1.0	my/t
Labour inputs other production activities:	LOP	500	my
Net income other production activities:	INOP	2000	Ksh/my
Indirect labour multiplier:	ILM	1.5	–

Land Resources. Kanjira sub-location comprises an area of 19 square kilometres. A considerable part of the area is occupied by small fields, averaging 0.4 hectares, while the homesteads on average occupy an area of 0.2 hectares. The remaining area consists of grazing land, small open water ponds, small pieces of uncultivated land, wooded areas, an airstrip, market centres, and some (dirt) roads.

TABLE 3 *Projects and scenarios considered in IRDEM*

1. Water-related projects:
   - wells;
   - roof catchments;
   - piped supply schemes.

2. Land-related projects:
   - reclamation;
   - change of land use patterns;
   - soil conservation;
   - intensification of crop land use.

3. Projects related to production activities:
   - increase agricultural inputs;
   - change cropping patterns;
   - increase livestock production;
   - increase fish production;
   - increase (organic) fuel production;
   - increase small-scale production activities.

4. Projects related to public facilities:
   - medical centres;
   - educational facilities;
   - electrification of rural centres.

5. Scenarios:
   - population growth;
   - changes in areas available productive soils (urbanization) and declining soil fertility (soil erosion and/or environmental degradation).

TABLE 4 *Model output variables*

Food (protein) available per capita
Net income per capita
Number of (active) labourers
Labour excess
Labour shortage
Water use per capita
Average distance to water source (people)
Labour involved in domestic water supply
Weighted quality index of domestic supply
Average distance to water source (livestock)
Available organic fuel per capita for people without electricity
Electricity supply service coverage
Average distance to school
Educational service coverage
Average distance to medical centre
Medical service coverage

Agricultural production. The type of farming practised is mainly at subsistence level. The climatological conditions of the sub-location usually sustain only one crop a year due to inadequate rainfall (around 1000 millimetres per year). The major crops grown are maize, sorghum, beans, groundnuts and cotton. Sunflower and sisal are also grown by a few individuals. Yields per crop type per hectare depend on the soil type and fertility, and the availability of rain, and are typically low.

Most of the farmers grow more than one crop type at any time, the most common combination being maize and sorghum, beans and maize, and beans and sorghum. Irrigation does not take place in the area, as this is not feasible in most places (except perhaps near the border of the lake). Nearly all families keep one or more animals.

Income situation. As the farming activities are mainly at subsistence level, the cash income from agricultural production is limited. Other sources of income are hardly available in the area. The average income is therefore low. According to the socio-economic survey that was held in South-Nyanza in relation to the shallow wells project, the annual income per capita is less than 500 Ksh for well over 50% of the population, whereas the income for the remaining part of the population is between 500 and 1000 Ksh per capita per year.

Demographic information. The present population is estimated at about 7500 people (some 1250 households). The majority of the people are farmers though others are engaged in activities like trading, carpentry, masonry, fishmongering and a few other skilled trades. The population seems to be growing at a rate which cannot be supported by the available arable land. Some farmers are engaged in a trade to supplement their income, the main trade being the sale of agricultural products like vegetables, eggs and groundnuts.

Public facilities. Kanjira has eight primary schools and one youth polytechnic. No public health centre exists in Kanjira. There are two private, poorly equipped and staffed, clinics at Otaro and Pala market centre and a mobile clinic providing occasional medical services. The nearest public dispensary is located at an average distance of 6 kilometres and is poorly equipped. There are three market centres, which are all located close to the borders between Kanjira and the neighbouring sub-locations.

Illustrative Results

The illustrative analysis

Problem statement. How to maintain, and possibly improve, the living conditions of the people in Kanjira in terms of available food, income, water supply, health situation, etc.

Objectives. To identify feasible sets of measures (strategies) to meet a number of selected criteria regarding socio-economic conditions.

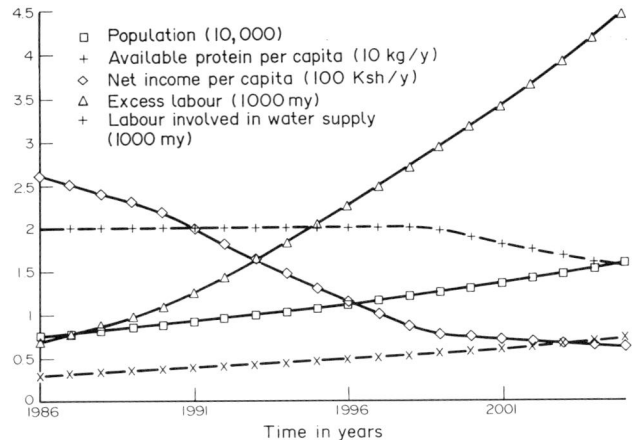

FIGURE 2   Socio-economic variables – no projects

System boundaries. The subject of the study is the total area of the Kanjira sub-location, including the people and animals in it, its land and water resources, the production activities, and the man-created facilities and institutions in the sub-location (roads, houses, medical centres, schools, etc.).

Time horizon. The period of study will be from 1985 till the year 2005, covering a range of 20 years.

Within the constraints of available land and water resources the analysis should concentrate on visualizing the socio-economic and environmental status of the rural system, its autonomous developments and the changes brought about by the various possible measures and combinations thereof. An evaluation of measures should take place by a comparison of alternative strategies in terms of costs and effects according to the adopted criteria. The presentation of alternatives should facilitate a clear insight into the merits of the various courses of action to enable a selection of the "optimal" strategy based on social and political preferences.

The input data used for the illustrative analysis (in the situation without projects) were presented above. The inputs used by IRDEM created the results as shown in the diagrams of Fig. 2 and 3. The following observations can be made on each of these figures.

Observations on Fig. 2

In the course of 20 years we see the population gradually increase to about 17,000. Average net income drops quite dramatically from about 260 Ksh to

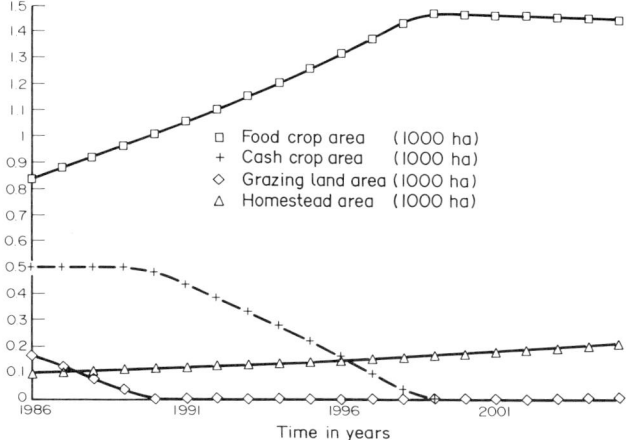

FIGURE 3 Land use variables – no projects

some 70 Ksh per capita, due to the lack of income generating production activities and the increasing population. For the same reason, the excess labour shoots up from about 700 to 4500. After 1997, the total amount of food produced falls short in providing the population's basic needs, and we see the available protein per capita drop from 20 to about 15 kilogrammes/ year in 2005. Increased population and no additional (near) water sources will cause the amount of labour involved with the water supply (walking up and down to seasonal ponds and Lake Victoria) to rise from about 350 to 700 manyears.

Observation on Fig. 3

The amount of food crop land shows a sharp increase to keep up with the needs of the growing population. This increase takes place at the expense of grazing and cash crop land. In 1998, the total agricultural area of Kanjira is occupied by crop land and the cash crop area will be zero. Grazing land already becomes zero in 1990. In addition, total agricultural area decreases because of the increase of homestead area, due to rising population. After the food crop area reaches its maximum, the available food per capita starts decreasing as could be observed in the previous figure. The fact that cash crop area decreases and eventually becomes zero is a major factor causing the drop in net income per capita.

With growing population and no additional water sources, the average distance to the water source increases as relatively more people have to rely on seasonal and permanent distant sources. Two effects emerge from that. The weighted water quality index goes down, as the seasonal ponds and Lake

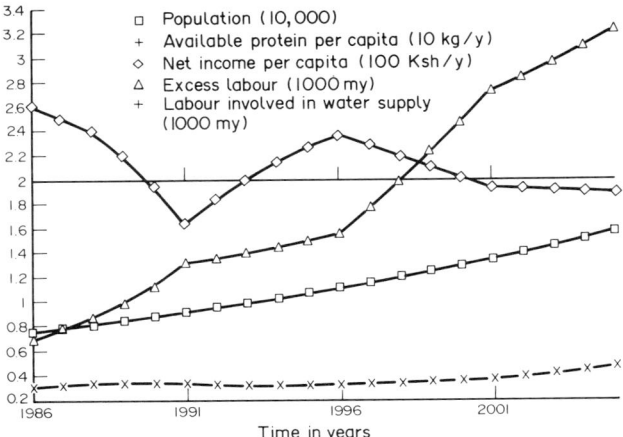

FIGURE 4 Socio-economic variables – no projects

Victoria have a low water quality index (to express the poor quality of these sources relative to groundwater wells, roof catchments and piped supply). Also, the water use per capita per day goes down, as people tend to use less water if they have to go further to get it. The average distance to the water source for livestock remains the same, as they only use the seasonal ponds and the Lake and no changes occur to these sources.

Effects of measures

As an illustration of the effects of possible measures to improve the conditions of the rural population, a number of different situations were simulated with IRDEM. These simulations produced the diagrams as shown in Figs 4 and 5. The following observations can be made regarding these figures.

Observations on Figs 4 and 5

A comparison should be made with Figs 3 and 4, respectively. The following measures to improve the situation were simulated.

- A gradual increase of food crop production per hectare, rising from 1.5 ton/ha to 2.5 ton/ha in the course of 10 years (1991 to 2000).
- A gradual increase of cash crop production from 2,000 to 3,000 Ksh/ha in the course of 10 years (1988 to 1997).
- An increase in the number of manyears involved with "other" production activities (generating net income worth of 2,000 Ksh per manyear). Increase is 100 manyears each year during the period 1991 to 1995 and 2001 to 2005.

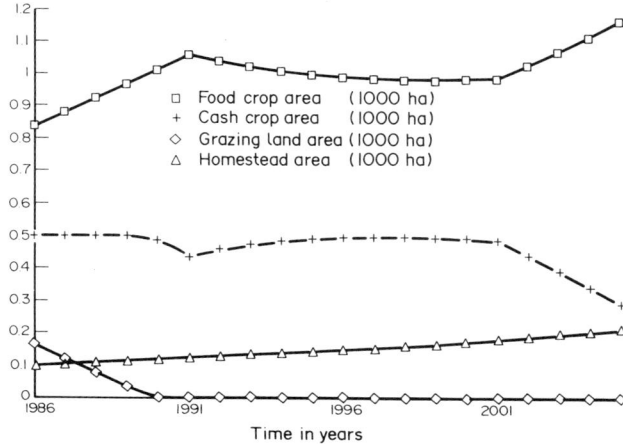

FIGURE 5 Land use variables – with projects

From Fig. 4 it can be observed that available food per capita now keeps up with the needs. Income per capita varies somewhat over time, and finally decreases to less than 200 Ksh in 2005. Excess labour now increases to about 3200 in 2005. No changes occur to population. Manyears involved with water supply go down due to some water projects. From Fig. 5 it appears that food crop area increases quite less dramatically now (to about 1150 ha in 2005). Cash crop area goes down somewhat, increases again and finally goes down to a little under 300 ha in 2005. Grazing land still goes to zero in 1990, as this is sacrificed to the increasing food crop area first. No changes occur to the development of the homestead area.

Session Review

The session started with an introduction of the model, explaining its background, logic behind its structure, and advantages and drawbacks, presented by the first author. The session participants subsequently explored the model (one person per computer). The design of the session included two hands-on sessions: (a) to explore the model; and (b) to test it through evaluation of a particular strategy. For two reasons the two sessions became one big session in practice. First, the model proved to be quite complex to understand for persons without some background in planning and development issues so that the first session took more time than expected. Second, because of the small group and the luxury of one computer per person, it was possible to explore and test the model at their own pace in a less structured manner than required for group work.

Even though the model is fairly complex, the participants became quite interested and succeeded to get to a level where they were interested in testing their own specific ideas (as opposed to merely taking in the structure and data already in the version they were presented with). It became clear, though, that the model can be more satisfactorily used for workshops of several days (for which it was developed) rather than sessions of a few hours, to allow the participants to achieve a higher level of control over the model.

In the discussion that concluded the session several interesting remarks and suggestions were made. They can be summarized as follows.

- It would be useful to have a facility to save specific strategies and their model results, for easy access to successful strategies afterwards.
- The automatic scaling of output graphs employed by the model confused some of the participants.
- Several participants found it frustrating that (given the population growth scenario in the model, using current Kenyan population growth rates) the model results invariably gave a very high unemployment, virtually regardless of the projects in the agriculture sector devised by the participants. That is a very real-life problem. It was suggested that the model should therefore pay more attention to modelling the employment opportunities in other sectors or outside the modelled area.

Conclusions and Recommendations

The modelling concept for IRDEM was specifically developed for the case study in the database project. A first version of the model has been programmed in a spreadsheet using LOTUS-123. This version, reflecting the above concepts, is now operational.

However, the concept as presented above should be regarded as a preliminary version. In discussions within the project team and with a number of senior planners within LBDA, some useful suggestions were made for future extensions of the modelling concept. In view of further developments aiming at a more general application of this concept, a short summary of these suggestions will be given below.

1. A major comment is that the above approach mainly deals with the socio-economic indicators that can be quantified. In addition, there may be a number of social and cultural variables that are very important for the final judgement on the feasibility and desirability of certain courses of action. Especially the acceptability by the local population, and their capability to deal with imposed changes is a decisive factor for the success of any proposed measure. The suggested approach does not provide this kind of information. The implication is that the approach can only be successfully applied in combination with a thorough social and cultural appraisal of the feasibility of alternative development options.

2. The model in its present version, reflects a more or less closed system. The basic assumption is that the study area should provide the basic needs of its population. In reality, all kinds of interactions exist with the surrounding areas. Especially when the situation becomes worse in future due to population pressure and the limited production capacity of the system, this is likely to have a major impact (people migrating or trying to exploit sources outside the system considered). It should be noted, however, that assumptions regarding these kinds of responses could be incorporated in the modelling concept in a fairly simple way.

3. In the present set-up, certain elements of the socio-economic and natural resources system have been schematized in a simplified way. Possible extensions in order to create a more powerful and more generally applicable tool are:

   - adding a demographic model (including e.g. fertility, mortality, migration, age structure);
   - extending the types of areas to be considered, e.g. forests, marshes, mountainous areas etc. that have different production potential from the areas presently considered;
   - including a more complete energy balance that includes all relevant potential sources, e.g. hydropower, fossil fuels, wind energy, etc.

4. A final version of the model should also include the cost functions of the various measures considered so that a full report of costs and benefits (in terms of the socio-economic indicators) can be produced for each strategy.

In trying to include more detailed relations in the model, the danger exists that the model as a whole will soon become too complicated to enable a proper interpretation of model results. Therefore the most logical further development is the creation of a number of alternative model versions that contain more detail with respect to some specific parts only. Depending on the amount of detail required in the description of specific relations and phenomena, the appropriate model version can then be selected.

Note

1. The project referred to, in the framework of which the IRDEM model was originally conceived, is a consulting assignment of DELFT HYDRAULICS for LDBA, funded by the Dutch development aid (DGIS).

References

Baarse, G. and Rijsberman, F. R. 1986. Ecology and tourism: protecting the coast of the Dutch Island of Texel. *Project Appraisal* 1:2.

Baarse, G. and Rijsberman, F. R. 1987. Policy Analysis. *In* Wind (1979).
Euroconsult. 1986. *The Programmatic approach to rural development*. Arnhem: Euroconsult.
Grigg, N. S. and Bryson, M. C. 1975. Interactive Simulation for Water Supply Dynamics. *Journal of the Urban and Development Division, ASCE* 101 (UP1).
Grigg, N. S. 1985. *Water Resource Planning*. New York: MacGraw-Hill.
Horsey, H. R. 1987. *Computer Decision Support for Water Supply Development*. Ph.D. Dissertation. Ft. Collins, CO: Colorado State University.
Lai, J. and Haiden, M. 1984. *Economics of African Agriculture*. Halcrow, UK: Longman.
Morrs, E. R. and Gow, D. 1983. *Implementing Rural Development Projects: Lessons from AID and World Bank Experiences*. Boulder, CO: Westview Press.
Organization of American States. 1984. *Integrated Regional Development Planning: Guidelines and case studies from OAS experience*. Washington, DC:OAS.
Rijsberman, F. R. 1987. Development of Strategies for Planning for Improvement of Irrigation System Performance. *Research Report T329.02*. Delft: Delft Hydraulics.
Wind, H. G. (ed.) 1979. *Impact of Sealevel Rise on Society*. Rotterdam: Balkema Publishers.

CHERNOBYL: a game of negotiations under stress

Claude Bourlès

Université Catholique de l'Ouest, Angers, France

ABSTRACT: CHERNOBYL is a negotiation game that can serve as an introduction to "Black Games", i.e. games dealing with uncomfortable objects. It is a norm-oriented game dealing with the behaviour of regional governments while experiencing a catastrophe. Using a pack of cards, a dice and a set of rules, players of the game can decide how to use the tools they have in order to minimize the damage. The game involves making decisions between helping the population directly or fighting the radio-active emissions on the reactor site, thereby helping the population indirectly. Some rules that are being used try to simulate the atmosphere of an authoritarian administration.
KEYWORDS: Negotiations, totalitarism, environment, black games, ethics.
ADDRESS: IPSA-UCO, BP 808, 49005 Angers Cedex, France.

Improving dental planning through computer simulation

E. M. Bronkhorst[a], G. J. Truin[a], J. H. G. Klabbers[b] and
A. J. M. Plasschaert[a]

[a]*University of Nymegen, The Netherlands*
[b]*Utrecht University, The Netherlands*

ABSTRACT: Recently a simulation model of supply and demand of the dental health care system in The Netherlands has been developed. The model includes major demographic, pathological, psychological, sociological and economical processes comprising the demand side. The supply side covers the availability of dentists, dental hygienists and factors related to care delivery. The main goal of the model is to provide government and local dental health planners as well as insurance companies and the dental profession with a tool to assist them in programme and policy planning. The model will be used to study an extensive variety of policy options related to major aspects of the dental health care system with a time-horizon of 40 years. Such simulations reveal short-term and long-term implications of a particular trend, policy or programme by tracing changes in critical dental variables.
KEYWORDS: Dental health care; computer-simulation; planning; large-scale modelling.
ADDRESS: E. M. Bronkhorst, Department of Cariology and Endodontologie, Dental School, University of Nymegen, Postbox 9101, 6500 HB Nymegen, The Netherlands.

Introduction

During the past decades the imbalance between demand and supply of dental health care in The Netherlands has been a matter of serious concern. In attempting to tune supply and demand, policy makers have to deal with a complex socio-political system and consequently with a dynamic, non-equilibrium system (Forrester, 1969; Mesarovic, 1970; Klabbers *et al.*, 1980).

Adequate policy formation requires not only research findings from different scientific disciplines, but especially implies suitable support systems for handling different dimensions of complexity. During the last decades it has become increasingly clear that both the systems-approach and computer simulation offer favourable perspectives for a more adequate integration of policy-options in complex social systems.

For the sake of surveying and investigating the structure and behaviour of the dental health care system more systematically, system dynamic computer-simulation models of the dental health care in The Netherlands have been developed.

History of the Model

As has been pointed out in the introduction, mainly due to imbalance between supply and demand of dental health care, the development of computer-simulation models to study the important characteristics of this imbalance seemed appropriate. In the late seventies a first relatively global model was developed. The relative simplicity had some disadvantages. For instance, no difference was made between age-groups and only one "treatment" was considered. This model has been embedded in an interactive computer-simulation game (Klabbers et al., 1980). In the game three groups of actors were considered: government, parliament and the dental profession. Each group had only access to partial information and different policy measures. The experiences during the game sessions were twofold. Firstly, the players were enthusiastic and expressed to have gained valuable insight in the complexity of the issue. Secondly, especially the dentists felt the need to look into the problem area in much more detail, focusing on age-groups, pathology etc.

As a result of this experience a model more detailed in the description of the dental health care system was developed and implemented on a mainframe computer. That configuration made it quite difficult to use it as an interactive model suitable for gaming purposes. However, as an instrument for performing scenario analyses concerning dental health care, it proved to be very suitable.

The availability of a new generation of powerful microcomputers made it possible to start the construction of a new simulation model combining the advantages of the first model with regard to its feasibility for gaming, with the level of desaggregation of the second model. Moreover, the construction of a new model offered the opportunity to make the model-structure more flexible. This was very important since the dental health care system in the Netherlands is momentarily in a great flux and some major changes in the financing structure of dental health care are bound to take place. Such a change would effect the whole dental health care system in the Netherlands. This third computer-simulation model of the dental health care system has not been completed yet. The model presented at the conference is a preliminary version. Completion is expected to take place early 1989.

Description of the Model

As has been stated above the model is of considerable size and complexity. Therefore, a full description of the model cannot be presented here. Instead an overview of the most important features of the model is stated below, providing an overall image.

Size of the model. A dominating characteristic of the model is its size. It contains about 300 state-variables and about 900 parameters, most of them time-dependent. The size of the model is also reflected in the various aspects included in the model: the model contains major demographic, pathological, psychological, sociological and economical processes.

Model Structure. An off-line population model is acting as a driving function for the main model. It generates data for a population divided into seven age-groups. In the main model every age-group is divided into two or more socio-economical groups. These groups for instance may represent people with different socio-economical status (SES) or may reflect differences between people taking part in different insurance-programmes. Each socio-economical group in its turn is divided into three groups: people visiting a dentist regularly, people visiting a dentist irregularly and people without any dentition (edentulous). The population structure thus obtained is the backbone of the model. Notwithstanding the links existing between the different subgroups (induced for instance by demographic processes or changing in dental attendance behaviour) the overall model contains a complete and self-contained submodel for each subgroup. As a result of this for each subgroup a set of state-variables (e.g. average number of decayed teeth per person, or percentage of people suffering from calculus) and a set of parameters (ranging from the sugar-consumption to the price-sensitivity for dental treatments considered, for that subgroups) is provided. An impression of the model structure is given in Fig. 1. All cells have the same internal structure. For obvious reasons the cells containing edentulous people have a much simpler structure.

Technical characteristics. The model is a non-interactive discrete-time model. In the future, after completion, it will be embedded in an interactive gaming-environment. The model is implemented in Pascal on a Macintosh II computer. Despite the considerable speed of the Macintosh II, the complexity of the model has its consequences for the run-time of the model. A run covering the 40-year period 1980–2020 has a run-time of about 2.5 minutes. Software for post-processing the data yielded by the model is available.

Session Review

The intention of the session was to give the participants an idea on how the model can be used to improve the competence of decision makers in the field of dental health care. In accordance with the spirit of the conference this was done by means of a gaming set-up. Since the model is not completed yet, a simple game was designed for this occasion. Participants not being familiar with the field of dental health care, were given a short introduction of the model explaining the most fundamental notions of it. The following three topics have been discussed:

1. Construction of the model with emphasis on the possibility to use various scenarios in analysing policy planning concerning dental health care.

Improving dental planning through computer simulation **329**

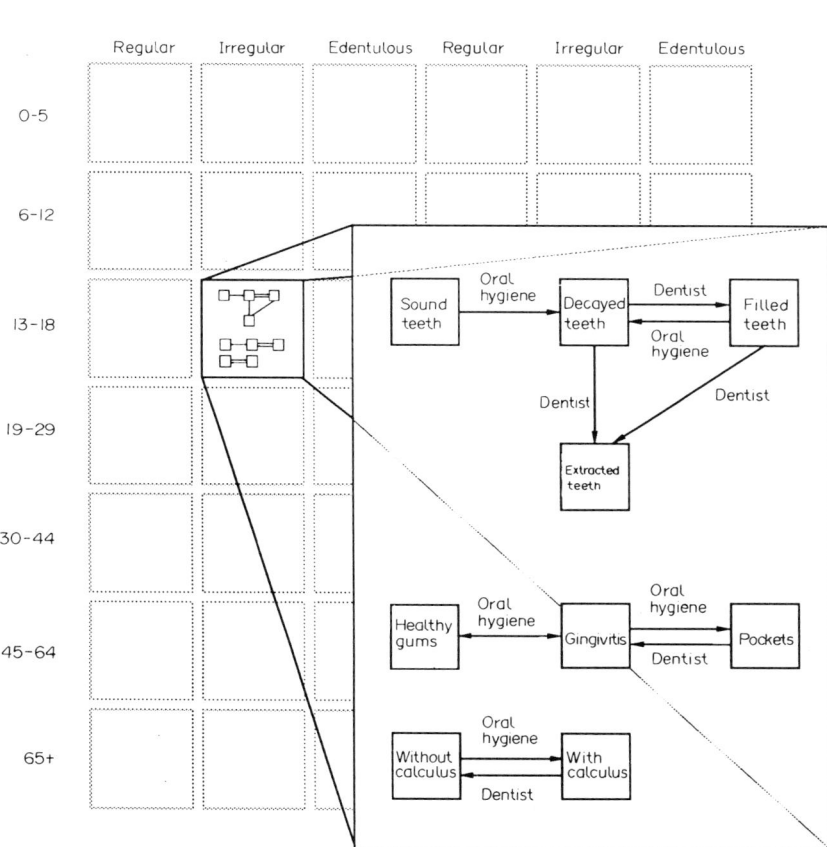

FIGURE 1   Overall structure of simulation model

2. Discussion of the long-term results obtained with the model, including scenarios made by participants in the workshop.
3. Implications of the use of the model. For instance, is the model a suitable tool for policy makers? Or, is the model flexible enough to serve as the core for a gaming environment?

The Game

During the game two players were invited to play the role of an imaginary adviser of the Ministry of Public Health. Their task was to allocate a number of restricted resources to a range of different options. The objective of the game was to find a package of measurements which was most effective in improving the oral status situation of the Dutch population. In order to avoid the game being to technical, the range of options consisted of relatively familiar actions as reducing the sugar-consumption, stimulating preventive measures such as the use of fluoride-tablets or toothpaste with fluoride in various age-groups, or a global cut-down on expenses for dental health care.

Its Results

Despite the rather simple nature of the game, it proved to be rather appealing to the players. During the session, while constructing the package of options, serious discussions about the expected effectiveness took place. A problem mentioned by almost all players was the feeling to be forced to make decisions without having the knowledge to oversee the consequences of their decisions. The occurrence of this problem supports our opinion that the model developed is feasible for getting people familiar with the problems encountered when steering a complex social system. Another promising experience has been acquired with the presentation of data. The model itself only generates the numerical data; the processing of the data is decoupled from the model run, giving the users the opportunity to manipulate the data in a very flexible way. The presentation can be adapted to the requirements for a particular occasion. During the conference, software for converting the numerical data into graphic representations was successfully used. After only a little explanation the players, although not familiar with many variables included in the model, were able to understand rapidly the differences between the results of the scenarios developed. Moreover, the possibility to compare the results of the various scenarios numerically and converting the differences into graphic representations enabled the players to judge for *themselves* about strengths and weaknesses points of their scenario compared to other scenarios. This certainly enforces the individual and collective learning process that gaming is aiming at.

Conclusions

Because the model is not completed yet, conclusions about its overall feasibility can not easily be drawn. Experiences with the preliminary version of the model during testing and during the conference have given us the opportunity to draw some preliminary conclusions:

- Powerful microcomputers like the Macintosh II provide the means to construct models of complex social systems which can combine features from two different fields i.e. simulation (a highly detailed numerical model aiming at an accurate description of the system) and gaming (an ambience of collective meaning processing).
- Software developed for post-processing the data generated by the model, proved to be useful in making the model results easily understandable for the players.

References

Forrester, J. W. 1969. *Urban Dynamics*. Cambridge: M.I.T. Press.
Klabbers, J. H. G., Hijden, P. P. van der; Hoefnagels, K. *et al.* 1980. Development of an interactive simulation game: A case study of the development of Dentist. *Simulation & Games* **11**:1
Mesarovic, M. D., Macko, D. and Takahara, Y. 1970. *Theory of hierarchical, multilevel systems*. New York: Academic Press.

Physiotherapist's dilemma

Louwrens ten Brummeler and Cor van Dijkum
Utrecht University, The Netherlands

ABSTRACT: In this paper a simulation game "Therapist's Dilemma" is described which have been developed for physiotherapists, family doctors, patients and representatives of health insurance companies. The game has to produce insight into the quality of treatment and related cost of physiotherapy. The game is interactive: (1) the participants are making joint decisions; (2) the consequences of these decisions are calculated with a computer program; (3) results are returned to the participants and are evaluated by them; (4) new decisions are made for the next cycle. The game is played in three or four cycles and is completed with a debriefing. Special interest is given to the introduction of the game with the help of a hypercard tutorial.
KEYWORDS: Quality, cost, physiotherapy, health care centre, interactive, causal map, vicious circles, hypercard.
ADDRESS: Faculty of Social Sciences, Utrecht University, Heidelberglaan 1, 3584 CS Utrecht, The Netherlands.

Introduction

Physiotherapists in the Netherlands are confronted with a financial system stressing productivity. Rigid rules by financiers leave little room for adequate quality control. The game we shall discuss here focuses on financial as well as organizational aspects of physiotherapy. It is based upon an evaluation study concerning quality and cost of physiotherapy in health care centres in Amsterdam. The game represents causal relations between various aspects of quality and cost. The purpose of the game is to help physiotherapists gain insight into chains of cause and effects. In this way they will be able to integrate theory into their practice and improve the quality of their work.

The Problem Situation

In our evaluation study (see Bertels, 1985) we showed that it is possible to improve the quality of physiotherapy and at the same time reduce the cost.

Physiotherapy in health care centres is a case in point. For example we demonstrated that physiotherapy can be a suitable substitute for the more expensive and specialized treatment of the medical specialist and family doctor.

In the study we were specifically interested in factors which determine quality and cost of treatment. "Recovery of daily functions" and "gaining insight into the nature (cause) of the complaint" appeared to be characteristic for the quality of treatment. This quality is determined by the kind of activities the physiotherapist has chosen and of course by the characteristics of both the complaint and the patient. The physiotherapist's approach itself (quality as a process: for example by "systematic transfer of physiotherapeutic knowledge to the patient") could indirectly effect the results. Overall cost of treatment depends both on the family doctors referral policy as to the physiotherapist's choice of treatment.

However the different actors in the field (e.g. physiotherapists, family doctors, medical specialists, the association of doctors, insurance companies, government, etc.) act from different perspectives on health care and use different instruments to control quality and cost. For example: government and doctors have different views on quality. Monitoring the quality is the result of negotiations between all actors concerned and is usually incomplete. In consequence quality is in most cases the outcome of *ad hoc* policy. Many options for improving quality as well as reducing costs are not utilized.

It is important that the participants are able to identify this basic problem in the game. Most important is that they can reconstruct the laborious process of negotiation, in order to see where it goes wrong and to become aware of various vicious circles (see Dijkum, 1988) which obstruct improvement of quality and reduction of costs and which are generated by actors pursuing their particular goals. To reconstruct these vicious circles one first has to reconstruct the causal chains which constitute physiotherapist's performance. These chains could be visually represented in a map as follows:

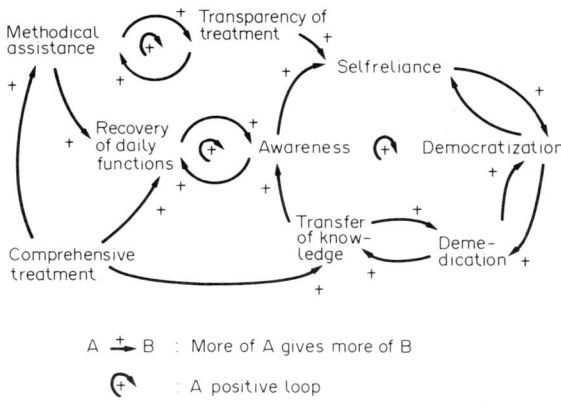

FIGURE 1

Put into Words

Comprehensive treatment implies that physiotherapists take into account circumstances of work, housing and social situation, to recover daily functions. This enables patients to become more aware of their lifestyle, as a potential cause of their complaints. Such an awareness may speed up the recovery process. Comprehensive treatment by the physiotherapist stimulates the transfer of physiotherapeutic knowledge about the complaint, which may improve the patient's understanding of his/her illness. Comprehensive treatment demonstrates the physiotherapist's skill as well, affecting favourably positively methodical treatment as such. This stimulates the transparency of the treatment, which permits the patient to increase his self-reliance, which triggers a positive loop consisting of participation (democratization), demedication, transfer of knowledge and awareness. In this way the effect of comprehensive treatment is amplified.

Both positive and negative loops are possible in these cause and effect relationships. In this way vicious and obstructing circles originate. For example: a decrease in transparency will cause a lower level of awareness, and subsequent reduced self-reliance. This is the problem the game has to deal with.

The Purpose of the Game

During the game players will reflect on questions like: "Which factors determine quality and cost of physiotherapy? Which vicious circles obstruct the actor's attempts to improve quality and reduce the cost?

In the game two treatment settings are relevant. First of all physiotherapy carried out in health care centres. Secondly, physiotherapy performed by therapists and family doctors working in private practice. In the game these two settings are compared with regard to quality and cost.

Patients are important in the game. The whole game concentrates on the treatment of patients. Patients' representatives play an especially important role in the discussion of the quality of treatment. The other actors are: physiotherapists and family doctors in health centres, physiotherapists and family doctors in private practice and representatives of financiers and health care policy makers.

The game should give these actors insight into the process of producing quality and making costs and into the influence they have on that process. In order to do so they have to discover their own particular role, the role of the other actors and discuss each other's position in order to share an understanding of each other's position. Via these communications and interactions we are aiming at a process of convergence towards mutual understanding i.e. metasubjectivity.

The Simulated System

The evaluation study provides the game with a metalevel of empirical knowledge. The results of the study are used as hypotheses to be questioned during the game. A computerprogram as part of the game reflects the empirical facts from the research. It represents the factors which determine the quality and cost of physiotherapy. A linear model has been deduced by means of numerous multivariate-analysis and programmed in a spreadsheet programme. In this model cost and quality are the dependent variables. The independent variables are: categories of complaints, the doctor's policy of referral and the physiotherapist's types of treatment. In the game this model is a starting point for a discussion on quality and cost of physiotherapy. In the game players become familiar with the results of the evaluation study. They are in a position to confirm or refute them and consequently gain new knowledge.

In our game then, it is simple to quantify costs that have to be paid for treatment. It is also easy to specify the way the costs are determined by the measures of doctors and physiotherapists.

It is more difficult to quantify quality. Quality in the game is calculated as a score, on a scale. Those calculated scores are compared between the physiotherapists working in the health care centre and the physiotherapist working in the private practice. The results are interpreted using qualitative considerations, resulting from our field study i.e. opinions and experiences of patients concerning the quality of treatment. In this way the objective and subjective aspects of the problem situation are simulated. The computer program presents the objective and quantitative aspects, the actors in the game introduce the personal, subjective and qualitative aspects. It is in this way that the game generates the causal chains of the problem situation, see Fig. 2.

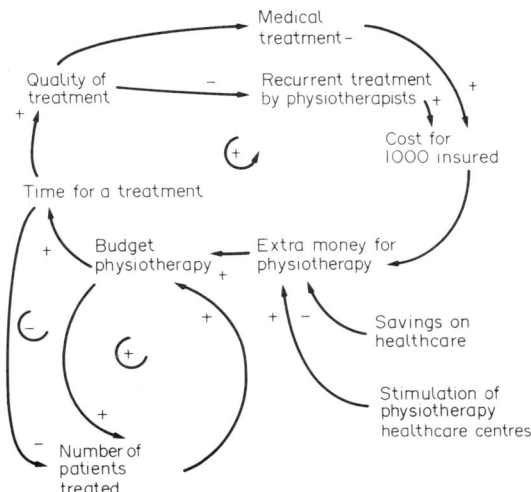

FIGURE 2

In certain situations a higher quality will induce reduced cost and that will induce increased quality. The reverse can also occur. In this way a vicious circle can originate which leads to raising of the cost and lowering of the quality. The actors in the game will try to resolve this vicious circle, but as you can see in the diagram one not only has to look at the quality of each treatment, but also at the number of patients who are treated. This complicates the issue of quality and cost considerably. The players have to resolve it to be successful.

The Game Itself

To start the game we have chosen an unfamiliar approach. We are using a tutorial (programmed instruction with the help of a computer) which guides the players through the game. In this way the participants get a quick overview of it before they start playing.

The game starts with a client seeing the family doctor. The players are confronted with realistic cases. We will illustrate this with the help of the hypercard tutorial[1].

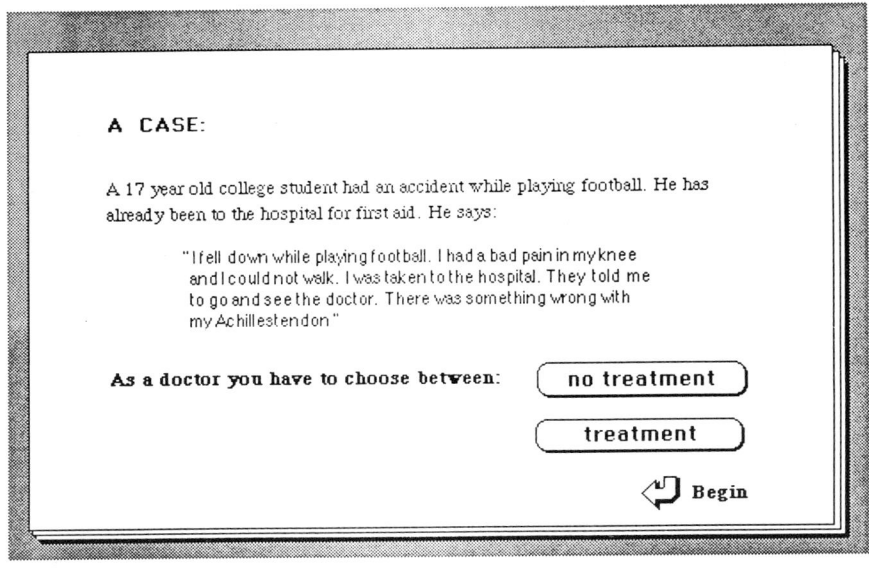

FIGURE 3

The doctor has to make several choices: 1. Is treatment necessary? 2. If so: by whom? By himself/herself or by a specialist, and if so which specialist? Or by a physiotherapist? When the doctor sends the patient to a physiotherapist, that physiotherapist has to decide: 3. What kind of treatment will be necessary? 4. How much time he/she has to spend on this patient? 5. How many visits will be needed? The physiotherapist can, in general, choose between more or less technical activities. For example he/she can choose: massage, exercises, electrical treatment, explaining to the patient the origin of the complaint. According to the insurance companies all these activities take different amounts of time. Each activity takes a standard amount of time and requires specific equipment. The physiotherapist has to find an acceptable balance between quality and cost in terms of equipment, number of visits and length of a visit. Anyway the physiotherapist has to decide about:

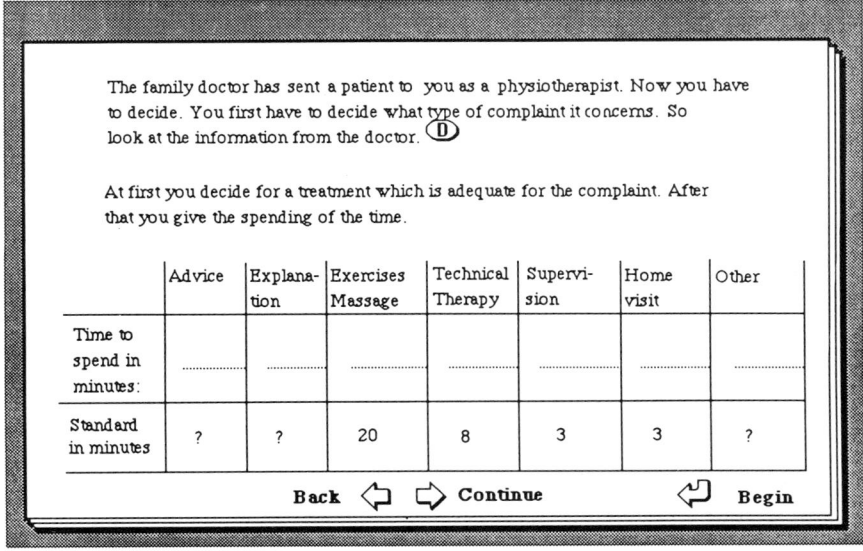

FIGURE 4

The patients' representatives receive the description of the hypothetical cases too. However, they receive more information. In particular they are informed about the actual treatment. On the basis of the evaluation study this involved information on the effect of treatment, both from the point of view of objective measurement, and from the patient's own opinion. In this way the patients' representatives are able to compare and to judge. They are able to compare the way a given complaint was actually treated with the way the physiotherapists treated the complaint in the game. With the help of this comparison the patients' representatives can judge what quality could be expected in the game. They are asked to express their expectation in a score on a scale running from 1 to 10.

After their qualitative sorting out of considerations for treatment the doctors and physiotherapists are asked for their planning for a period of one month. They are now confronted with a whole cohort of patients. These

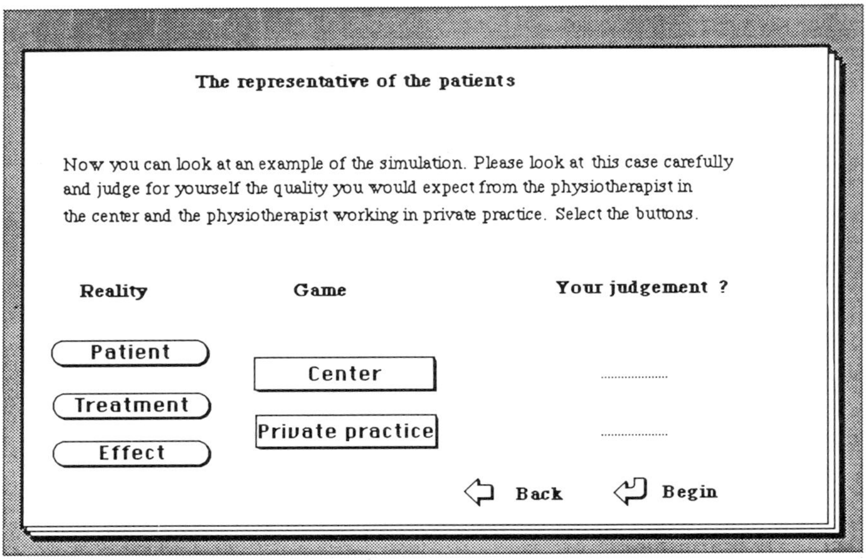

```
┌─────────────────────────────────────────────────────────────┐
│              The representative of the patients              │
│                                                               │
│   Now you can look at an example of the simulation. Please   │
│   look at this case carefully and judge for yourself the     │
│   quality you would expect from the physiotherapist in the   │
│   center and the physiotherapist working in private practice.│
│   Select the buttons.                                         │
│                                                               │
│      Reality           Game            Your judgement ?       │
│                                                               │
│     ( Patient  )    ┌─────────┐                               │
│                     │ Center  │        ............           │
│     ( Treatment )   └─────────┘                               │
│                     ┌──────────────┐                          │
│     ( Effect   )    │Private practice│      ............     │
│                     └──────────────┘                          │
│                                   ⇐ Back    ⇐ Begin           │
└─────────────────────────────────────────────────────────────┘
```

FIGURE 5

patients are not described in detail, but in general, requiring the doctors and physiotherapists to plan treatments concerning groups of patients. This is not an easy problem for the physiotherapists. In any case the plans and judgements of patients' representatives are fed into the computer program subsequently. The resulting cost and quality are calculated. These figures are shown to the doctors, physiotherapists, patients' representatives and insurance companies. After which it becomes clear what has gone wrong.

For example insurance companies are confronted with costs which are too high. In any case it is obvious that in the plenary meeting, which follows the first round of the game, the insurance companies will present compelling financial limits. Probably the health care centre will be told to revise their treatment policy, otherwise there will be a financial disaster. The physiotherapists will argue that they provide better quality than the physiotherapists in private practice. They will also draw attention to the low cost of treatment in the centre, in relation to the cost of the private practice. They will look for support from family doctors, doctors working at the centre and patients' representatives.

The game is now on the way and can safely run for four or five rounds.

Summary

So far, while running "Therapist's Dilemma" we have noticed that actors gained essential insight in the planning of treatment. Options and opportunities are discussed that otherwise would be tackled much less fruitfully.

The hypercard tutorial is a promising way of introducing the game. It is even possible to put the whole game session into a hypercard program. As face to face discussions of the participants during the game are very important for its success, we wonder however whether we will lose important features of the game if we put it all in hypercard. We shall continue to experiment on the use of such tutorials.

Note

1. In the program hypercard (on an Apple Macintosh) a tutorial for a game can be programmed easily. People who are interested can write to us and receive a copy of the hypercard stack.

References

Bertels M., Brummeler L. ten, Dijkum C. van, Giebels R. and Van der Mannen J., 1985. *Tijd voor kwaliteit*. Amsterdam: Siswo.

Dijkum C. van 1988. *Spelen met onderzoek: naar een nieuwe methodologie*. Meppel: Boom.

Diplomatic games

Paul W. Meerts

Netherlands Institute of International Relations "Clingendael"
The Netherlands

ABSTRACT: The author tries to answer the question: are games useful as a means to train people for the diplomatic service? If yes, what kind of games are most suitable? How to build them, practice them, evaluate them?
KEYWORDS: Training of skills concerning negotiation processes, enlargement of insights in international affairs, balance and dynamics of diplomatic games.
ADDRESS: Clingendael Institute, Clingendael 7, 2597 VH The Hague, The Netherlands.

Training Diplomats

To train a professional diplomat, is to prepare him/her for a career with an abundance of value and interest conflicts, uncertainty (in public and private affairs), and complex relationships between people and institutions.

As a diplomat is a representative of a country or a group of countries, the essence of his work is defending the interests of those he represents. It is needless to say that these interests are very complex and cover a broad scope of issues. This means that a diplomat must be competent, actually, he cannot be competent enough. Of course diplomats are not necessarily the most competent people around (Renaissance men are hard to find and do not necessarily join the diplomatic service), if they were it would be an advantage.

In view of the complexity of a diplomat's work it is worth while to enhance his competence by training him/her to deal with difficult situations involving a broad variety of issues and factors like politics, trade, development, security, culture, law and human rights.

There is a good reason for training them in skills in order to cope with uncertainty in situations where the ties with their power-base are weak. In a crisis for example, they may have to act without the support from their colleagues at home because time is running out, or they do not react adequately because they are not able to, or do not want to.

Besides the above-mentioned value conflicts will occur in abundance: different political and social cultures, clashing cultural values. Nevertheless: the gap must be bridged, so knowledge and skills are at least useful tools in managing the crisis.

The Context

Conference diplomacy in the form of a simulated multilateral meeting is the game most suitable for the training of future, and even experienced diplomats. Because it prepares them for a highly complex situation, where uncertainty can be diminished by setting strict rules, and value conflicts play a prominent role as a factor obscuring the negotiations that are going on. Actually, everything could be summarized as a negotiation between many parties in a situation where one tries to avoid uncertainty by the creation of rules governing a process plagued by clashes between completely different value systems. Therefore: the multilateral game is an outstanding forum to deal with the improvement of skills and knowledge concerning complexity, uncertainty and value conflicts.

There is yet another reason why multilateralism is a good means to reach the goals mentioned before: it is rather easy to apply it. First of all one does not need all kinds of machines and computers to set up a multilateral game, so the role-playing can be executed at any time at any place (nearly). Secondly there is an abundance of documentation on all kinds of political, economic and other issues that can be used in the game. Thirdly these issues are changing day by day, so the game receives incentives throughout the process.

This makes it very interesting for the participants, because they have the impression they act in a real situation, not in an artificial one (but the game remains a model of course, a simplification of diplomatic practice). Last but not least there is a large pool of multilateral fora: from the Security Council and the General Assembly of the United Nations to a meeting of the ministers of Foreign Affairs of the countries of the European Community or the Atlantic Alliance. A variety of negotiations at a diplomatic level (diplomats), to talks at a political level (politicians: ministers, heads of states).

One of the multilateral fora that can be used is the meeting of the 35 states participating in the "Helsinki Process" for European détente: the "Conference on Security and Cooperation in Europe" (CSCE). Simulating parts of this process has many advantages:

- the CSCE meets every 3 years and issues a continuing stream of rules and documents, that can be used as a basis for the game to be executed (moreover: there is a basic set of rules: the Helsinki Final Act);
- a lot of research has been done in order to analyse the negotiation process within the CSCE-context, so the game can be made as realistic as possible as far as the institution and the process are concerned;
- the issues involved cover nearly all aspects of inter-European relations: from security to economics, from environment to human rights; moreover: the issues are officially interconnected and this complicates the negotiation process extremely;
- there is a wide variety of countries with very different interests and nevertheless allied to one of three "blocs": the East, the West, or the

Neutrals. A fierce value and interest conflict between East and West, combined with a potential compromising force: the Neutral states;
- the level of negotiation can be adapted to the interest of the participants because of the parties within CSCE: there can be negotiation between junior diplomats, senior diplomats or even ministers;
- the number of participants can be varied by command: one can sample out the countries that are most involved, or one can organize a meeting of all states with full delegations;
- the combination of negotiations within caucusses with negotiation between caucusses brings in a very interesting dynamism: the delegates are caught between their loyalties to their own state, their caucus and the necessity to ensure security and co-operation in Europa.

The Case

The dynamism of the game depends mainly on the problem that has to be solved and the context in which this has to be achieved. The question(s) at stake should be complicated enough to keep everybody busy for some time. On the other hand the gap between the positions must not be so wide that compromises are impossible. Why should one start to negotiate if there is no way to bridge the gap? In order to build a balanced case, there should be a dichotomy with some common ground between the poles. Preferably this common ground must be heterogenious. In other words: if three parties are involved dealing with issues, the common ground must vary for each party. A and B can agree on point I, but have some trouble to agree on II, while they are adversaries on III. On the other hand B and C agree on II and more or less on III, while C and A have a joint interest in III and share some interest in I.

This balance between opposing and converging interests is the key to the success of the game. If there is a bias, the game-master is obliged to interfere in the process, introducing some new elements. There are games where outside interference is useful. But the really good game must have enough dynamics and surprises built in, so the convenor can stay outside the game. This is very rewarding for the participants: they have the feeling to be autonomous. It is their game now, so their involvement is as high as possible. Maybe they will reach agreements, but the most important thing is the process. That must be interesting and useful. In such a simulation the role of the game master is threefold: he is responsible for a good introduction at the start of the session, his task is to be present for technical advice during the session, and most important of all: he chairs the evaluation. The evaluation of the game is the decisive moment of learning. At that time the experiences during the simulated process are compared with the premises given in the introduction and with the situation in real life.

At the Clingendael Institute three cases have been developed in order to simulate a multilateral conference within the context of the Helsinki Process.

First of all there is a real-time game, dealing with the actual topics at the CSCE. If new decisions are reached in reality, or new events influence the real meeting, the game changes its course. The initial instructions given by the staff change automatically as a consequence of the new information. Such a diplomatic game can go on for many days and needs many weeks of preparation. The attention tends to be directed towards the subjects of the game, its content.

The second game deals with an international crisis around the non-existing country of Turghanistan in Transcaucasia, between the Soviet Union, Iran and Turkey. New developments in day-to-day life play no part, but the game master can make the problems big or small as he wishes. No fear for early compromises in CSCE that distort your simulation.

The third variant combines the real time and the fantasy games. It is about a national crisis in the Socialist Federal Republic of Yugoslavia, spilling over into international tension.

The case is about an economic crisis, with a strong political background, leading to problems in the fields of human rights and military security. These problem areas fit into the three "Baskets of Helsinki": the main "chapters" of the Helsinki Final Act. The Conference on Security and Cooperation in Europe has to manage the crisis. In order to do so the three "caucusses" (Atlantic Alliance, Warsaw Pact, Neutral and Non-aligned countries) meet for preparatory discussions.

On the basis of short instructions for every country, they try to reach consensus in order to be ready for the second round of negotiations: between the three caucusses in three working-groups (the three "Baskets": Security, Economy, Human Rights).

After the second round they return to the caucus (third round) in order to set the strategy and tactics for the final negotiations in the "Baskets" (fourth round). In between they convene in plenary sessions and have informal talks in the corridors.

The CSCE Yugoslavia game offers the opportunity to develop skills in multilateral negotiation, to deepen one's knowledge about inter-European relationships and problem areas, and it gives the participant more insight in the political culture of other countries. The players can improve their competence to diminish their uncertainties in dealing with complex issues linked to sharply diverging value patterns.

The Background

Yugoslavia is in a profound economic crisis. This is partly a consequence of the economic crisis in the world and partly of the inability of the post-Titoïst structure of the Yugoslavian State and Society.

The structure of the federal state, with its rotation of leaders at all levels in order to prevent one man dominating the political scene, results in giving priority to the interests of the various Yugoslavian provinces and is a major obstacle to an effective economic policy.

The president of the Republic, Milovan Servic, also head of the Presidential Council (a body-politic representing the major national entities) decides to declare a state of emergency. He is supported by the General Staff of the Peoples' Army and by the leaders of the Yugoslavian Communist Party in the provinces of Serbia, Montenegro and Bosnia (all provinces with a Serbian majority).

Consequences of the new situation:

- the Presidential Council is suspended;
- the Council of Ministers must obey the orders of the president only;
- all provincial governments have to obey direct orders of the president;
- the press is under full censorship;
- all political activity is forbidden;
- five hundred political opponents are arrested;
- the army is in a state of preparedness;
- imports of luxury goods are prohibited.

The president flies to Moscow for a short visit to Mr. Gorbachev. A statement is issued, saying Yugoslavia will receive Russian oil at a very competitive price and that trade relations between Yugoslavia and the Comecon countries will become more intense in the near future. Back in Belgrade, the president sends messages to the capitals of the Western powers asking for loans up to $5.000.000.000 at an interest rate of 5%. A declaration is issued containing a phrase, saying that Yugoslavia might become a member of Comecon if the loans cannot be awarded.

The NATO countries condemn the state of emergency, the Neutral and Non-aligned (NNA) countries await developments and are afraid that Yugoslavia will join the Socialist Bloc and lose its NNA-status, the Warsaw Pact countries welcome the new situation.

Meanwhile the situation is discussed at the Third Follow-up Meeting of the Helsinki Final Act, that is in progress for some time in the Austrian capital Vienna.

The US send ships of their Mediterranean Fleet to the Adriatic, the Soviets, Hungarians and Bulgarians start manoeuvres near the Yugoslavian borders.

The Roles

Each one of the seventeen participants of the workshop represents a country of his own and receives an individual instruction of one page on his position in general and on the issues of security, economics and human rights.

These countries are: the Soviet Union, Hungary, Poland, the German Democratic Republic, Bulgaria and Rumania (together the caucus of the Warsaw Pact); the United States, United Kingdom, France, the Federal

Republic of Germany, Italy and the Netherlands (the Atlantic Alliance or NATO-caucus); Austria, Switzerland, Sweden, Finland and Yugoslavia (the caucus of the neutral and non-aligned, or NNA countries).

In other cases Malta is represented too, belonging to the NNA group. The Soviet Union has the instruction to pull Yugoslavia in the direction of a more intense co-operation with the socialist countries. Bulgarians and Germans support the USSR, but Hungarians, Poles and especially the Rumanians try to prevent a closer co-operation. They fear the economic burden of this and favour a neutral Yugoslavia.

The United States condemn the state of emergency in Yugoslavia, and especially the consequences for the implementation of human rights. The United Kingdom favours this standpoint, the Federal Republic and Italy oppose, and the Netherlands and France take an intermediate position. The NNA countries are also divided. The Finns want to approach the crisis in a moderate way, the Swedes and Swiss are rather critical, and the Austrians play a broker's role. They need to do this as they are the chairmen of the CSCE-conference.

The Yugoslavs, finally, try to gain as much as possible from all sides. In a way they blackmail the others: "if you do not support me, I will support your opponents".

The Process

The workshop starts with an introductionary phase of 90 minutes: a lecture on Yugoslavia as it is in the real world (10 minutes), on the history and structure of the CSCE (10 minutes), on the case (10 minutes), and on the principles of multilateral negotiation (30 minutes). After this lecture the participants have 30 minutes to read the case, their individual instruction, and to prepare themselves for the negotiations: what do I want, how do I want to achieve it, whom do I need?

After this first period, there is a second one (60 minutes): the parallel negotiations in the three caucusses. In every caucus decisions must be taken in order to reach a consensus concerning the issues at stake, the tactics and strategy to be carried out collectively, the representation of the caucus in the negotiations with the other ones: which country will represent us on which issues? This is a necessity as the third period consists of a plenary session (15 minutes: the chairman of every caucus delivers a policy statement) and a round of preliminary discussions in the three parallel working groups: one on security, one of economics, one on humanitarian affairs (30 minutes).

In the fourth period (30 minutes) the caucusses convene again in order to discuss the information they got out of the preliminary talks with the representatives of the other two groups. In the following fifth period they must reach consensus on a concluding document, linking all subjects to each other. This round consists of alternating meetings of the working groups and the caucusses, with bilateral and trilateral negotiations in the corridors (60 to 90 minutes), concluded by a plenary session (15 minutes) in which the final agreement, if any, is communicated to all participating countries.

Finally the game is wrapped up in an evaluating session (fifth period thirty up to a hundred and twenty minutes). The game is over now. Staff and participants review the simulated conference on process and content. If the meetings are video-taped, 2 hours are a bare minimum.

The Workshop

Playing the CSCE Yugoslavia game ("Crisis in Yugoslavia") with seventeen participants during a whole day, several conclusions can be drawn concerning the use of this exercise *vis à vis* the improvement of competence in dealing with complexity, uncertainty and value conflicts.

As far as the complexity is concerned, the players tried to cope with it by drawing up a clear agenda, nominating chairmen and secretariats, setting rules for the negotiation, etc. They made lists of issues, trying to reach agreement on each item individually, warning their colleagues that they withheld their final approvement, until they were satisfied about the agreement as a whole.

As there were seventeen countries, having three items in each of their instructions, they had to cope with more than fifty issues. Thanks to good time management, furthered by the structure of the multilateral process in separated phases, they reached a rather substantial compromise.

To avoid uncertainty as much as possible, the participants operated cautiously. They tried to prevent attacks by keeping their positions veiled as long as this seemed possible. They looked for support inside and outside their caucus. They tried to link difficult issues to issues more agreeable to others. Working step-by-step, channelling demands through an intermediate consensus in their own bloc to a full consensus in the working group, they diminished the risk to be exposed to threats alone. Of course the Yugoslav delegation seemed to be in a very unstable situation, but as everyone tried to win the Yugoslavs over to their own side, their position became very comfortable. Actually, there was a win-win situation in the end. The Yugoslavs went home with a lot of financial support and Europe contained a crisis that threatened its security.

Value conflicts in abundance: differences of political cultures led to many misunderstandings and "ideological" struggles. Of course this could not be as serious as in reality, as all participants were Dutch. But it was amazing to see people getting acquainted to their roles so quickly. As the day progressed, they identified themselves more with their country. In this respect the character of the participating people made of course a lot of difference, but the position of their country also played an important part in determining the intensity of the value conflicts.

On the whole, the neutral countries were less involved in these conflicts than countries belonging to the blocs of East and West. This was furthered by the fact that the neutrals had to preside over the meetings of the working groups. Here too Yugoslavia was an exception: it could not chair any meeting and could not choose its working group. Yugoslavia had to participate in the economic "Basket".

What did the participants learn? They got a better insight in the handling of a situation in which different outlooks play a dominant role, in which a multitude of issues are at stake, in which they are exposed to threats from many sides. Furthermore they got a better understanding of East–West relations and the items in it, they enhanced their understanding of the importance of diplomatic conference as a means to manage a crisis, and they learned to watch negotiation processes more carefully.

Maybe they will understand now why it takes such a long time to reach such a rather vague consensus, and at the same time why it is – nevertheless – so important to use these diplomatic means if they are at hand. In other words: maybe they realized that diplomats are doing a more valuable job than they supposed before they started playing the game.

Competency in gambling games

Danny Saunders and Dave Turner
The Polytechnic of Wales, UK

ABSTRACT: Aspects of gambling are linked with general features of game playing and game structures. Central concepts are emphasized: such as risk, uncertainty, strategy, competition, and information. It is argued that the distinction between skill and chance based gambling activities proves problematic when individual perceptions of causality for game payoffs and outcomes are considered. This recognition has implications for the selection of choices by players for each move of the gambling frame, and in turn for any general description of "competence". Four levels of skill can be associated with competence; these being labelled intellectual, social, emotional, and personal. The discussion concludes with observations of incompetent gambling, and asks whether competency can be suitably defined.
KEYWORDS: Gambling, game, simulation, risk, uncertainty, strategy, competence, incompetence, bet, information, competition, zero-sum.
ADDRESS: Department of Behavioural and Communication Studies. The Polytechnic of Wales, Pontypridd CF37 1DL.

Introduction

Whilst applications of games and simulations to educational contexts have enjoyed recent success, we should perhaps remember a broader commercial significance: what Smith and Apt (1984) have called the business of games. The reader might initially think of corporate and business training when this phrase is used, but here we refer to the relatively neglected issue of gambling. After all, much gaming involves risking material wealth or social reputation, and we should not forget the gigantic multi-billion dollar leisure and entertainment industries associated with money games. In this chapter we explore some of the major concepts which bring together games and gambling, and focus upon issues of competence and incompetence when the individual is involved with risk taking and competition.

Gambling and Competition

Gambling is defined by Scarne (1975) as "risking something one possesses in the hope of obtaining something better". With this statement, all games involve gambling, and indeed so do most everyday pursuits. As Scarne says, "life itself forces us to make bets on Dame Fortune". The central component therefore is *risk*, which involves some degree of uncertainty. If we were to adopt a more blinkered attitude, gambling could be more narrowly defined as risk taking with a currency recognized by all members within a participating economy. At the most tangible levels this means money, but we should also be aware of other invisible currencies (such as affection) which are associated with social behaviour and which result in the definition of such vague entities as "relationships".

Within the game context, and based on strategy formation which is confined within a given rule system, we provide a kernal statement which allows us to define gambling. A gamble is characterized by a fundamental disagreement between players, which results in the staking of material wealth on specific options or choices. Because these are associated with a move that has other alternatives, there is uncertainty about the direction of the outcome or payoff.

The next recognition involves the staking of (for example) money, so that there is a pool of wealth which is made up of all of the stakes (or bets) of players. Those involved will therefore be competing for a part, or all, of the aggregate sum.

Some gambling games can now be described as predatory and zero-sum, in that one player openly wins at the expense of everyone else. Other games can be more closely linked with a "them and us" situation wherein all players – especially teams and syndicates (Scott, 1968; MacDonald, 1979) – can win, at the expense of the administrator of the game. Thus poker involves the most direct face-to-face competition between players who openly try to bankrupt one another, whilst the game's administrator (at least for official games) is guaranteed a share of the kitty. By contrast, betting on races involves punters versus bookies: in betting shops, there will even be a physical barrier between the two in the form of a counter and security screen (see Newman, 1972). To summarize, with poker there is only one winner, but with racing there can be any number of winners on one race.

Consequently the focus of competition in gambling can be diverse, and depends on the rule systems associated with the overall game framework. The reader may want to consider similar foci for games which are not traditionally linked with gambling (see Christopher, 1987; Cudworth, 1988) but where participants are either involved in intense rivalry (e.g. STARPOWER) or hostility towards both other players and the facilitator (e.g. ME, THE SLOW LEARNER). At this point in our discussion, we should recognize these broader links which have been made between gaming and gambling by some central theorists within both areas.

Key Concepts: Games, Gambling and Play

The concept of "game" is usually applied to gambling, and so it is worthwhile providing a satisfactory definition:

> the term game is applied to those simulations that work wholly or partly on the basis of player's decisions . . . players who have roles to play, goals they seek to achieve, activities to perform, constraints on what can be done, and payoffs (positive and negative) as a result of their actions

and the actions of other elements in the system (including chance). In a gaming simulation, the game roles, goals, activities, constraints and consequences, and the linkages between them are patterned from real life, or, in the term used above, they simulate those elements of the real-world system. (Greenblat, 1988; p14)

Ultimately these notions of payoffs based on decision making take place within a competitive environment where performance is measured on at least an ordinal – and preferably an interval or ratio-scale. In many games, players' positions can be defined in terms of not only having "more than" or "less than" other players (in which case positions can be ranked); the differences can also be measured according to a scale of equal units wherein there is often room for an absolute zero – such as having no points, or being bankrupted, or even (in certain role-play games) being "killed".

The above quotation neatly links the "game" concept with that of "simulation". The distinction between games and simulations has been discussed at length elsewhere (see Crookall *et al.*, 1987); perhaps the best summary advice comes from Jones (1986):

However, the main areas of difference between games and simulations can undoubtedly be observed in the thoughts and behaviour of the participants. (p. 4)

Interpreting this statement, we can perhaps argue that simulations require an act of faith on the part of participants as regards believing their relevance to what some non-existentialists call "the real world" out there. Games can be a reality in their own right, and usually a more competitive one. People are said to "play" at games but to "participate" in simulations. The problems in distinguishing between the two appear with Greenblat's subtle recognition of gaming simulation, where the game outcomes and payoffs have very real relevance even though it is located in some fantasy like and seemingly peripheral location. Here the game is perceived of as a self-contained reality and as symbolic of "real life".

This is where our interest in gambling appears. People can become totally involved within, and committed to, the playing of a game. But when that game involves the risking of wealth (usually money), and when the same scale of wealth is used in everyday life, *then the activity has the potential of becoming both a game and a simulation.* A losing streak, for example, may not just be viewed as a series of coincidental game outcomes; it may also be seen as *"typical"* or *"symbolic"* of that player's general bad luck and incompetence. The relevance of such a losing run becomes even more profound when significant sums of money are involved. In this instance the gambling "frame" is at its most powerful: it is an abstract game activity, yet (as a simulator) it has acquired personal relevance, and it has also provided very real financial consequences.

Perhaps this is why some major theorists view gambling as the most dramatic form of gaming. One of the most well-known treatises on the subject comes from Caillois (1961), who identified four forms of play in games, and who illustrates some of these with gambling scenarios. In the form of *Agon*, for example, play is combative and players at the outset usually stand a near equal chance of winning. The point of the game is for each player, through skilled performance, to have his or her superiority recognized: it is left to the contestants to sort out the winners and losers within the given rule system. With the *Agon* category, many sporting activities are linked with betting and gambling – including horse racing. Another form of play is labelled *Alea*, which means "dice" in Latin, and which refers to games where decisions are made independently of the player: hence crap, roulette and lottery games become useful illustrations of Alea games where Caillois assumes – somewhat generally – that people surrender to destiny. These concepts of play have implications for the labelling of players as "competent" in specific situations; an issue which will be returned to shortly.

A more powerful analysis of game characteristics and episodes is provided by Von Neumann and Morgenstern (1953), who offer a brilliant insight into the game of poker, whilst identifying a series of crucial game concepts. Thus a game is defined as "simply the totality of the rules which describe it" (p.49), whilst a play involves the behavioural dimension wherein the game becomes active through people following (and even breaking) those rules. Play can be sub-divided into an entire sequence of moves, which correspond to an abstract occasion of a choice between various options. The actual choice which is made by (or for) a player constitutes the concrete rather than abstract play. Finally (and of most interest to economists, mathematicians and psychologists) there are strategies adopted by the players – these are open to choice and should be freely determined decisions within the confines of the overall rule framework. Strategies ultimately refer to plans of action based on each player's belief in control and perception of *causality*. We are now recognizing a psychological dimension to game playing, which adds further depth to Caillois' game taxonomy. It is even possible for Alea to be viewed as Agon by certain players and spectators.

All gambling activities can be described in terms of rules, play, choices and strategies; but what interests us most is this linking of an event with a cause. If Caillois is literally interpreted, some (Alea) games can be characterized as "chance" based, wherein the moves and choices are determined by some external agency which "randomly" decides an outcome or payoff. These are referred to as "decision machines" by another major game theorist, Erving Goffman (1967). Dice, a pack of cards, a roulette wheel and a "one armed bandit" are all decision machines; so too are animals used for racing or fighting.

By contrast, other (Agon) games are described as "skill" based, wherein the determination of an outcome or move can be based more on the players control of the situation, and depend more heavily on negotiation. But this is a theoretical and objective dichotomy between skill and chance based activity:

life is fortunately not so simple! When the concept of "perception" is applied, there is the recognition of knowledge affecting people's interpretations of a game's outcomes. Those interpretations in turn affect beliefs and attributions concerning the influences and actions of decision machines. With gambling, some players may think that the wheel is weighted, and devise a strategy wherein only certain numbers are backed at certain times. Others may have a confidence in divine intervention and back a horse which has the name of a dear long lost relative. Yet others will have a special ritual for throwing dice in the belief that this action imposes skill on a chance event.

Much of this kind of cognition is based on superstition (Skinner, 1953) and appeals to notions of magic, fate or luck – but it should be remembered that some decision machines are far from perfect when it comes to random action. When this happens it is much easier to succeed in the search for skill-based strategies of control – and to demonstrate gambling competence.

Perceptions of skill or chance factors operate at still more complex levels. Players also reach conclusions, or have suspicions, about how other people who are playing the same game view the process of causality. Strategies of play are therefore concerned with our own views about the operation of the decision machine, and also our interpretation of others' views. The situation can spiral: player A perceives player B's interpretation of the game moves, but also of player B's perception of player A's interpretation of player B!

This process of "meta-perspective" is crucial for many Agon game contexts where players are trying to conceal things from one another. In certain forms of gambling it is the essence of successful strategy formation when no external decision machine is in operation. It is not usually necessary to hide one's feelings and strategies when gamblers are playing blackjack or roulette: the odds are set by the game organizer, irrespective of what customers do. But consider the case of poker wherein the essence of "good play" is to disguise one's hand. As von Neumann and Morgenstern (1953) emphasize, such deception involves bluffing. We should "play down" four aces and a king and put on a show of calmness or even sadness: if we were to dramatically up the stakes with wild ecstasy many players, if not all, would back down. By contrast:

> ... a player with a weak hand may occasionally obtain a gain against a stronger opponent by creating the (false) impression of strength by a high bid, or by overbid – thus conceivably inducing his opponent to pass (Von Neumann and Morgenstern, 1953; p 188)

Clearly, the intensity and style of interpersonal communication varies considerably with the perception of chance versus skill based events. We argue that the more skilful the game is perceived to be, in terms of deciding each move's outcomes, the greater the degree of conflict and competition between players. The competent player resolves such competition with ingenuity and enjoyment, and with a relatively moderate degree of success (which might nonetheless entail an overall loss). We now turn to a more detailed discussion of "competence" as applied to gambling, and emphasize that when competence is defined by *experience*, then many experienced gamblers have reached the stage of Agon play even when Alea game structures are involved.

Competency in Gambling

So far we have discussed some key concepts associated with both gaming and gambling, and identified the central issues of competition and risk taking in such uncertain situations. Before turning to issues of competency it is vital that a bridging concept be considered; this being information. If people are taking risks in games (and are going to stake something that is of value in order to demonstrate confidence in their decisions) then they are by definition acting on partial information. It follows that the more comprehensive the information, the less risk and uncertainty. When attributing causes for uncertain events, the player turns to a variety of official and unofficial sources in order to *predict* future outcomes.

It is argued that *the securing and digesting of information, along with the making of a decision based on that material, has implications for the subsequent reputation and wealth of the "competent" gambler*. This statement assumes that we can define what competency means, which is problematic when the relativity of the concept is remembered: competency in one culture may be seen as incompetency and stupidity in another. To be more perverse, competency might even be defined as that which is not judged as incompetent! For now we conform to the standard dictionary definitions, which emphasize ability, sound judgement, qualified status, experience, fitness, and comfortable performance. Whilst competency does not imply excellence, it does mean a demonstration of reasonable ability which exceeds some minimum standard.

Given these interpretations, a secondary difficulty emerges: nobody has ever defined a minimum standard which has to be achieved by "competent gamblers"! We are therefore unsure of our parameters and yardsticks, and ask for the reader's patience when we suggest that the following criteria be respected:

(i) that gambling should not prove to have catastrophic financial consequences.
(ii) that the player's ability to predict future game outcomes is viewed with respect by informed observers.
(iii) that the gambler does not bet money that they (or others) cannot afford to lose.
(iv) that there is sound understanding of the rule system within which the gamble is operational.
(v) that there is an historical knowledge of the game being played.

(vi) that the social skills of the player conform to group norms and standards – should these be applicable.

These recommendations are based on our own qualitative and quantitative observations of experienced gamblers, (Saunders and Turner, 1987; Turner and Saunders, 1988), but can be easily applied to other game scenarios. Other criteria have been considered but finally dismissed: for example, that gambling should not interfere with social relations, and that gambling should be enjoyed as a true form of play. However, both of these statements are irrelevant to definitions of competence within the game scenario when it comes to issues of prediction, decision making, and self-control (although they do of course have relevance to broader considerations).

Five levels and styles of competency emerge; all based on the processing of information. It is emphasized that these categories are not exhaustive or exclusive of one another.

Intellectual

Here we refer to an ability to handle concepts and assimilate a large amount of knowledge in order to reach an informed judgement or decision. This can even imply a decision not to play, as well as the more usual involvement in a gamble. A useful illustration of intellectual competency involves an understanding of how to calculate a book for a horse or dog race, and of what various odds as offered by bookmakers mean. Dowie (1980; p.65) explains the statistical mechanism involved in interpreting the odds for just one runner:

> ... to translate the odds against an outcome happening into the probability of it happening, replace the top (first) number in the odds by the sum of the top and bottom (second) numbers, leave the bottom number as it is, and then turn the ratio upside down. (Multiply by 100 if the percentage probability is wanted). For example:
>
> $\frac{6}{4}$ against $-> \frac{6+4}{4} = \frac{10}{4} \frac{4}{10} \times 100 = 40\%$

Imagine how much more intricate the process is when calculating and interpreting odds for a 20 horse race! Understanding odds is only a part of the intellectual competency in gambling: with the above example of horseracing, all that they do is reflect the popularity of the runners as perceived by other punters. In addition to odds, there is the studying of "form", and possibly the securing of more pertinent information about a race from more informal sources. Intellectual competence also refers to historical knowledge about the game – including stories of great moments in gambling, and of legendary characters.

Social

The above possibility reflects *social* as well as intellectual competency, for the playing of many gambling games involves peer groups, social networks and contacts. Many gambling groups involve friendship as well as rivalry, and when members have known one another for years then it is inevitable that ritualistic codes of conduct develop. Furthermore, when gambling involves socializing, then conversation often revolves around stories of past exploits involving competent and even "professional" players. Such communication relies on knowledge about the history of a game and its players, and whilst socially based it also involves the remembering of information.

Whilst the gambling environment may be described as economically competitive, the perception of that scene by participants may vary considerably – especially when the consequence of risk taking are small. Occasionally, for example, players may share their winnings with the rest of the gambling circle. This also constitutes social competency, and it is even possible for there to be a norm of generosity within that circle of players – much to the annoyance of academic theorists who assume that all gambling is motivated by individual profit!

Emotional

This refers to the ability to handle and control oneself when observing and even participating within the actual decision of a game outcome. It is here that Goffman's (1967) identification of "span of play" becomes relevant – and one which complements those earlier concepts from von Neumann and Morgenstern's analysis of games. To begin with there is squaring off, which involves the intellectual competency of securing relevant information for decision making. This is followed by deliberation, where the bet is made or a commitment to risk is taken. Disclosure and settlement is the (usually) brief phase wherein the gamble is decided and the game's decision machine is thrown into motion: this is especially relevant to emotional competency, because high degrees of euphoria or dysphoria may be experienced when the risk consequences are significant. Goffman (1967) provides some fascinating descriptions of public presentations of self by gamblers: for example, the respect that is earned by the player who maintains calm and dignity when winning or losing large amounts of money. What Goffman ignores, however, is self-control associated with emotionally competent gamblers who resist the temptation to either chase their losses, or reinvest their winnings.

Manual

In certain games a degree of manual dexterity distinguishes the more experienced players from the novices and beginners. With gambling, the best examples are set by card players, who can perfect a virtual performance which is characterized by speed of action, accuracy and delicacy. Obvious examples include shuffling methods, dealing, and the holding or viewing of a hand of cards. The issue of manual competence is important to Goffman's discussions about self presentation – especially when *style* is considered when players play games. Furthermore, it is especially in this area that such action distinguishes the professional player from the amateur, and the game administrator from his or her customers. Consider those cases where "card sharps" can demonstrate their skills, or where croupiers can speed up the action so that span of play is shortened. A further possibility which develops from such competence concerns the ability to cheat, provided other players are less knowledgable about such methods.

Personal

The last level in our list is personal competence – a rarely discussed possibility, but one that refers to the perception of the self by the self (rather than by others). In effect the individual compares the self-at-play with the "ideal self" and with previous performances. A judgement of "improvement" – even though there may still be an overall loss in gambling – can therefore be interpreted as a move towards greater "self-esteem" and competence. As noted by Scott (1968) and Saunders and Turner (1987), some players follow a career path wherein an "apprenticeship" phase is expected and wherein consistent losses teach the necessary gambling and gaming skills. In this context an incompetent performance is viewed as a temporary but necessary and even enjoyable phase of game playing.

Conclusion: Incompetent Gambling

These are therefore levels of competency which should be considered whenever observing competitive games where a degree of wealth is risked in uncertain situations. It is hoped that our discussion has provided a bridge between the normally separate worlds of gaming and gambling. The isolation of both concerns is somewhat surprising at times – especially when the actual titles of regulatory bodies in the gambling world are remembered: for example, in Britain we have "The Betting and Gaming Board" which observes and regulates the casinos. The other side of the coin involves educational games which use decision machines that are normally associated with gambling: dice being the best example. The gambling dimension is further enhanced when prizes are offered as an incentive for playing such games.

The issue of competency in game playing is a crucial one for observers of problematic gambling behaviour. If competency has been previously defined as ability, sound judgement and fitness, then incompetence implies an inability, illness and lack of judgement. But this is a very limiting and circular style of definition: are we really content to say that competent gaming is that which is not abnormal – and vice versa? Whilst gambling incompetence is a favourite issue within existing literature (as reflected by numerous analyses of "compulsive" gambling), the major domain of controlled and enjoyable financial involvement in games of chance and risk remains a neglected area of discussion. Perhaps a closer understanding of "addictive" (Kusyzyn, 1972) or "pathological" (Moran, 1970) gambling behaviour will be gained through discussing the more general aspects of playing games for leisure.

As a final comment, we would like to return to Greenblat's (1988) recognition of game-simulations, where the activity has relevance to real life and constitutes a reality in its own right. With certain cases of incompetent gambling this is not applicable, because the game activity actually excludes everything else – as illustrated by the professional gambler (and gaming executive) Mike Goodman:

> When I was a very young boy, I was shooting crap for pennies with the other youngsters. Suddenly a cop turned the corner and we were caught. Everyone else made a run for it. I stood there and pleaded with the policeman, "Just let me see if I can make the four". (1963, p.89; and quoted in Smith and Abt 1984).

With other forms of incompetent gambling the game scenario is not only taken too seriously, it also symbolizes that player's entire life. It is especially here that Greenblat's concept becomes useful in our analysis of gambling. In such cases the game simulates something else, and the over-involved player differs from most other players in perceiving such symbolism and meaning. This is not a sudden process, but is more a career progression in game playing wherein the minority of compulsive players were, in earlier times, more controlled and possibly more competent in their game activities. The previously mentioned level of personal competence becomes crucial when compulsive gaming is considered: there may be such a negative self-image that the individual actually wants to *lose* in order to punish themselves, and to publically demonstrate their incompetence. Such a possibility has been extensively discussed in psychoanalytical interpretations of gambling, and especially by Freud (1928) when observing the development of guilt within the individual.

The next step is to establish the conditions which allow for a transition from gambling competency to incompetency, and to investigate whether specific individuals, groups or societies are more prone to crossing such boundaries. It is hoped that such investigation will provide a more extensive and detailed definition of competence within games involving risk and uncertainty.

References

Caillois, R. 1961. *Man, Play and Games*. New York: Free Press.
Christopher, E. 1987. *Leadership Training: through gaming*. London: Kogan Page.
Crookall, D.; Oxford R. and Saunders, D. M. 1987. Towards a reconceptualization of simulation: from representation to reality. *Simulation/Games for Learning* 17:4.
Greenblat, C. 1988. *Designing Games and Simulations*. Beverly Hills: Sage.

Dowie, J. 1980. *The World of Monetary Risk*. Open University (Block U201). Milton Keynes: Open University Press.
Freud, S. 1928. Dostoevsky and parricide. *In*: *Collected Works of Sigmund Freud* (ed. Strachey J. 1959). New York: Basic Books.
Goffman, E. 1967. *Interaction Ritual*. New York: Anchor.
Goodman, M. 1963. *How to Win*. Los Angeles: Holloway House.
Jones, K. 1986. *Designing Your Own Simulations*. London: Methuen.
Kusyzyn, I. 1972. The gambling addict versus the gambling professional. *International Journal of the Addictions* **7**.
McDonald, D. 1979. Heads they win, tails you lose. *New Statesman*. 31 August.
Moran, E. 1970. Varieties of pathological gambling. *The British Journal of Psychiatry* **116**.
Newman, O. 1972. *Gambling, Hazard and Reward*. London: Athlone Press.
Von Neumann, J. and Morgenstern, O. 1953. *Theory of Games and Economic Behaviour*. Princeton: Princeton University Press.
Saunders, D. M. and Turner, D. 1987. Gambling and leisure: the case of racing. *Leisure Studies* **6**.
Saunders, D. M., Coote A., and Crookall D. 1988. *Learning From Experience through games and simulations*. Loughborough University: Sagset.
Scarne, J. 1975. *Scarne's New Complete Guide to Gambling*. London: Constable.
Scott, M. B. 1968. *The Racing Game*. Chicago: Aldine.
Skinner, B. F. 1953. *Science and Human Behaviour*. New York: MacMillan.
Smith, J. F. and Abt, V. 1984. Gambling as play. *In*: Frey J. and Eadington, W. R. (eds) *Gambling: views from the social sciences*. Beverly Hills: Sage.
Turner, D. and Saunders, D. M. 1988. Gamblers Anonymous: an ethnographic interpretation. *Society for the Study of Gambling Newsletter*.

Game References

ME, THE SLOW LEARNER by Thatcher, D. and Robinson, J. 1988. *In*: Saunders, D.; Coote, A. and Crookall, D. (op. cit.).
STARPOWER by Shirts, G. 1969. Western Behavioural Sciences Institute: La Jolla, California.
THE SURVIVAL GAME by Cudworth, A. 1988. *In* Saunders, D.; Coote, A.; and Crookall, D. (eds) *Learning From Experience: through games and simulations*. Loughborough University: Sagset.

SURVIVAL: a case study in risk taking

Alan L. Cudworth
Trent Polytechnic, Nottingham, UK.

ABSTRACT: Survival is a simulation-game illustrating the procedures adopted in traditional tendering operations for building work. Groups of participants compete with each other to obtain profitable building work. Each group is offered contracts to tender for and presented with identical estimated costs for all labour, material and plant required to complete the work. The basic requirement of each group is that they add to the basic cost a percentage to cover their overheads and profit. A random adjustment or gamble is added to/deducted from both the tenders submitted by each group and to the estimated contract cost to illustrate the uncertainty that is experienced in reality. The winning group is the one that makes the most profit after the recovery of their overheads.

The simulation game illustrates the risk taking or gamble experienced by many firms and how necessity becomes the driving force to the estimator. The debriefing, an essential part of the exercise will explore the motives and emotions behind the risk taking, the chances of success and the effects of failure on the firm and its operatives.
KEYWORDS: Simulation-game; risk taking; building; tendering.
ADDRESS: Department of Surveying, Trent Polytechnic, Burton Street, Nottingham NG1 4BU.

> Education is an admirable thing,
> but it is well to remember
> that nothing that is worth
> knowing can be taught
>   Oscar Wilde 1894

The perception of Oscar Wilde in this quotation was reflected in the presentation of the paper and the subsequent participating workshop. It is arguable that basic estimating procedures can be taught by conventional techniques such as lectures, seminars and personal reading but I believe that the essence of tendering is in the uncertainty of obtaining the work, the urgent need to provide sufficient work to keep the operatives in full employment and the need to recover overheads and profit. These may be vital to the survival of the firm.

This simulation-game was devised in order to develop those skills and attitudes and to enable the participants to experience real-life emotions in a protected environment. It provides a vehicle for the participants to experience the roles of building company directors making decisions which will ultimately affect the performance of the firm and possibly affect its survival in a very competitive world. The simulation enables the participants to develop a tendering strategy by carefully controlling the amount of work on offer and forcing the participants to make decisions regarding the recovery of their overheads and profit and in some cases to take risks in order to try and secure work for their firm. The eventual consequences of these actions were developed as the simulation progressed and recovery was found to be very difficult in some cases, as it certainly is in reality.

The part played by the "random factor" was a very minor one in that it was used only to adjust the tenders submitted and the actual estimated cost to ensure that there were no identical tenders submitted (as no two estimators would ever arrive at exactly the same figure in real life) and to illustrate that the estimated cost of carrying out the work would vary beyond the control of the contractor (due to material changes, variations made by the architect, and inclement weather). The simulation is played in rounds each representing 3 months of the year and over a total simulated period of 2 years.

General Introduction to the Simulation

Each group represents a building company the details of which follow this introduction. A group leader was elected to represent the managing director of the firm and to be also responsible for the overall co-ordination of the work of his/her group. All deadlines given to the participants were to be met and the following objectives of the exercise were to be fulfilled.

1. To use the firms resources to their maximum effect by obtaining profitable construction work.
2. To obtain the maximum amount of profit for the company after allowing for overhead and running costs.

Each group (as a building company) had to tender for ALL work presented to it and were provided with certain similar assets which they had to use in order to maximize their profits.

Extracts from the Simulation Instructions (Extended Version)

The Building Company

You are a medium-sized building contractor operating in the East Midlands. Your work varies and is made up of both private and public undertakings. You have a good relationship with local architects and generally if work is to be tendered for in your area you would be one of those firms asked to quote.

You have financial resources of £300,000 which you are able to "turnover" 10 times i.e. a £3,000,000 limit for each 12 months of trading. Any increased turnover will require you to "borrow" additional finance in units of £100,000 to enable you successfully to carry out the work. Fortunately with your high standing locally this borrowing should not present any problems but you can only approach the Bank with a maximum of two requests. You may plough back 50% of profits made (50% tax) in lieu of bank borrowing.

TENDER SUBMISSION SHEET
(To be handed to Exercise Controller before applying the random factors)

GROUP NO.

Signed Group Leader	
No.	Estimate
1	3,000,000 – 24 months
2	150,000 – 3 months
3	1,500,000 – 18 months
4	500,000 – 12 months

FIRST QUARTER

FIGURE 1

Your normal mark up (overheads and profit) is set at 8% of your required £3m turnover per annum, i.e. £240,000.

However, if you have to borrow money because of your higher turnover you will be required to obtain a 10% mark up on the additional turnover, to cover the additional cost of borrowing.

The value of existing contracts already obtained by your company are as follows and each can be assumed to include 8% to cover overheads.

1st Quarter	*2nd Quarter*	*3rd Quarter*	*4th Quarter*
£600,000	£400,000	£200,000	—

The bidding simulation is played in rounds with each "round" representing 3 months of the year, over a simulated period of 2 years.

During each "round" you will be given a number of estimates of possible contracts together with their individual contract periods that you must tender for.

Each group should discuss their tendering strategy and complete form B recording their thoughts and reasons for the proposed mark up.

These estimates shown on Fig. 1 represent the cost of all labour, material and plant required to carry out the work and in practice will be the figure arrived at by the estimator and submitted to the board for consideration. You

Form B

Group No. _____

LOG SHEET

Contract No.	Estimate	Proposed Mark up %	Comments (Enter in this column your reason for the mark up stated)

FIGURE 2

represent this board of directors and are required to add to the estimate a "mark up" to cover overheads and profit and complete the form.

Each series of tenders will be submitted to the exercise controller and a random mark up/down of between ±1% in increments of 0.1% will be applied individually to each tender submitted. This represents a discrepancy factor for human error in the preparation of the estimate.

The lowest tender (including random factor) submitted will be awarded the contract. Each estimate as submitted to the groups will also be subjected to a random factor of ±3% in increments of 0.5%. This represents the *actual cost* of doing the work assuming that the ultimate solution is applied to all resources in all situations i.e. this represents the most economic method of carrying out the work.

The difference between this figure, arrived at by way of applying the random factor and the lowest tender (including factor) received, represents the profit or loss to the firm on that contract.

At the end of each series there will be a period of time to enable each group to assess the outcome of the bidding and complete all documentation. It also allows the group time to discuss future strategy when the next series of contracts are presented.

When a contract has been obtained it must be entered on the contract

SURVIVAL: a case study in risk taking

BIDDING SIMULATION DATE   CONTRACT PROGRAMME

CONTRACT	January/February/March	April/May/June	July/August/September	October/November/December
EXISTING CONTRACTS	600,000	400,000	200,000	
Turnover Per Month/Quarter				

Total Turnover For Year

programme commencing at the beginning of the next quarter and its value divided by the contract period to obtain the monthly value of work executed (including overheads and profit).

For the purposes of the workshop a simplified version was used, simply to illustrate the risk taking or gamble. It was not necessary to complete all the documentation, just sufficient to identify the groups trading position and provide information to "calculate" the next gamble.

General Observation and Conclusions from the Development Operation and Evaluation of "SURVIVAL"

There are perhaps two kinds of general observations to be made. Firstly there are those lessons learned from running the "Exercise" which contribute to its improvement for future use. Many of these, of course, contributed to the formative evaluation and development of the project. Secondly, there are those observations that may well be more generally applicable in the wider field of education and training through games and simulations. The after game discussion highlighted several of these important points and although the participants had no prior knowledge of tendering procedures it was evident that an understanding of the philosophy of tendering had been achieved. The observations recorded here are taken not only from this particular workshop but also from the many exercise runs within the college environment. The following is a necessarily selective list of some of these more generally applicable observations.

(i) *High motivation.* In common with almost all studies of experiential learning there was a high motivation by the participants and students and a strong commitment by students to an "Exercise" they saw as being directly related to the profession they were training for.

(ii) *Realism and accuracy.* It was necessary to provide realistic and accurate information in the initial scenario. Any small error or artificiality was quickly observed and produced the most dissatisfaction from the students (in the initial development stages).

(iii) *Physical environment.* The provision of some form of realistic surroundings at college to work in, has also added an element of realism. In the "Exercise" run, each group was given a number of large display screens which provided both a measure of privacy and an area to display drawings and notices. The whole "Exercise" was undertaken in a large open area designed to resemble an open plan office. This kind of provision was not possible to achieve in this particular workshop.

(iv) *The group leader.* A noticeable fact in all "Exercise" periods was the influence of the elected group leader. A group with a highly motivated group leader tended to produce work of a higher standard and a more creative outlook. Consequently there could be merit in paying more attention to the selection and special briefing of the group leader.

(v) *Winning*. The simulation-game is designed to produce a clearly defined "winner", an absolute requirement in any competitive game. The participants and students were clearly motivated by the thought of success and as work became more scarce and difficult to obtain so the competitiveness of the participants and students increased.

(vi) *Debriefing*. The debriefing sessions with the tutor helped the students and participants to understand where their actions or decisions represented good or bad practice and why! There was a need to explain the reasons for any assessment made by the tutor. It is unreasonable to expect participants of this level to accept a form of assessment without explanation or the opportunity to discuss and criticize.

Review of the Workshop

The debriefing provided an opportunity for the participants to reflect upon the motives of the manager in submitting tenders for building work and how the risk taking or gamble is part of business life. The gamble in this environment could range from a few hundred pounds to millions depending upon the size of the contract and the participants commented upon their emotions and motivation in submitting their tenders particularly high sums over 3 million pounds.

The consequences of failure were there to be seen as the "luck" went away so some groups struggled to find success and consequently the gamble or risk taking for them increased.

The simulation highlighted the fact that people are prepared to take enormous gambles, often with other peoples money and livelihoods at stake, as the firm's survival becomes more and more under threat.

Before any innovations of this kind can be successfully devised the innovator must be clear about his objectives and what he hopes to achieve. Curriculum theory and curriculum models provided the method by which the innovation described earlier could be devised and tested and showed how conclusions could be presented. The model chosen provided the framework within which the surveying curriculum could be examined in detail and provided a similar framework for devising and testing the innovation. Curriculum theory aided the illumination process, identifying components and interactions for further study.

The Contribution of Experiential Techniques to the General Intellectual Development of the Student or Participant

Many students enter higher education believing important issues in reality are clear cut, easily identified and can be seen as "black or white". In practice this is not the case and normal student development in higher education entails a move from this position to a more relativistic position as described by Perry (1970) who explains that:

> the student after a period recognizes some kinds of knowledge as *Contextual*. That is, the answers depend on so many circumstances that individual judgements have to be made in each context. Not everyone's judgements are as good as anyone else's. The quality of individual judgements will depend on the experience and openness and insight of the individual. Some people are better qualified than others to make judgements in certain areas.
>
> *Later on* the student discovers that more and more of the knowledge he previously considered to be Absolute is in fact Contextual.
>
> The student then comes to recognize that the personality, attitudes and skills appropriate in a relativistic world are different from those sustained by a simple belief in certainty.

There is some evidence from observing and talking to students and participants that exercises such as "SURVIVAL" make a significant contribution to that development.

Each student had entered the "Exercise" with some basic knowledge gained from "formal" teaching techniques. Many students believed the information given to them through formal teaching was "clear cut" or "absolute". However, as the simulation progressed, many students found their beliefs and attitudes changing in much the same way as discussed by Perry. Although the participants in this workshop had not had the benefit of any formal teaching it was clear to them that many decisions were made in a contextual environment.

In the development of this simulation-game an attempt has been made to afford each student the opportunity to reflect upon personal judgement, attitudes and skills and to consider realistic, contextual solutions to the problems that have been set.

Any perspective on possible and desirable developments of techniques such as case studies, games and simulations must face the reasons why teachers in higher education have such a strong attachment to the lecture which remains the dominant form of instruction in most areas of higher education.

Gibbs (1982) suggests a number of reasons for this including ignorance, overwork, lack of resources, time and attitudes. However, there are signs that teachers are becoming more adventurous in their use of alternative methods.

Special subject groups have been established such as the Association for Study of Medical Education. There has also been the development of education sections in many of the professional institutions such as the Royal Institute of Chemistry, the Institute of Biology and of course the Royal Institution of Chartered Surveyors which have encouraged the use of alternative strategies.

New developments in teaching take a long time to become accepted by the majority of teachers, but as teaching in higher education becomes less *individual* and *private* and more co-operative and public so it seems more likely that worthwhile innovations will spread more rapidly.

References

Gibbs, G. I. 1982. *Twenty terrible reasons for lecturing.* SCEDSIP.
Perry, W. G. 1970. *Forms of Intellectual and Ethical Development in the College Years: A Scheme.* New York: Holt, Rinehart and Winstons.

Gaming, gambling, risk taking, addictions and a developmental model of a pathology of man-machine relationships

R. Iain F. Brown

University of Glasgow, Scotland, United Kingdom

ABSTRACT: Links and interactions between gaming, gambling and risk taking were explored and discussed. In particular the phenomenon of excitement or arousal experienced by the participants of all these activities was examined and a demonstration of it was given in the workshop by recording the heart rate of participants as they played well-known games. The phenomena of addictions in general and gambling addictions in particular were described and discussed. A study on videogame addictions among adolescents was then reported and discussed and compared with studies of adolescent machine gamblers in the US and the UK. A developmental model was proposed outlining one possible path along which the development of machine gambling addictions could stem from an early pathology of man-machine relationships.

KEYWORDS: Gaming; gambling; risk taking; addictions; pathological; machine-gamblers; video games; amusement with prizes; development; and adolescents.

ADDRESS: Department of Psychology, Adam Smith Building, University of Glasgow, Glasgow G12 8RT, Scotland, United Kingdom.

DATE DUE

NOV 30 2002			
AUG 04 2003			

HIGHSMITH #45230   Printed in USA